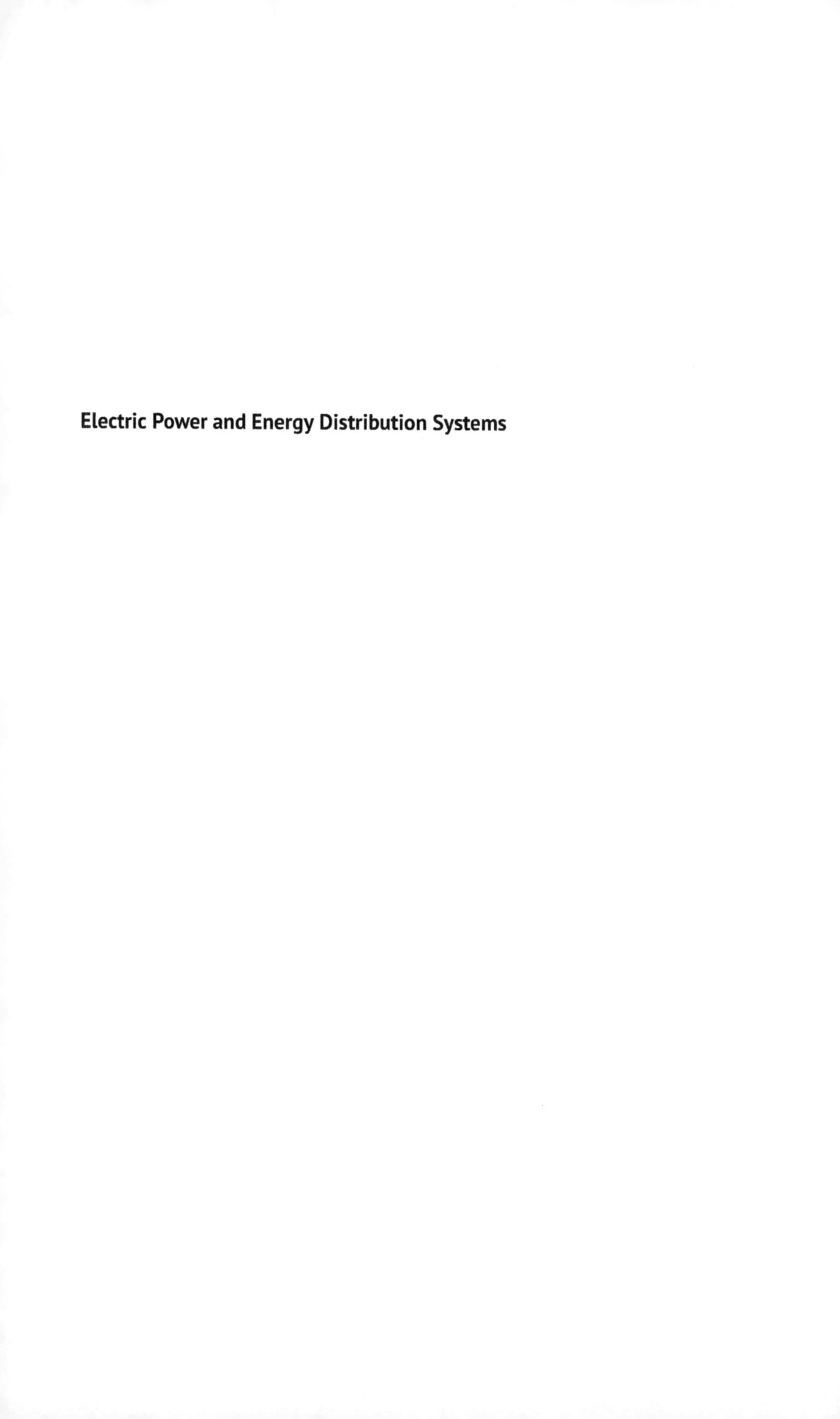

Electric Power and Energy Distribution Systems

IEEE Press
445 Hoes Lane
Piscataway, NJ 08854

Electric Power and Energy Distribution Systems

Models, Methods, and Applications

Subrahmanyam S. Venkata
Venkata Consulting Solutions, LLC
Tuscon, Arizona, US

Anil Pahwa
Kansas State University
Manhattan, Kansas, US

Published by John Wiley & Sons, Inc., Hoboken, New Jersey.
Published simultaneously in Canada.

Library of Congress Cataloging-in-Publication Data Applied for:

Hardback ISBN: 9781119838258

Cover Design: Wiley
Cover Image: Courtesy of Anil Pahwa

Set in 9.5/12.5pt STIXTwoText by Straive, Chennai, India

Dedicated to our spouses
Padma and Mukta

Contents

Biography

Subrahmanyam S. Venkata is Affiliate Professor of Electrical and Computer Engineering with the University of Washington (UW), Seattle, Washington, where he has taught since 1979. He was Dean and Distinguished Professor of Wallace H. Coulter School of Engineering at Clarkson University, Potsdam, New York, during 2004–2005. In 2003, he was Palmer Chair Professor of Electrical and Computer Engineering Department at Iowa State University, Ames, Iowa. From 1996 to 2002, he was Professor and Chairman of the department at ISU. Before joining ISU, he taught at the University of Washington, Seattle, West Virginia University, and the University of Massachusetts, Lowell, for 25 years. He received his B.S.E.E and M.S.E.E. degrees from India, and his Ph.D. degree from the University of South Carolina, Columbia, in 1971. In 2016, he received the Robert M. Janowiak Outstanding Leadership and Service Award from ECEDHA. In 2015, he received the IEEE PES Douglas M. Staszesky Distribution Automation Award. He received the IEEE Millennium Award in 2000. In 1996, he received the Outstanding Power Engineering Educator Award from the IEEE Power Engineering Society.

He is President of Venkata Consulting Solutions LLC. He was with GE Power/Alstom Grid Inc. from January 2011 to September 2017. Dr. Venkata is a Life Fellow of the IEEE. At the IEEE level, he served as a member of the IEEE Fellows Committee for six years. He also served on the PES Board as Vice-President, Publications PES, during 2004-07.

Anil Pahwa is University Distinguished Professor at Kansas State University and holds the Logan-Fetterhoof Chair in Electrical and Computer Engineering. He received the B.E. (honors) degree in electrical engineering from Birla Institute of Technology and Science, Pilani, India, in 1975, the M.S. degree in electrical engineering from University of Maine, Orono, in 1979, and the Ph.D. degree in electrical engineering from Texas A&M University, College Station, in 1983. He has been a faculty member in Electrical and Computer Engineering Department at Kansas State University since 1983. The National Academies selected him for the Jefferson Science Fellowship in 2014. He served as a Senior Scientific Advisor in Economic Policy Office of East Asian and Pacific Affairs Bureau of the U.S. Department of State as a part of the fellowship. He served as Program Director for power and energy at the National Science Foundation from 2018 to 2020. He has served in several officer positions in IEEE Power and Energy Society (PES) including chair of Power and Energy Education Committee from 2012 to 2013, and an editor of *IEEE Transactions on Power Systems* from 2010 to 2015. He has received several awards over his professional career including the IEEE PES Douglas M. Staszesky Distribution Automation Award in 2012, IEEE PES Prize Paper Award in 2013; Distinguished Researcher Award in 2020, Snell Excellence in Undergraduate Teaching Award in 2017, Frankenhoff Outstanding Research Award in 2012, and Erickson Public Service Award in 2011 from Kansas State University College of Engineering; and Outstanding Alumni Award from Birla Institute of Technology and Science in 2014. He is a Life Fellow of IEEE. He is presently serving as a member of the IEEE Fellows Committee. His research focuses on reliability, automation, and optimization of power distribution systems. His research has provided innovative and practical solutions for application of advanced communication and cyber technologies for automation of electricity distribution to customers and large-scale integration of renewable energy resources in the system.

Preface

Electric power systems have seen enormous changes over the past 15 years due to the growth of renewable resources and the integration of information technology. Power and energy distribution systems have had the most changes. They are at the lowest end with less than 69 kV and close to customers. In the past, they were completely passive with very little information available beyond the substation. Now they are becoming active with consumer-owned rooftop solar generation, electric vehicles, energy storage, and advanced metering infrastructure. It is imperative to integrate cyber technologies to automate operation of the emerging distribution systems for higher efficiency, reliability, resiliency, and flexibility. We can expect higher deployment of distributed energy resources (DERs), more electric vehicle ownership, and consumers becoming more engaged in the future. As a result, the planning and operation of power distribution systems becomes more challenging.

Power distribution systems did not receive much attention until about 15 years ago. With the introduction of the Smart Grid concept and related investment by industry in distribution systems for integration of advanced technologies, interest in distribution systems grew. Increased deployment of distributed resources and microgrids further fueled interest in the subject. Many universities now offer a course in distribution systems. We expect that this book will encourage more faculty to offer courses, as the lack of a suitable book is often a deterrent to offering a new course. We also expect that parts of this book may be used as a part of other courses in power systems. As we have expertise and combined experience of about 90 man-years in distribution systems, several colleagues asked for our help and urged us to write a textbook. This book will be a valuable resource for professors teaching a course on distribution systems and for students to prepare for their future careers. In addition, it will be a comprehensive reference for professionals who want to expand their knowledge about distribution systems.

The book provides a comprehensive treatment of various aspects associated with distribution systems including modeling, methodologies, analysis, planning, economics, distribution automation, reliability, grounding, protection, power quality, DERs, and microgrids. Modeling and analysis provide rigorous modeling of distribution system components including distribution transformers, feeders, and load followed by system analysis. Planning and economics include topics such as load characteristics, feeder layouts, and optimal selection of devices. Distribution automation focuses on operation functions and their analysis. Case studies of distribution automation are provided. Reliability, grounding, protection, and power quality topics deal with legacy issues as well as address issues relevant to emerging distribution systems. The chapter on DERs and microgrids discusses various types of distributed resources and challenges associated with operation and planning of distribution systems with increased proliferation of these resources, and design and operation of microgrids.

The book is designed to provide comprehensive treatment of various topics relevant to classical and emerging electric power and energy distribution systems. The topics address legacy as well as contemporary issues supported by rigorous analysis and practical insight. The book provides a good blend of theory and practice. It is ideally suited to teach both senior and first-year graduate courses at any university. We are assuming that the students have taken a basic course in power systems. The material can be covered in two semesters, but the instructors can select topics for a one-semester course based on the emphasis of the course. Some topics are more relevant for a graduate-level course.

Organization of the Book

The book consists of 13 chapters.

Chapter 1 introduces power distribution systems with a historical perspective and looks into the future, guided by the new developments. Differences in terminologies as relevant to distribution systems in the United States and the rest of the world are included. Descriptions of various devices in distribution substations and feeders are provided.

Chapter 2 presents different types of transformers used in distribution systems and associated standards. Models for steady-state analysis and performance measures, such as efficiency and regulation, are discussed. Various schemes and associated analyses for transformer connections both for single- and three-phase transformers for distribution of electricity to customers are included.

Chapter 3 presents a rigorous analysis for modeling overhead lines and underground cables. Effects of unbalances including mutual coupling between phases and effects of ground are included in models for series impedance.

Chapter 4 includes models for sources and loads. DERs are introduced, and their integration with the system as per IEEE Standard 1547 is discussed. This is followed by a source-load iteration method for power flow analysis of unbalanced three-phase power distribution systems. Voltage regulation is explained with simple examples. The chapter concludes with methods for computing different unbalanced faults both in the phase domain and the sequence domain.

Chapter 5 focuses on distribution system planning while comparing traditional vs. modern approaches for planning to achieve optimal designs. Effects of load forecasting, load demand characteristics, and standards on system design are discussed. Load characteristics include coincidence and effects of temporal and spatial aggregation of loads. Topological designs for the substation, primary feeders, and secondary feeders are presented. An introduction to cold load pickup and its effects on planning are discussed.

Chapter 6 introduces fundamental economics concepts, such as present worth, and annuity, relevant to power distribution systems. Applications of these concepts for optimal selection of conductors for feeders and transformers are presented. Tariffs for selling electricity to consumers of different types are discussed.

Chapter 7 focuses on distribution automation for the operation of distribution systems. Basic ideas on communication infrastructure needed for automation are provided. Details of various common operation functions are discussed along with benefits of their automation. Cost–benefit evaluation of distribution automation with examples of cost and benefits is presented. Case studies of automating various functions in urban and rural systems are provided.

Chapter 8 provides an in-depth analysis of various operation functions presented in the previous chapter. Details on novel approaches with mathematical formulations for outage management, voltage and var control, system reconfiguration, and systems restoration including cold load pickup are presented. This chapter relies heavily on integrating the results of the authors' research on these topics.

Chapter 9 introduces the concept of distribution system reliability supported by various indices that are used by industry. Mathematical approaches for component modeling are discussed. Analytical methods as well as Monte Carlo methods for reliability evaluation of distribution systems are presented. This is followed by discussion on regulations for reliability.

Chapter 10 provides the definition and need for system grounding for distribution systems. The effects of soil resistivity and frequency on neutral grounding are discussed. Applicability of National Electric Safety Code for primary system grounding and the National Electric Code for secondary system grounding are presented.

Chapter 11 focuses on philosophy and architecture of protection with relevance to distribution systems. Selection of fuses, reclosers, and overcurrent relays and their coordination for overcurrent protection are discussed. Equations for relay characteristics are included. Students can use these and also standard characteristics of other devices provided by the manufacturers for computer-aided design of distribution system protection.

Chapter 12 introduces various indices used in power quality evaluation of distribution systems. This is followed by approaches for harmonic analysis. Effects of motor starting and its effects on flicker in the system are discussed. Voltage sag and swell and behavior of sensitive loads, such as computers, which are vulnerable to transients in the system, are presented.

Chapter 13 defines DERs with a focus on wind, solar, and battery storage. This is followed by interconnection issues with a focus on the role of all applicable standards including IEEE 1547 and other standards addressing the interconnection issue. Architecture, control, and protection of microgrids along with the methods for their performance evaluation are presented.

While there are various other topics that are relevant and several issues still evolving, we feel that the book provides a comprehensive coverage of the most relevant topics. We will address additional topics as necessary in the next edition of the book.

Acknowledgments

First and foremost, we thank our parents for nurturing us and providing us the right opportunities for education and inspiration for professional success. We thank our respective spouses, Padma Venkata and Mukta Pahwa, for their unrelenting support throughout our professional careers. Without their support and encouragement, writing this book would not have been possible. We thank our respective children Sri and Harish (Venkata) and Samir and Mrinal (Pahwa) and their families for their love and respect and guidance during preparation of the book. We thank Padma Venkata for her proof reading and editing. Several of our professional colleagues reviewed our book proposal and provided valuable comments, which allowed us to improve the quality of the book. Thanks to H. Lee Willis, whose books on power distribution planning inspired the authors to write some of the chapters of this book. Thanks to Dr. Chanan Singh (Texas A&M University) for his help in preparation of the book proposal. Special thanks to Dr. Sukumar Brahma (Clemson University), who suggested several enhancements to the book and served as a sounding board during the writing of the book. Thanks to Dr. Ned Mohan (University of Minnesota) for his general advice on book writing. Thanks to our graduate students from whom we learned a lot. Although we have not listed individual names, the work of several graduate students has contributed to portions of the book. We want to thank Eaton Corporation, S&C Electric, ABB, Schweitzer Engineering Laboratories, Power Standard Laboratories, and the Information Technology Industry Council (ITI) for providing diagrams and product images for inclusion in the book. Finally, we thank the staff of Wiley and IEEE Press for their help throughout the course of writing this book.

Subrahmanyam S. (Mani) Venkata
Anil Pahwa

About the Companion Website

The book is accompanied by a companion website which has data sheets, supplementary files, additional problems, and problem solutions. Problem solutions are accessible only by instructors, but the rest of the material is accessible to both instructors and students.

www.wiley.com/go/Pahwa/ElectricPowerDistributionSystems

1

Introduction

1.1 Prologue

The development of distribution systems poses new challenges in the changing world, where levels of electrification need to be increased and electricity served reliably for sustainable, economic, and social development. Technological development and adequate regulations are required at the distribution level to respond to new energy challenges and the restructured environment. The need for a change in the way distribution systems are designed, planned, operated, and managed is a must for both developed and developing countries. All changes should ultimately ensure optimal and economic service to the consumers of electricity. While the basic parameters remain the same, the challenges to be met are substantially different. The past, present, and the future of the distribution systems are reviewed.

1.2 The Past

Towards the end of the nineteenth century, direct current (dc) distribution systems came into existence. Recognizing the value of electric energy and the need for development of economic sources of electricity, our forefathers wisely replaced dc with alternating current (ac) during the early part of the twentieth century. Subsequently, ac systems grew enormously, making the development of the electric power system the greatest achievement of the century. Unfortunately, several persistent problems with distribution systems have lingered on for many decades. They did not receive the attention they deserved under the regulated environment, when compared to generation and transmission. Very little attention was paid to the planning, design, operation, and management of these nonbulk systems. Performance optimization for efficiency, regulation, and other measures were not adequately addressed.

Electric Power and Energy Distribution Systems: Models, Methods, and Applications, First Edition.
Subrahmanyam S. Venkata and Anil Pahwa.

Companion website: www.wiley.com/go/Pahwa/ElectricPowerDistributionSystems

1.3 The Present

The way the business was regulated contributed to the current situation in many distribution systems worldwide. While analyzing just efficiency we find that many electric companies in several countries, both public and private, are still experiencing extremely high system losses, in the range of 30–50%. In addition, voltage regulation at the customers' premises sometimes is very poor, placing undue stress on the loads at this end. In many developing countries this has an added complex socioeconomic dimension: the need to have access to affordable electricity supply as a basic human need. The cumulative benign neglect of the past is now coming home to roost in the form of aging distribution infrastructure, that is still operational though it has far exceeded its intended life span. Its ability to survive natural disasters is more because of chance than design. For example, there are underground cables and overhead poles installed in the 1930s that are still in service. Aging conductors of inadequate capacity (from the current demand point of view) are still supplying power, but with poor performance. It is amazing that these major components have survived this long. The conservativeness of the design and operation may have prolonged their life expectancy. The question that naturally arises is: How long will they survive and at what cost to the utilities, or for that matter, to the world citizenry?

1.4 The Future

The outlook for the future is not all doom and gloom. Many positive changes have been rapidly occurring during the past decade, perhaps due to the deregulation (or reregulation) of the industry. We have become increasingly dependent on electricity being a necessity for our existence. It is also the backbone for future economic development if we are committed to improving the quality of life for all mankind. We have witnessed electric power systems becoming larger and more complex in the past 60 years due to the unprecedented growth in the demand for electricity coupled with the population growth and with the higher standards demanded by society. Distribution systems are no exception. Globally, these lower voltage power systems are facing intense competition, with tremendous challenges to cover the ground of past neglect and to deliver cost-effective electric supply, while meeting ever-increasing customer expectations. Globalization is yet another factor to keep in mind when designing, planning, and operating distribution systems of the future.

1.5 New Developments

On the technology front, the penetration of new technologies and materials for efficient distribution systems, including distributed generation, and the availability of efficient computation and analysis tools provide the encouragement and the impetus to make the distribution systems of the future more efficient and effective. As a result, monitoring, control, protection, and automation of these systems in real time are becoming a reality. Demand management at the consumer level to match the availability of supplies to lower costs is a distinct possibility in such an environment. The distribution community should work now to make these lower voltage systems safer, more secure, and more reliable while meeting the ever-increasing demand with the highest possible performance. The asset management intended to prolong the life of the existing equipment while integrating the new technologies, is receiving increased attention. The overall risk management of resources, including finances, will assist the utilities to utilize them wisely and effectively. Many optimistic trends are emerging as we started our journey into the twenty-first century.

1.6 Epilogue

Distribution systems require regular upgrading and modernization to continue providing quality service to consumers. In countries where the demand for electricity has reached a plateau, systems suffer from aging infrastructure and reliability issues. In countries where the power demand is high, the extension of an upgraded electricity infrastructure in urban areas is becoming a necessity and this will require investment. Reliability of supply in many cases is compounded by the shortage in power supply and by inadequate power delivery systems. The need to stimulate efficiency in investment and operation is a must. The tools to evolve the solutions for the problems and the technology for implementation are available. The cost of implementation will, however, be substantial. Therefore, solutions must be structured for phased implementation to ensure acceptability. Challenging, but interesting, times are faced by engineers contributing to these developments. For further insight into any of the topics, readers are encouraged to go through the recent literature and the reference cited at the end of the chapter. While the basic principles for distribution of electricity are the same throughout the world, there are some differences between North America and the rest of the world with respect to topology and terminology. Some aspects of these differences are highlighted in Table 1.1.

Table 1.1 Common terminologies [1].

North America	Rest of the world
Distribution systems	Distribution networks
Primary distribution	Medium-voltage, high tension
Secondary distribution	Low-voltage, low tension
Consumption: kWh	Units
Topology: radial tree structure	Radial with primary and/or secondary selective
Primary feeder protection: reclosers, fuses	Circuit breakers

1.7 The Electric Power System

Electric power systems have three main building blocks: generation, transmission, and distribution. In terms of capital expended, generation systems have approximately 40%, transmission systems have 20%, and distribution systems have 40% of the total. It is the last block and subsystem, which is closest to the consumers, that is the main focus of this textbook. Table 1.2 shows comparison of transmission and distribution systems based on various characteristics. Figure 1.1 shows the layout of a typical distribution system. The starting point for a distribution system is a distribution substation that steps down the power flowing through it from a transmission (or a subtransmission) level, say 115 kV, to a primary distribution level between 4.16 and 34.5 kV level. The power-handling capability of a distribution substation usually varies from 5 to 25 MVA. A substation may feed two to eight three-phase primary feeders. Several three- or single-phase laterals branch

Table 1.2 Characteristics of distribution and transmission systems [1].

Characteristics	Distribution	Transmission
Topology	Radial	Network or loop
Power	100 MVA or below	Bulk (100–1000 MVA)
Voltage	<69 kV class	>120 kV class
No. of phases	Both 1 and 3	Only 3
Load	Distributed	Concentrated
Unbalance	20–30%	5%
No. of components	10 times more	10 times less
Capital outlay	40%	20%

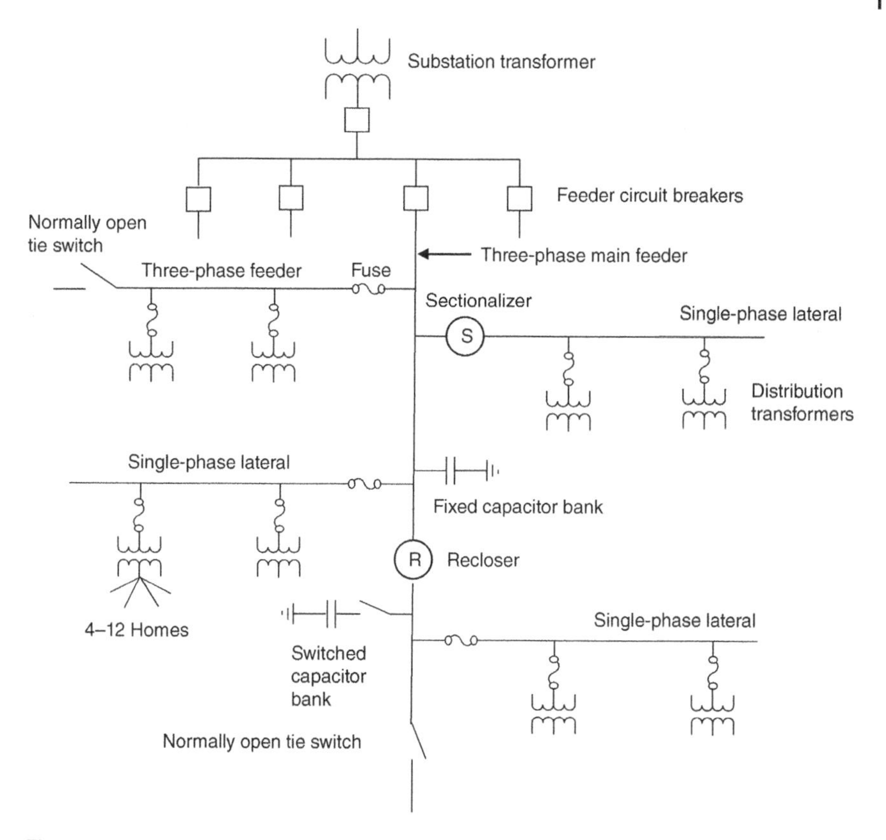

Figure 1.1 Typical distribution system layout.

off from the primary feeders. Depending on the nature of the load, the power is further stepped down to a secondary distribution level of 600 V and less via distribution transformers. The secondary lines feed residential loads, whereas the commercial and industrial loads are predominantly three phase in structure and may be fed directly from the primary level, depending on the connected load and power. Most of the distribution systems around the world have overhead feeders due to cost considerations, but underground feeders have become popular due to aesthetics and higher reliability.

1.8 Distribution System Devices

Distribution systems have a substantial number of devices, all the way from the substation to the service entrance at customer locations. The devices include transformers, switchgear, compensating devices, protection equipment, and control and monitoring devices. In this chapter, we explore these devices.

1.8.1 Substation Devices

Distribution substations are the link between the transmission system and the distribution system. Figures 1.2 and 1.3 show a distribution substation. The right-hand side of Figure 1.2 shows the 115-kV side including the incoming feeder, switching equipment, and busbars. The left-hand side of this figure shows the 12.47-kV side. A power transformer to step down the voltage is in the middle of the figure. Figure 1.3 shows an enlarged view of the 12.47-kV side. Substations are typically located on the periphery of cities, but they can also be inside the cities. Sometimes the distribution substation can be part of a large transmission substation. Distribution substations can have air- or gas-insulated equipment. Modern substations are typically gas insulated because gas reduces the size of equipment and provides additional advantages. Sulfur hexafluoride (SF_6) has been used for substation equipment for many years. However, there are some environmental concerns associated with SF_6, which is prompting scientists to look for alternatives.

1.8.1.1 Power Transformers

Power transformers are large transformers that receive power from the transmission system and reduce the voltage for distribution of power to consumers. These transformers have multiple power ratings, such as 15/20/25 MVA OA/FA/FOA. The rating implies that the transformer will handle up to 15 MVA with cooling provided by convective flow of oil through fins (OA), up to 20 MVA with additional cooling provided by fans circulating air through fins (FA), and up to 25 MVA with cooling aided by forced air as well as forced oil circulation (FOA). These transformers typically have a load tap changer (LTC) on the secondary side to change the low-voltage side voltage up or down, depending on the load on the system.

Figure 1.2 A 115-kV to 12.47-kV step-down distribution substation. The right side of the picture shows the 115-kV equipment, and the left side of the picture shows the 12.47-kV equipment.

Figure 1.3 The 12.47-kV side of the substation showing the transformer, circuit breakers, and the rest of the equipment.

1.8.1.2 Switchgear

Switchgear includes all the devices that are used for opening or closing an electrical path. The most important of them is the circuit breaker, which is designed to open under fault conditions. A circuit breaker can have air, oil, vacuum, or gas as the media. Modern circuit breakers at high voltages are SF_6 based, but most of the circuit breakers in the 15-kV class use vacuum-based interruption mechanism. Load break switches are designed to interrupt a circuit, but their capacity is limited to the maximum expected load on the circuit. They cannot interrupt a circuit under fault conditions. Disconnect switches are purely manual devices that are used to isolate a circuit or component that has already been deenergized.

1.8.1.3 Compensating Devices

Compensating devices are used to adjust the voltage or reactive power flow. The Regulator is like an autotransformer, which can change the output voltage by moving the tap up or down. Capacitors provide reactive power. They can be fixed or switched. Switch capacitors are switched on to provide reactive power in response to voltage, reactive power flow, or temperature, or they can have a fixed time-based switching schedule. Similar to capacitors, reactors can be installed in substations to absorb reactive power. Usually, reactors are not used in distribution substations, but they are sometimes used in transmission substations to compensate for the reactive power of high-voltage transmission lines under light load conditions.

1.8.1.4 Protection Equipment

These are the equipment that respond to abnormal conditions, such as high current due to faults and high voltage caused by circuit switching or lightning. Inverse time overcurrent and instantaneous relays are the most common types of relays deployed in distribution systems. They monitor the current flowing on the distribution feeders emanating from the substation and send trip signals to associated circuit breakers to trip if the current exceeds the threshold values. High-voltage fuses are sometimes used on the high-voltage side of power transformers to protect them from faults. Other protective equipment include surge arresters for limiting voltage on equipment by discharging or bypassing surge current created by switching or lightning. In addition, the substations have static wires, which are at the top of the poles bringing an overhead transmission line into the substation or an overhead distribution line going out of the substation. These wires protect the substation equipment during lightning storms by diverting the lightning surges to the ground.

1.8.1.5 Control and Monitoring Devices

These devices include current transformers (CTs) that reduce the current flowing on the lines to lower values for meters and relays. Similarly, voltage transformers (VTs) or potential transformers (PTs) reduce the high voltage to a lower value for metering and protection. Traditional voltage transformers use the inductive principle to reduce the voltage, but capacitive principle can also be used for reductive voltage. The devices that use this principle are called capacitive voltage transformers (CVTs). Substations also have various transducers to measure different quantities such as ambient temperature, oil temperatures of transformers, and dissolved gases in transformer oil. Since modern substations substantially integrate cyber technology, they have computers and communication links. Microwave, fiber optics, and radio are some of the options for communications. In addition, the substations have remote terminal units (RTUs), which collect information and convey this information to the supervisory control and data acquisition (SCADA) system in the control room through the communication link.

1.8.2 Primary System Components

The primary system consists of feeders that emanate from the substation and go all the way to the distribution transformers. Figure 1.4 shows an example of an overhead primary distribution feeder along a city street. These feeders can have different configurations and conductor types. They also have different associated devices. In this section, we present a brief overview of these devices.

Figure 1.4 A 12.47-kV primary feeder.

1.8.2.1 Feeders and Laterals

Feeders and laterals can be overhead or underground. Overhead feeders have bare conductors mounted on poles. These conductors can be copper, aluminum, or aluminum conductor steel reinforced (ACSR). ACSR conductors are the most used bare wires for overhead distribution feeders. Underground feeders use insulated cables with copper or aluminum conductors and are insulated with ethylene propylene rubber (EPR) or cross-linked polyethylene (XLPE) polymeric insulation. A third option for feeders are the tree wires, which are copper or aluminum conductors coated with insulation for overhead feeders. They are used in areas with dense vegetation with high likelihood of contacts with trees.

1.8.2.2 Switches

Primary feeders, which are mainly three phase, have different types of switches installed on them to provide operating flexibility. Manual switches are installed to isolate a part of the feeder for maintenance. They can only operate on a feeder that is deenergized. Sectionalizing switches or sectionalizers are devices that isolate parts of a feeder under fault conditions. They do not have fault-clearing capabilities but operate in conjunction with reclosers, which have fault-clearing capabilities. Sectionalizers have a counting mechanism to count the number of times the recloser has operated, and after a predetermined number of operations they open when the recloser is open to disconnect the downstream part of the feeder. In automated systems, advanced sectionalizing switches with communication capabilities are being deployed. Communication capabilities allow the automated switches to precisely locate the fault to isolate the faulted part from the rest of the system. Three-phase main feeders also have a switch at the end of the feeder, which is called the tie switch, which is normally open. This switch can be manual or automated. In legacy systems, these switches were typically manual; but in modern systems, they are automated to provide higher flexibility for system reconfiguration.

1.8.2.3 Compensating Devices

Similar to substations, feeders also have compensating devices. The most common are capacitors and voltage regulators. Capacitors are installed at strategic locations to inject reactive power to maintain proper voltage on the feeders under changing load conditions. Capacitors can be switched based on local control using time, temperature, voltage, reactive power, or a combination of them, or they can be switched in coordination with other capacitors and devices in the system. Regulators are like autotransformers and are used on exceedingly long feeders to boost voltage under heavy load conditions.

1.8.2.4 Protection Equipment

Fuses are the most common protection equipment used on distribution feeders. Every lateral branching off the main feeders has a fuse to protect it. Also, the main feeder can have fuses in certain situations. They are also used to protect distribution transformers. Fuses are very inexpensive and have provided reliable protection for over a century. A disadvantage with them is that they must be manually replaced. One can argue that they should be replaced by automated protective devices. However, the cost advantage they offer outweighs any benefits an automated protective device would provide. Hence, they will continue to be used for protection of downstream portions of distribution systems. Reclosers are like circuit breakers and have fault-clearing capabilities. Unlike circuit breakers, which depend on separate relays to initiate operation during faults, reclosers have their own fault-detection mechanism. However, reclosers are smaller in size and are mounted on top of poles in overhead feeders. They also have a reclosing feature, which allows them to reclose a selected number of times before locking out. Since many faults in distribution systems are temporary, this feature permits fuse saving for such faults. The feeders also have surge arresters and static lines, which have the same functions as those for substations.

1.8.2.5 Control and Monitoring Devices

Legacy distribution systems had very little control and monitoring devices on the feeders. But emerging distribution systems with automation have larger proliferation of such equipment. Monitoring devices include current transformers (CTs), voltage transformers (VTs), transducers, and RTUs. Automated distribution systems also have an overlay of communication network for communication between different devices and the control center to make optimal operating decisions in real time.

1.8.2.6 Distribution Transformers

Distribution transformers are at the end of the primary distribution system. They reduce the voltage to utilization level for distribution to the customers. They are pole mounted for overhead systems and pad mounted for underground systems. They are single phase for residential customers but can be three phase for commercial and industrial customers, depending on the size of the load. A single-phase transformer typically feeds one to eight residential customers through the secondary part of the system.

1.8.2.7 Types of Primary Systems

Primary systems can be public or private. Typically, large consumers have their own primary distribution system, which is connected to the local utility. For example, the process industry, such as Boeing at Everett, WA, has a load higher than 10 MVA, and University of Washington in Seattle, has load higher than 40 MVA. While the focus on this book is not on terrestrial power systems, ships, aircrafts, and space use power systems that are different. Ships use three-phase distribution at 60 Hz, but aircraft and space systems use 400 Hz.

1.8.3 Secondary System Components

Secondary systems connect the distribution transformers to service entrance in homes and businesses. For an overhead system, triplex cable provides this connection, and for an underground system, aluminum cable is used for this connection. Service entrance has a meter to record energy consumption. Smart meters, which are prevalent now, allow remote metering capabilities with the ability to meter energy consumption over a 15-minute period. They can also have capability to report loss of power to the utilities. This feature is useful for locating outages when enough meters report loss of power. In addition, the customers are installing their own devices for generation and storage of energy. For residential customers, rooftop solar photovoltaic (PV) and battery are a viable option. Industrial and commercial customers can typically have co-generation (1–25 MW): solar sources (100 kW to 25 MW), wind parks (100 kW to 25 MW), batteries (1–25 MW), and fuel cells (1–25 MW).

1.9 Frequently Asked Questions on Distribution Systems [1]

While we explore many issues related to distribution systems in this book, it is worth considering the following questions:

(1) Should secondary distribution systems be single or three phase?
(2) What voltage should be used for primary distribution?
(3) Should a unit or modular substation or a gas-insulated substation (GIS) be used instead of a conventional substation?
(4) How do we judge the economics of installing voltage regulators, capacitors, and automating the system?
(5) Can distribution systems be designed optimally?
(6) Can distribution systems be designed, planned, and operated automatically with the help of computers?
(7) How can distribution systems be protected effectively?
(8) How can distribution systems be planned and operated with dispersed storage and distributed energy resources (DER)?
(9) How are energy losses evaluated in a distribution system?
(10) What are the effective methods for load forecasting?
(11) How can the highest level of power quality be delivered to the customers?
(12) What will be the impact of electric vehicles (EV) on distribution systems?
(13) Can real-time pricing and electricity markets be implemented in distribution systems?

Reference

1 Venkata, S. S., Pahwa, A., Brown, R. E. and Christie, R. D. "What Future Distribution Engineers Need to Learn," *IEEE Transactions on Power Systems*, Vol. 19, No. 1, February 2004, pp. 17–23.

2

Distribution System Transformers

2.1 Definition

Transformers with a rating of 1000 kVA or less are classified as distribution transformers. Those with larger than 1000 kVA are grouped as power transformers. In a distribution system, distribution transformers are those that are located close to the loads and are used to reduce voltage to the utilization level, such as 120/240 V. Power transformers are those that are located in substations. In some situations, for large industrial or commercial loads, power transformers can be deployed close to the loads.

2.2 Types of Distribution Transformers

One way to classify distribution transformers is based on the medium employed for cooling and insulation. In a broader sense, they are either the dry type or liquid filled. The dry type uses air as the medium, and is most commonly used for commercial or light industrial applications. Liquid-filled transformers use oil. They are commonly used for residential service. They are pole mounted for overhead feeders and pad mounted or underground for underground feeders.

2.2.1 Overhead Transformers

A majority of transformers deployed in overhead distribution feeders are conventional or do not have any built-in protection. They may have a fuse on the primary side for protection against overload and faults. Completely self-protecting (CSP) transformers have fault, lightning, and overload protection built in them. Another variety of transformers called completely self-protecting banked (CSPB) have all the protection features but also allow secondaries to be paralleled.

Electric Power and Energy Distribution Systems: Models, Methods, and Applications, First Edition.
Subrahmanyam S. Venkata and Anil Pahwa.

Companion website: www.wiley.com/go/Pahwa/ElectricPowerDistributionSystems

2.2.2 Underground Transformers

Low-cost residential transformers for underground feeders are conventional with no protection. Subway transformers, which are designed for installation in vaults, can be either conventional or current protected. Transformers that are in locations with high flooding risk are designed with submersible capabilities. Network transformers, which are used in secondary network systems, have built-in protection.

2.3 Standards

Temperature rise in a transformer is dependent on the amount of current it carries and the time duration of that current. Transformers have a long life if the load on them does not exceed the rated load. However, they can carry current up to twice the rated current, but currents over the rated current accelerate aging and cause loss of life. Therefore, overload is permitted only for short durations under emergencies. The larger the current, the higher the loss of life. Hence, the allowed time for overload is reduced with increase in load to keep the oil and winding temperature rise within allowable limits. Name plate ratings typically specify temperature rise rating based on these standards in addition to volt-ampere (VA) rating, voltage ratings, and percent impedance.

2.3.1 Loading of Transformers

Transformers are assigned a kVA rating. According to standards, the transformer can continuously carry a current corresponding to this kVA without exceeding an average winding temperature of 55 or 65 °C [1, 2] above the ambient temperature. Insulation for older transformers was designed with 55 °C temperature rise, but the new transformers typically have insulation suitable for 65 °C temperature rise. The nameplate rating also serves a useful commercial purpose by specifying the kVA at which guaranteed losses and regulation must be met. In service, however, a distribution transformer is rarely loaded continuously at its rated kVA but usually goes through a daily load cycle characterized by a short-time peak load.

The primary consideration of loading which determines the life of a particular transformer is the deterioration of insulation during its service life. The rate of deterioration is greatly influenced by the temperature to which the insulation is subjected. For insulation used in liquid-immersed transformers, the standards require hot-spot temperature (the highest temperature on the windings) to be limited to 110 °C for 65 °C temperature rise transformers and 95 °C for 55 °C temperature rise transformers for normal life expectancy [1]. One solution to the cyclic loading problem would be to limit the peak load of the transformer to nameplate rating. However, this would result in uneconomical use of the transformer loading capability.

There are two primary characteristics of the transformer that permit short-time peak overloads to be carried without decreasing the expected life. The first one of these is the relatively long thermal time constant of the transformer. While the load on the transformer can increase very rapidly, the oil temperature increases much more gradually along an exponential curve with a time constant in the order of a few hours. The temperature differential between winding and oil increases to its ultimate value much more quickly, but the total winding temperature is held down by the oil. The magnitude and duration of the overload which can be carried without exceeding 110 or 95 °C degree hot-spot temperature depend on the ambient temperature, initial loading, loss ratio, etc. Procedures for calculating the temperature transients are given in the literature [3]. The second factor that aids in carrying peak overloads is the thermal aging characteristics of insulation used in distribution transformers. Temperatures considerably above 110 °C (95 °C) can be carried for short periods of time without decrease in normal life expectancy, if this condition is offset by extended operation at temperatures below 110 °C (95 °C). In other words, the elevated temperatures do not cause failure of insulation but only increase the rate at which deterioration occurs.

2.3.2 Types of Cooling

Usually, all distribution transformers are self-cooled. Forced-oil and forced-air cooling are invariably used for power transformers, which are used in substations. The basic types of cooling are referred to by different designations.

2.3.2.1 OA – Oil-Immersed Self-Cooled

In this type of transformer, the insulating oil circulates by natural convection within a tank having smooth sides, corrugated sides, integral tubular sides, or detachable radiators. Smooth tanks are used for small distribution transformers, but because the losses increase more rapidly than the tank surface area as kVA capacity goes up, a smooth tank transformer larger than 50 kVA would have to be abnormally large to provide sufficient radiating surface. Integral tubular-type construction is used up to about 3000 kVA and in some cases to larger capacities, though shipping restrictions usually limit this type of construction for the larger ratings. Above 3000 kVA, detachable radiators are usually supplied. Transformers rated 46 kV and below may also be filled with Inerteen fire-proof insulating liquid, instead of oil. The OA transformer is a basic type and serves as a standard for rating and pricing other types.

2.3.2.2 OA/FA – Oil-Immersed Self-Cooled/Forced-Air Cooled

This type of transformer is basically an OA unit with the addition of fans to increase the rate of heat transfer from the cooling surfaces, thereby increasing the permissible transformer output. The OA/FA transformer is applicable in

Table 2.1 OA and FA ratings of transformers.

OA rating (kVA)	FA rating (kVA)
2500 kVA and below	1.15× kVA (OA)
2501–9999 kVA single phase or 11,999 kVA three phase	1.25× kVA (OA)
10,000 kVA single phase or 12,000 kVA three phase and above	1.333× kVA (OA)

situations that require short-time peak loads to be carried recurrently, without affecting the normal expected transformer life. This transformer may be purchased with fans already installed, or it may be purchased with the option of adding fans later. The higher kVA capacity attained by the use of fans is dependent on the self-cooled rating of the transformer and may be calculated as shown in Table 2.1. These ratings are standardized and are based on a hottest spot winding temperature rise of 65 °C.

2.3.2.3 OA/FA/FOA – Oil-immersed Self-Cooled/Forced-Air Cooled/Forced-Oil Forced-Air Cooled

The rating of an oil-immersed transformer may be increased from its OA rating by the addition of some combination of fans and oil pumps. Such transformers are normally built in the range of 10,000 kVA (OA) single phase or 12,000 kVA (OA) three phase and above. Increased ratings are defined in two steps, 1.333 and 1.667 times the OA rating, respectively. Recognized variations of these triple-rated transformers are the OA/FA/FA and the OA/FA/FOA types. Automatic controls responsive to oil temperature are normally used to start the fans and pumps in a selected sequence as transformer loading increases.

2.3.2.4 FOA – Oil-Immersed Forced-Oil Cooled with Forced-Air Cooled

This type of transformer is intended for use only when both oil pumps and fans are operating. Under these conditions the transformer may carry any load up to full rated kVA. Some designs are capable of carrying excitation current with no fans or pumps in operation, but this is not universally true. Heat transfer from oil to air is accomplished in external oil-to-air heat exchangers.

2.3.2.5 OW – Oil-Immersed Water Cooled

In this type of water-cooled transformer, the cooling water runs through the coils of a pipe which are in contact with the insulating oil of the transformer. The oil flows around the outside of these pipe coils by natural convection, thereby effecting the desired heat transfer to the cooling water. This type has no self-cooled rating.

2.3.2.6 FOW – Oil-Immersed Forced-Oil Cooled with Forced-Water Cooled

External oil-to-water heat exchangers are used in this type of unit to transfer heat from oil to cooling water; otherwise, the transformer is similar to the FOA type.

2.3.2.7 AA – Dry-Type Self-cooled

Dry-type transformers, available at voltage ratings of 15 kV and below, contain no oil or other liquid to perform insulating and cooling functions. Air is the medium that surrounds the core and coils, and cooling must be accomplished primarily by air flow inside the transformer. The self-cooled type is arranged to permit circulation of air by natural convection.

2.3.2.8 AFA – Dry-Type Forced-Air Cooled

This type of transformer has a single rating, based on forced circulation of air by fans or blowers.

2.3.2.9 AA/FA – Dry-Type Self-cooled/Forced-Air Cooled

This design has one rating based on natural convection and a second rating based on forced circulation of air fans or blowers.

2.3.3 Terminal Markings and Polarity

As shown in Figure 2.1, usually the high-voltage terminals of a transformer core are marked as $H_1, H_2, \ldots$, and low-voltage terminals are labeled as $X_1, X_2, \ldots$ Note the dot marking (•), which specifies the polarity of voltages on the windings.

2.3.4 Insulation Class

The insulation class of a transformer defines the dielectric test the unit can withstand, which is usually defined as a number in kV. The number corresponds to the

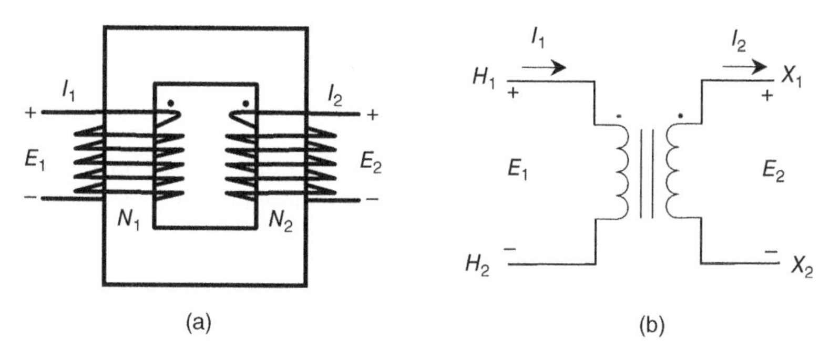

Figure 2.1 Two-winding diagram (a) and schematic (b) of a single-phase transformer [6].

maximum-rated voltage between terminals for phase-to-phase connection of the highest rated voltage that falls within the particular insulation class. For example, a transformer for connection in either wye or delta on a 69-kV system would be in the 69-kV insulation class. An exception to the rule stated above applied to single-phase transformers with voltage ratings 8.66 kV and below. These are insulated for voltages corresponding to the wye connection. Hence, for transformers with delta connection, classification in one class higher is necessary.

Dielectric test consists of impulse, applied potential, or induced voltage tests. The basic insulation level (BIL) is the highest standard impulse wave (1.2 × 50 ms) that a transformer can withstand.

2.4 Single-Phase Transformer

Figure 2.1 shows two-winding diagram and schematic of a single-phase transformer.

The fundamental governing equations of voltage and current for ideal single-phase transformer are described next. The voltage ratio is directly proportional to turns ratio, or

$$^{N_1}/_{N_2} = {}^{E_1}/_{E_2} \tag{2.1}$$

This implies that volts per turn is constant, that is

$$^{E_1}/_{N_1} = {}^{E_2}/_{N_2} \tag{2.2}$$

For an ideal transformer, there is no power loss, and power invariance implies that

$$\boldsymbol{E}_1\boldsymbol{I}_1^* = \boldsymbol{E}_2\boldsymbol{I}_2^* \tag{2.3}$$

Hence, currents are inversely proportional to turns ratio

$$^{I_1}/_{N_1} = {}^{I_2}/_{N_2} \tag{2.4}$$

2.4.1 Model for a Single-Phase Transformer

Since practical transformers are not ideal, model of a single-phase transformer shown in Figure 2.2 accounts for various resistances and inductances associated with it.

where

$$R_s' = \left({}^{N_1}/_{N_2}\right)^2 R_s = a^2 R_s \tag{2.5}$$

$$X_s' = a^2 X_s \tag{2.6}$$

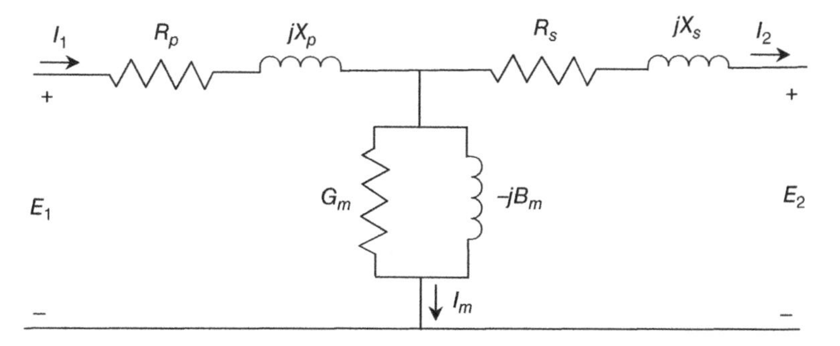

Figure 2.2 Model of a practical single-phase transformer.

Figure 2.3 Simple model of a single-phase transformer.

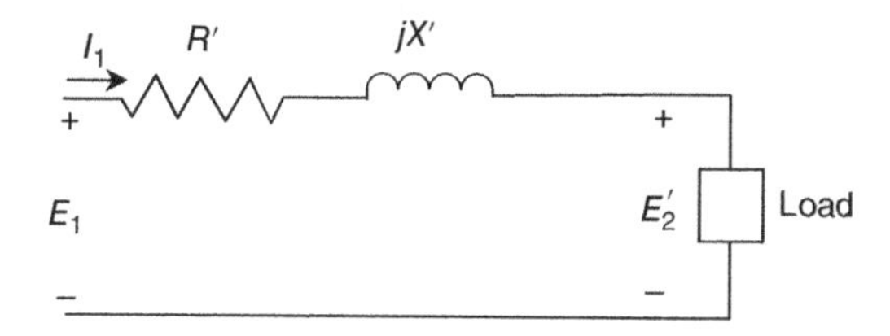

$$E_2' = aE_2 \tag{2.7}$$

$$I_2' = I_2/a \tag{2.8}$$

$$a = \left({}^{N_1}/_{N_2} \right) \tag{2.9}$$

where a is the turns ratio of the windings, and R_s' and X_s' are the resistance and reactance of the secondary side referred to the primary side.

The exciting branch can be neglected for the steady-state performance of transformers. Upon doing that and combining the series impedances, the model reduces to a simple series impedance model as shown in Figure 2.3.

In this model

$$R' = R_p + R_s' \tag{2.10}$$

$$X' = X_p + X_s' \tag{2.11}$$

and the short circuit impedance Z referred to primary is

$$Z = R' + jX' \tag{2.12}$$

The equivalent circuit referred to the secondary side can similarly be drawn. In per unit (pu), the equivalent circuit is the same referred to either primary or secondary.

The name plate rating usually specifies Z in % of the name plate ratings. Then, Z in ohms referred to primary (p) can be obtained

$$Z\,(\text{ohms}) = \frac{Z\,(\%)}{100}\,\frac{(\text{kV}_{\text{rated}_p})^2}{\text{MVA}_{\text{rated}}} \tag{2.13}$$

Westinghouse's Distribution Reference Book [4] lists the preferred voltage and kVA ratings and typical impedances for various distribution transformers. In percent or pu, these values are in a narrow range of 1–6.5% or 0.01–0.065 pu (see Appendix A for details on representing values of power system quantities in pu). The reference also lists no-load losses and total losses that could occur in distribution transformers.

2.4.2 Performance Analysis

Transformers have core losses or iron losses (Fe) and load losses or copper losses (Cu). The core losses are independent of the load and are called no-load losses. On the other hand, copper losses are proportional to the square of the load current.

$$\text{Cu} = I_p^2 R' \tag{2.14}$$

and

$$\text{Losses} = \text{Fe} + \text{Cu} \tag{2.15}$$

$$\text{Efficiency}\,(\eta) = \frac{\text{output}}{\text{input}} = 1 - \frac{\text{Losses}}{\text{input}} \tag{2.16}$$

Since no-load losses are constant and load losses are variable, maximum efficiency will occur only at one particular load. This happens when Fe = Cu. However, for typical transformers under normal operating conditions, the copper losses are 2.5–5 times higher than the iron losses. With loss ratio (r) defined as the ratio of copper to iron losses, the load at which transformer operates most efficiently can be determined.

$$L = \sqrt{\frac{\text{Fe}}{\text{Cu}}} = \frac{1}{\sqrt{r}} \tag{2.17}$$

2.4.3 Regulation

Consider the circuit in Figure 2.3 to find an expression of primary voltage in terms of secondary voltage, current, and transformers parameters.

$$\boldsymbol{E}_1 = \boldsymbol{E}_2' + \boldsymbol{I}_1 \boldsymbol{Z} \tag{2.18}$$

This equation can be depicted in a phasor diagram form as shown in Figure 2.4. It is considered that the load is inductive with current lagging the voltage by angle Ø.

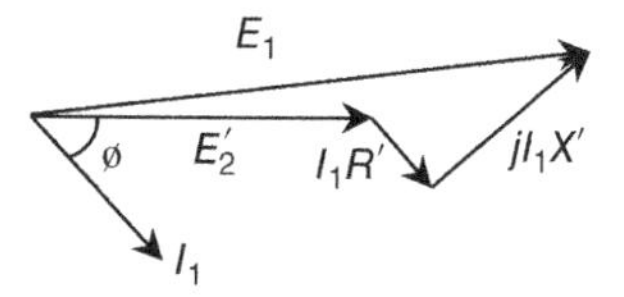

Figure 2.4 Phasor diagram of a simplified transformer model.

Percent regulation is defined as

$$\begin{aligned}\%\text{Regulation} &= \frac{|\boldsymbol{E}_1 - \boldsymbol{E}_2'|}{|\boldsymbol{E}_2'|} \times 100 \\ &= \frac{|\boldsymbol{I}_1\boldsymbol{Z}|}{|\boldsymbol{E}_2'|} \times 100 \\ &= \frac{\sqrt{(I_1R'\cos\emptyset + I_1X'\sin\emptyset)^2 + (I_1X'\cos\emptyset - I_1R'\sin\emptyset)^2}}{|\boldsymbol{E}_2'|} \times 100\end{aligned} \tag{2.19}$$

2.4.4 Taps

The output voltage of a transformer can be changed by changing the turns ratio between the primary and secondary windings, which can be done by tapping the primary winding at various points. Primary winding taps perform two main functions:

1. They can be used to compensate for the voltage drop caused by resistance (R') and leakage reactance (X').
2. They can be used so that a single transformer will operate on primary systems with slightly different voltage levels. For example, a transformer with properly selected taps will operate on either 2160 or 2400 V primary system.

Taps could also exist on the secondary side of a transformer performing similar functions as outlined above.

2.5 Distribution Transformer Connections

A single-phase distribution transformer can have different connections, depending on the type of load it serves. If all the loads on the secondary side have the same voltage, a simple two-winding transformer is used. However, homes in the United States have loads that operate at 120 or 240 V. Large-capacity loads, such as air conditioners and clothes washers, operate at 240 V, and the other loads in the same house operate at 120 V. To facilitate supplying both 120 and 240 V to consumers from a single transformer, single-phase distribution transformers are

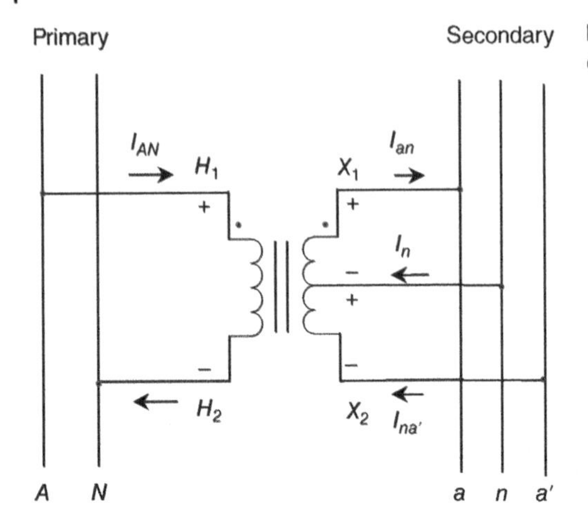

Figure 2.5 Three-wire secondary distribution transformer.

designed with the midpoint of the secondary grounded, as shown in Figure 2.5. This connection is called series, multiple secondary, or three-wire secondary; 7200 V primary to 120/240 V on the secondary is a good example of such a transformer. Typically, loads on the two circuits are distributed to create a balance between the two circuits. However, practically, it is impossible to have a balance under normal operation. In such cases, the current in the neutral is the difference between the currents flowing in the two circuits.

In certain applications, where variable secondary voltages are desired, a booster transformer is used. This transformer is similar to an autotransformer. It has low percent Z, low cost, and high efficiency. One disadvantage is that it has electrical connection between the two sides of the transformer [5]. To increase the capacity, two transformers can be operated in parallel under certain conditions. Specifically, they must have the same turns ratio and must be connected to the same primary phase. Also, to maximize the capability, they should have nearly equal impedances.

2.5.1 Example

Consider a transformer with 120 V loads connected, as shown in Figure 2.6. Under balanced conditions, no neutral current flows, but under unbalanced conditions, a neutral current would flow, sometimes substantial. In fact, the amount of neutral current is a direct measure of unbalance.

Now, consider the load across phases **a** to **n** to be 36 A at 0.95 pf lagging, and the load across phases **a′** to **n** to be 25 A 0.85 pf lagging. Determine the current flow in the neutral wire and the load supplied by each secondary coil of the transformer.

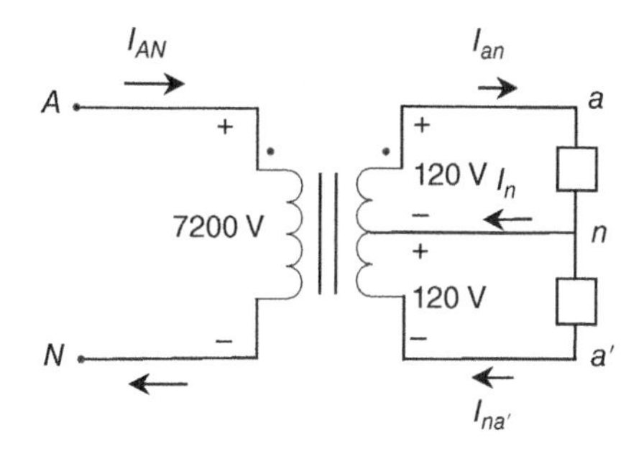

Figure 2.6 Three-wire secondary distribution transformer with 120 V loads.

Also, determine the kVA rating of this transformer such that the thermal rating of it is not exceeded.

Solution

$$\boldsymbol{I}_{an} = 36\angle -18.19 = (34.2 - j11.3)\ \text{A}$$

$$\boldsymbol{I}_{na'} = 25\angle -31.79 = (21.2 - j13.2)\ \text{A}$$

$$\boldsymbol{I}_n = \boldsymbol{I}_{an} - \boldsymbol{I}_{na'} = (34.2 - j11.3) - (21.2 - j13.2) = (13.0 + j1.9)$$

$$= 13.2\angle -8.3\ \text{A}$$

kVA supplied by a–n coil $= 120 \times 36 = 4330$ VA $= 4.33$ kVA

kVA supplied by a′–n coil $= 120 \times 25 = 3000$ VA $= 3$ kVA

The total kVA is 7.33 kVA. Rounding it off to the next higher available rating gives the transformer rating of 10 kVA.

2.5.2 Parallel Operation of Three-wire Transformers

Two single-phase three-wire transformers can be connected in parallel to supply increased load as shown in Figure 2.7, if the following conditions are met:

1. They must have the same primary and secondary voltage ratings and therefore the same turns ratio.
2. They must be connected to the same primary phase.
3. They must be connected to the same secondary phase.

However, they may have different kVA ratings and percent leakage impedance, though not too different from each other. To maximize the total load delivered without exceeding the thermal limits, they should have nearly equal impedances.

Using KVL gives $\boldsymbol{I}_1\boldsymbol{Z}_1 = \boldsymbol{I}_2\boldsymbol{Z}_2$ or $I_1Z_1 = I_2Z_2$, which gives

$$\frac{I_1}{I_2} = \frac{Z_2}{Z_1} \tag{2.20}$$

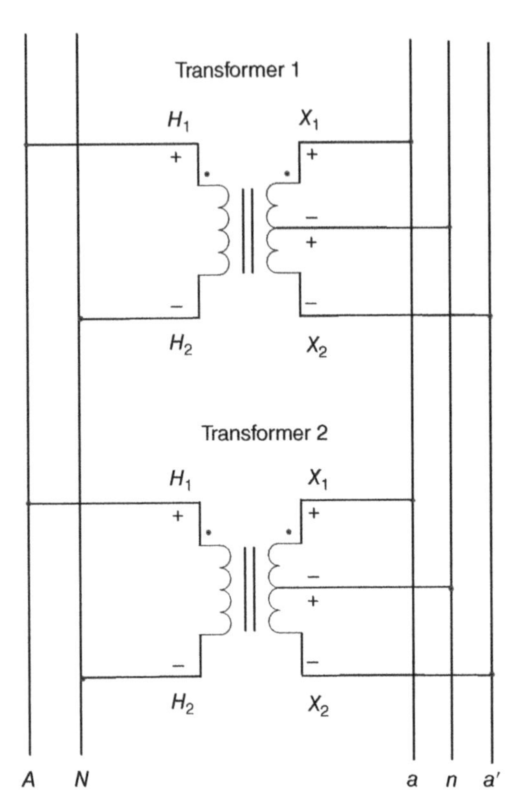

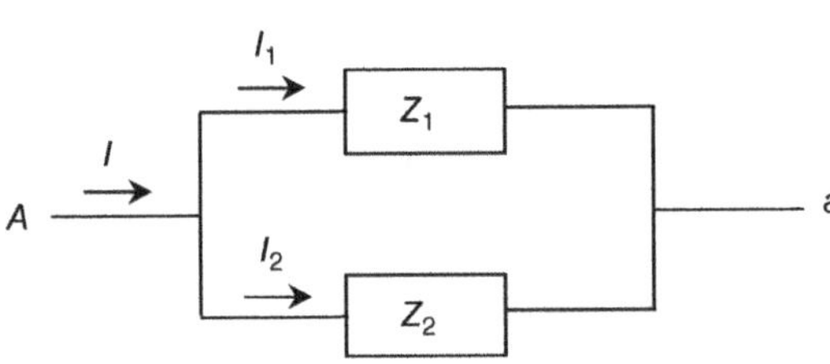

Figure 2.7 Two three-wire secondary distribution transformers connected in parallel.

where $\boldsymbol{Z}_1$ and $\boldsymbol{Z}_2$ are the impedances of transformers 1 and 2, respectively. Let Z_{b1} and Z_{b2} be the base impedances of these transformers computed based on their respective nameplate ratings S_{b1} and S_{b2}.

Therefore,

$$\frac{I_1}{I_2} = \frac{Z_2\%\ Z_{b2}}{Z_1\%\ Z_{b1}} \tag{2.21}$$

$$Z_{b1} = \frac{V_{b1}^2}{S_{b1}} \text{ and } Z_{b2} = \frac{V_{b2}^2}{S_{b2}} \tag{2.22}$$

Since $V_{b1} = V_{b2}$ are equal by requirement of parallel operation,

$$\frac{I_1}{I_2} = \frac{Z_2\%}{Z_{1\%}} \frac{S_{b1}}{S_{b2}} \tag{2.23}$$

Now,

$$\frac{I_1}{I_2} = \frac{I_1}{I_2} \frac{V}{V} = \frac{S_{L1}}{S_{L2}} \tag{2.24}$$

where S_{L1} is kVA supplied by transformer 1, and S_{L2} is kVA supplied by transformer 2.

Hence,

$$\frac{S_{L1}}{S_{L2}} = \frac{Z_2\%}{Z_1\%} \frac{S_{b1}}{S_{b2}} \tag{2.25}$$

Example

Consider that a 50-kVA transformer of 4% impedance is connected in parallel with a 50-kVA transformer of 3% impedance. Find the total load that these transformers can deliver.

Solution

Applying Eq. (2.25) gives

$$\frac{S_{L1}}{S_{L2}} = 0.75$$

if transformer is loaded to its full capacity of $S_{L1} = 50$ kVA and $S_{L2} = \frac{50}{0.75} =$ 66.7 kVA

Note that 66.7 kVA is higher than the second transformer's rating. In order to limit the secondary transformer's loading to 50 kVA, the first transformer's loading has to be limited to 37.5 kVA. Therefore, the total load that could be supplied is 37.5 + 50 = 87.5 kVA instead of 100 kVA. If identical transformers are not available, the best way to attain equal loading is to choose the two transformers with their impedance ratio equal to inverse kVA ratings, which can be verified from Eq. (2.25).

2.5.3 Single-Phase Autotransformers

Consider a two-winding transformer with $N : 1$ turns ratio, kVA rating S, and impedance $\%Z$. Now consider that its windings are connected in series to create an autotransformer to realize the turns ratio $N_A = \frac{N+1}{N}$, as shown in Figure 2.8.

$$Z_A\% = \frac{N_A - 1}{N_A} (Z\%) = \frac{1}{N+1} (Z\%) \tag{2.26}$$

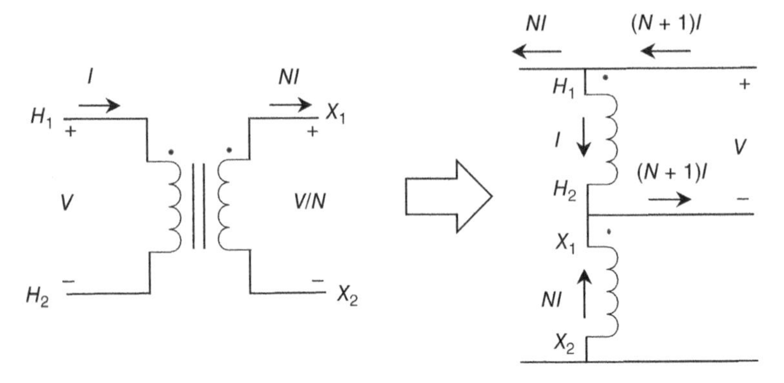

Figure 2.8 A two-winding transformer and its connections to create an autotransformer.

and

$$S_A = V\frac{N+1}{N}(NI) = V(N+1)I = (N+1)S \tag{2.27}$$

Losses in watts are the same for both the cases. However, the efficiency of the autotransformer will be higher, and the voltage regulation will be better.

2.6 Three-Phase Transformer Connections

The following connections are well known and widely used for power transformers in distribution systems:

- **Wye–Delta:** This connection is the most commonly used connection in transmission and distributions systems. It blocks the flow of the third harmonic from one side to the other. It also provides neutral for grounding, which is important for ground-fault protection. In transmission systems, this connection is used with wye connection on the high-voltage side, so as to step up the voltage from generation to transmission, for lower insulation cost. However, in distribution systems, the low-voltage side is connected in wye. This is specifically important because wye connection separates the three phases from one another, which allows distribution of power to consumers with one of the phases and the neutral.
- **Delta–Delta:** This connection is used in three-wire systems without neutral. It also blocks the flow of the third harmonic. Another advantage with this connection is that it can be operated at reduced capacity as an open delta with only two of the windings functioning. However, it lacks a neutral, which creates significant challenges for ground-fault protection.
- **Wye–Wye:** This connection can be used only if the neutral connection is available on both sides. It allows the flow of the third harmonic from one side to other, which is a problem. A delta-connected tertiary winding can solve this

problem. Other problems with it are poor voltage regulation for single-phase loads and neutral inversion (neutral shift outside of the line voltage triangle).

2.6.1 Analysis of Y/Δ Transformer with Unbalanced Load

Usually, such three-phase transformers are made of three, similar single-phase units to realize the desired type of connection. The aim is to analyze the three-phase transformer in the presence of a single-phase lighting type of load. Figure 2.9 shows a transformer with three-phase balanced load as well as a single-phase load connected across phases **a** and **b** on the low-voltage side. We consider an example to compute the high-voltage side currents and kVA rating of each single-phase transformer unit.

$\boldsymbol{I}_A$, $\boldsymbol{I}_B$, and $\boldsymbol{I}_C$ are not balanced due to the single-phase load. Therefore, kVA ratings of each single-phase transformer units are not identical. Since the neutral is not grounded, it is not at zero potential. However,

$$\boldsymbol{I}_A + \boldsymbol{I}_B + \boldsymbol{I}_C = 0 \text{ and } \boldsymbol{I}_N = 0 \tag{2.28}$$

If the neutral is not grounded, $\boldsymbol{I}_N$ will not be zero. Applying the superposition principle, the winding currents on the LV side can be obtained.

Example

Consider a transformer with line-to-line voltage to be 12.47 kV on the HV side and 240 V on the LV side. Let the three-phase load be 75 kVA at 0.9 pf lagging, and the single-phase load be 25 kVA at unity pf. Find kVA load supplied by each phase and pf of each phase from the primary side.

Solution

Three-phase load:

Choose $\boldsymbol{V}_{an}$ as the reference voltage, which gives $\boldsymbol{V}_{ab} = 240\angle 30°$. Now, we can draw a phasor diagram showing all the voltages and current due to the three-phase

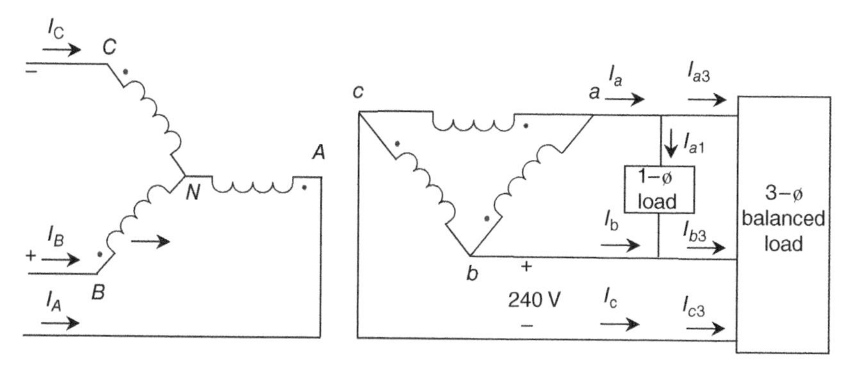

Figure 2.9 A three-phase wye–delta transformer with three-phase balanced and single-phase loads.

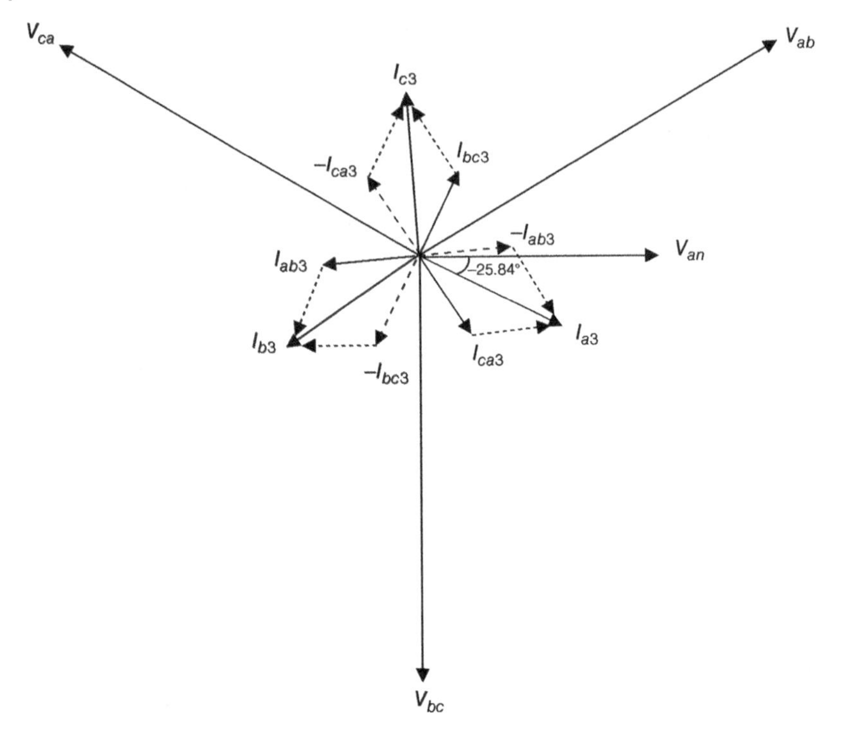

Figure 2.10 Phasor diagram due to three-phase load.

load as shown in Figure 2.10. Hence,

$$\begin{aligned} I_{a3} &= \frac{\text{kVA}_{3\emptyset}}{\sqrt{3}\,V_{ll}} \angle -\cos^{-1}(0.9) \\ &= \frac{75 \times 10^3}{\sqrt{3}\,(240)} \angle -\cos^{-1}(0.9) = 180.42\angle -25.84°\,\text{A} \end{aligned}$$

Due to the balanced nature of the three-phase load, we can obtain I_{b3} and I_{c3}:

$$I_{b3} = 180.42\angle -145.84°\,\text{A and } I_{c3} = 180.42\angle -94.16°\,\text{A}$$

Then, the winding currents can be obtained, or

$$I_{ca3} = \frac{I_{a3}}{\sqrt{3}}\angle -25.84 - 30 = 104.17\angle -55.84°\,\text{A}$$

$$I_{ab3} = 104.17\angle -174.84°\,\text{A, and } I_{bc3} = 104.17\angle 64.16°\,\text{A}$$

Single-phase load:

$$|I_{a1}| = |I_{b1}| = \frac{(kVA)_{1\emptyset}}{V_{ab}}$$

Therefore, $I_{a1} = -I_{b1} = \frac{25\times10^3}{240}\angle 30 = 104.167\angle 30°\,\text{A}$

This current flows into the transformer through terminal "*b*" and gets split into two parts while flowing toward "*a*": two-thirds of it flow directly to "*a*" from "*b*," and one-third of it takes the path "*b*"–"*c*"–"*a*" due to the current division. Therefore,

$$\boldsymbol{I_{ab1}} = -\boldsymbol{I_{ba1}} = -\frac{2}{3}\boldsymbol{I_{a1}} = -69.44\angle 30^\circ \text{A}$$

$$\boldsymbol{I_{bc1}} = \boldsymbol{I_{ca1}} = \frac{1}{3}\boldsymbol{I_{a1}} = 34.72\angle 30^\circ \text{A}$$

Applying the superposition principle, the secondary phase currents of the transformer can be obtained:

$$\boldsymbol{I_{ab}} = \boldsymbol{I_{ab1}} + \boldsymbol{I_{ab3}} = -69.44\angle 30^\circ + 104.17\angle -174.84^\circ = 169.39\angle -165.55^\circ \text{A}$$

$$\boldsymbol{I_{bc}} = \boldsymbol{I_{bc1}} + \boldsymbol{I_{bc3}} = 34.72\angle 30^\circ + 104.17\angle 64.16^\circ = 132.33\angle 55.81^\circ \text{A}$$

$$\boldsymbol{I_{ca}} = \boldsymbol{I_{ca1}} + \boldsymbol{I_{ca3}} = 34.72\angle 30^\circ + 104.17\angle -55.84^\circ = 112.16\angle -37.85^\circ \text{A}$$

Note that these currents are unbalanced. With the transformation ratio of $\frac{12{,}470/\sqrt{3}}{240} = 30$, the primary currents are

$$\boldsymbol{I_A} = \frac{\boldsymbol{I_{ca}}}{30} = 3.74\angle -37.85^\circ \text{A},$$

$$\boldsymbol{I_B} = \frac{\boldsymbol{I_{ab}}}{30} = 5.46\angle -165.55^\circ \text{A, and}$$

$$\boldsymbol{I_C} = \frac{\boldsymbol{I_{bc}}}{30} = 4.47\angle 55.81^\circ \text{A}$$

The kVA supplied by each phase is

$$S_A = V_{AN} I_A = 7200 \times 3.74 = 26.92 \text{ kVA}$$

$$S_B = V_{BN} I_B = 7200 \times 5.46 = 39.34 \text{ kVA}$$

$$S_C = V_{CN} I_C = 7200 \times 4.47 = 32.22 \text{ kVA}$$

The power factor of each phase as seen from the Y side is

$$\cos(\angle \boldsymbol{V_{AN}} - \angle \boldsymbol{I_A}) = \cos(-7.85^\circ) = 0.9906 \text{ lagging}$$

$$\cos(\angle \boldsymbol{V_{BN}} - \angle \boldsymbol{I_B}) = \cos(-15.55^\circ) = 0.9934 \text{ lagging}$$

$$\cos(\angle \boldsymbol{V_{CN}} - \angle \boldsymbol{I_C}) = \cos(-34.18^\circ) = 0.8272 \text{ lagging}$$

2.6.2 Analysis of Y/Y Transformer

This connection scheme is seldom used with isolated neutrals due to the instability caused by

(a) **Magnetizing currents:** Unbalances in magnetizing current could occur due to differences in the iron characteristics of the three single-phase transformer cores or in the shell type of core. If the transformers are of three-phase core type, then the unbalance is minimum. If the transformer neutral is isolated, the resulting unbalance is of minimum importance, especially since line-to-line voltages are unaffected by it. If the transformer neutral is grounded, the asymmetry is impressed on the line capacitance to ground and is enhanced when the line charging current is comparable to the magnetizing current of the transformer.

(b) **Third-harmonic currents:** Third-harmonic magnetizing current is suppressed if the neutrals are ungrounded. However, the third-harmonic voltages manifest across the line-to-neutral load on the secondary side, though they do not appear across line-to-line voltages. In single-phase transformers, these third-harmonic components could be as high as 50%. Thus, the resultant line-to-neutral voltage becomes

$$V_{ln} = \sqrt{1 + \left(\frac{1}{2}\right)^2} = 1.12\,\text{pu} \tag{2.29}$$

or 12% higher and distorted.

(c) **Line-to-neutral load:** Line-to-neutral or single-phase loads could cause neutral instability in ungrounded Y–Y transformer schemes. The line-to-neutral voltage effectively goes to zero on an ungrounded transformer scheme. If the secondary side is grounded, then excessive voltage drops could occur if single-phase or shell-type transformers are used, thus resulting in poor voltage regulation.

Example

Consider a transformer rated 500 kVA, 12.47 kV/208 V. There is a balanced three-phase load of 400 kVA at 0.94 pf lagging and a load of 100 kVA at 0.85 pf lagging across phases **b** and **c** on the low-voltage side of the transformer as shown in Figure 2.11. Find $\boldsymbol{I_A}$, $\boldsymbol{I_B}$, $\boldsymbol{I_C}$, $\boldsymbol{I_N}$, and $\boldsymbol{I_n}$.

Solution

We first find $\boldsymbol{I_a}$, $\boldsymbol{I_b}$, and $\boldsymbol{I_c}$ due to the three-phase load alone; then, we find these currents due to the single-phase load only. The resultant currents will be evaluated using the superposition principle.

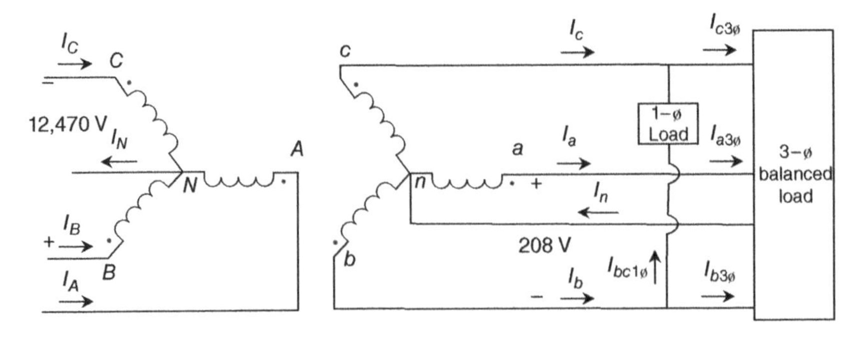

Figure 2.11 Y/Y transformer with unbalanced loading.

Three-phase load:

The magnitudes of the currents on the secondary side are

$$I_{a3\emptyset} = I_{b3\emptyset} = I_{c3\emptyset} = \frac{400 \times 10^3}{\sqrt{3}\ (208)} = 1110.26\ \text{A}$$

Assuming $\boldsymbol{V_{an}}$ to be the reference, or $\boldsymbol{V_{an}} = 208\ \angle 0°$, we can find the angles of currents:

$$\boldsymbol{I_{a3\emptyset}} = 1110\angle\cos^{-1}(0.94) = 1110.26\angle -19.95°\text{A}$$

$$\boldsymbol{I_{b3\emptyset}} = 1110.26\angle -139.95°\text{A}$$

$$\boldsymbol{I_{a3\emptyset}} = 1110.26\angle -259.95°\text{A}$$

Single-phase load:

$$I_{bc1\emptyset} = \frac{100 \times 10^3}{(208)} = 480.77\ \text{A}$$

Since $\boldsymbol{E_{bc}}$ lags $\boldsymbol{E_{an}}$ by 90°,

$$\boldsymbol{I_{bc1\emptyset}} = 480.77\angle(-90° - \cos^{-1}(0.85)) = 480.77\angle -121.79°\text{A}$$

Also, $\boldsymbol{I_{bc1\emptyset}} = -\boldsymbol{I_{c1\emptyset}} = 480.77\ \angle\ -121.79\,°\text{A}$

Therefore, the resultant currents are

$$\boldsymbol{I_a} = 1110.26\angle -19.95°\text{A}$$

$$\boldsymbol{I_b} = \boldsymbol{I_{b3\emptyset}} + \boldsymbol{I_{b1\emptyset}} = 1110.26\angle -139.95° + 480.77\angle -121.79°$$
$$= 1574.23\angle -134.49°\text{A}$$

$$\boldsymbol{I_c} = \boldsymbol{I_{c3\emptyset}} + \boldsymbol{I_{c1\emptyset}} = 1110.26\angle -259.95° - 480.77\angle -121.79°$$
$$= 1503.06\angle 87.71°\text{A}$$

$$\boldsymbol{I_n} = \boldsymbol{I_a} + \boldsymbol{I_b} + \boldsymbol{I_c} = 1110.26\angle -19.95° + 1574.23\angle -134.49°$$
$$+ 1503.06\angle 87.71°\text{A} = 0$$

The turns ratio of the transformer is 12470 : 208 or 59.95 : 1.

Hence, the currents on the primary side can be computed:

$$\boldsymbol{I_A} = \frac{1110.26}{59.95}\angle -19.95°\text{A} = 18.52\angle -19.95°\text{A}$$

$$\boldsymbol{I_B} = 26.26\angle -134.49°\text{A}, \boldsymbol{I_C} = 25.07\angle 87.71°\text{A}, \text{and } \boldsymbol{I_N} = 0$$

2.6.3 Three-winding Transformer

Example Three-winding transformers are used in special situation, such as auxiliary supply. For example, the transformer shown in Figure 2.12 has the secondary side rated at 13.09 kV (L–L) and the tertiary side rated at 2.4 kV (L–L). A single-phase load of 5 MVA of 0.95 pf lagging is connected across phase **a** and the neutral of the secondary winding. The goal is to determine the currents on the primary side as well as in the tertiary windings with $\boldsymbol{V_{an}}$ as the reference voltage.

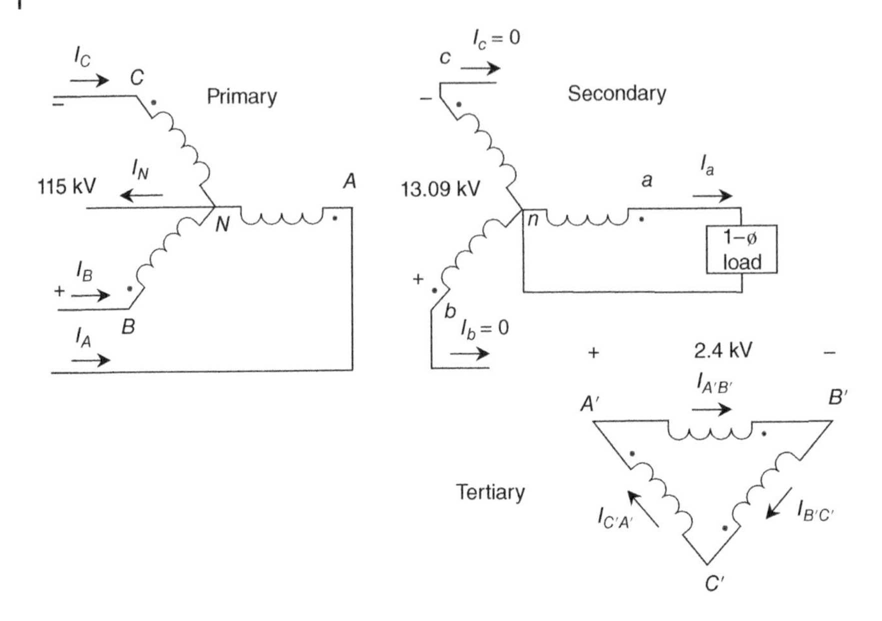

Figure 2.12 A three-winding transformer with single-phase load.

Solution

$$\boldsymbol{I_a} = \frac{5000\ \text{kVA}}{\left(13.09\ \text{kV}/\sqrt{3}\right)}\angle -\cos^{-1} 0.95 = 661.51\angle -18.19°\text{A}$$

$$\boldsymbol{I_b} = \boldsymbol{I_c} = \boldsymbol{0}$$

For flux balance

$$-N_s\boldsymbol{I_a} - N_T\boldsymbol{I_{A'B'}} + N_p\boldsymbol{I_A} = \boldsymbol{0} \tag{2.30}$$

or

$$-\boldsymbol{I_a} - \left(\frac{N_T}{N_s}\boldsymbol{I_{A'B'}}\right) + \left(\frac{N_p}{N_s}\boldsymbol{I_A}\right) = \boldsymbol{0} \tag{2.31}$$

Similarly,

$$-\boldsymbol{I_b} - \left(\frac{N_T}{N_s}\boldsymbol{I_{B'C'}}\right) + \left(\frac{N_p}{N_s}\boldsymbol{I_B}\right) = \boldsymbol{0} \tag{2.32}$$

and

$$-\boldsymbol{I_c} - \left(\frac{N_T}{N_s}\boldsymbol{I_{C'A'}}\right) + \left(\frac{N_p}{N_s}\boldsymbol{I_C}\right) = \boldsymbol{0} \tag{2.33}$$

All the currents circulating in the tertiary windings are equal. Therefore,

$$\boldsymbol{I_{A'B'}} = \boldsymbol{I_{B'C'}} = \boldsymbol{I_{C'A'}} \tag{2.34}$$

For the specified voltages in this example

$$\frac{N_T}{N_S} = \frac{2.4\ \text{kV}}{7.558\ \text{kV}} = 0.317 \text{ and } \frac{N_P}{N_S} = \frac{66.395\ \text{kV}}{7.558\ \text{kV}} = 8.7847 \tag{2.35}$$

Upon substituting the known numerical values into (2.31), (2.32), and (2.33), we get

$$-661.51\angle -18.19^\circ\ -0.3175\boldsymbol{I}_{A'B'} + 8.7847\,\boldsymbol{I}_A = \boldsymbol{0} \tag{2.36}$$

$$-0.3175\,\boldsymbol{I}_{A'B'} + 8.7847\,\boldsymbol{I}_B = 0 \tag{2.37}$$

$$-0.3175\,\boldsymbol{I}_{A'B'} + 8.7847\,\boldsymbol{I}_C = 0 \tag{2.38}$$

Therefore,

$$\boldsymbol{I}_B = \boldsymbol{I}_C \tag{2.39}$$

Since $\boldsymbol{I}_A + \boldsymbol{I}_B + \boldsymbol{I}_C = \boldsymbol{0}$, we get

$$\boldsymbol{I}_A = -\boldsymbol{2}\,\boldsymbol{I}_B = -\boldsymbol{2}\,\boldsymbol{I}_C \tag{2.40}$$

Substituting this in (2.36) gives

$$\boldsymbol{I}_{A'B'} = -13.8342\,\boldsymbol{I}_A \tag{2.41}$$

Further substituting this in (2.36) gives

$$\boldsymbol{I}_A = 50.25\angle -18.19^\circ\text{A, and}$$

$$\boldsymbol{I}_B = \boldsymbol{I}_C = -25.125\angle -18.19^\circ\text{A}$$

Also,

$$\boldsymbol{I}_{A'B'} = \boldsymbol{I}_{B'C'} = \boldsymbol{I}_{C'A'} = -694.50\angle -18.19^\circ\text{A}$$

Problems

2.1 In Figure 2.13, the load from **a** to **n** is 1.2 kW with 0.95 pf, from **a′** to **n** is 1.5 kW with 0.92 pf, and from **a** to **a′** is 5 kW with 0.9 pf. Find the current flowing in lines connected to **a**, **a′**, and **n**. Also, find the currents on the two windings of the transformer on the secondary side. What is the desired rating of the transformer for this condition?

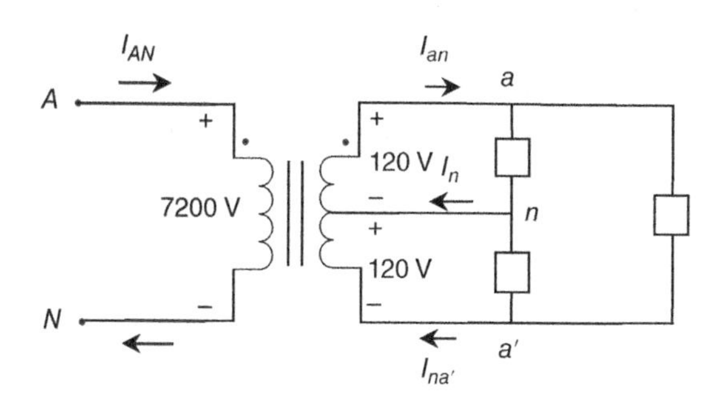

Figure 2.13 Distribution transformer serving single-phase loads.

2.2 A single-phase 25 kVA transformer of 4% impedance is connected in parallel with a 25-kVA transformer of 5% impedance. Find the total kVA load these transformers can supply without overloading any one of them.

2.3 Consider a three-phase wye–delta transformer shown in Figure 2.9. The three-phase line-to-line voltage is 12.47 kV on the HV side and 240 V on the LV side. The three-phase load is 100 kW at 0.8 pf lagging, and a single-phase load of 20 kVA at 0.9 pf lagging is connected between phases **b** and **c**. Find I_{ab}, I_{bc}, I_{ca}, I_A, I_B, and I_C, the kVA rating of each of the transformer windings, and the power factor of each phase on the primary side of the transformer.

2.4 Consider a three-phase wye–wye transformer rated 300 kVA and 12.47 kV/208 V as shown in Figure 2.11. There is a balanced three-phase load of 200 kVA at 0.9 pf lagging and a load of 100 kVA at 0.8 pf lagging across phases **c** and **a** on the low-voltage side of the transformer. Find I_A, I_B, I_C, I_N, and I_n.

2.5 A three-winding transformer shown in Figure 2.12 has a secondary side rated at 12.47 kV (L–L) and a tertiary side rated at 3.3 kV (L–L). A single-phase load of 3 MVA of 0.9 pf lagging is connected across phase **b** and the neutral of the secondary winding. Determine the currents on the primary side as well as in the tertiary windings, with V_{an} as the reference voltage.

References

1 ANSI/IEEE C57.91.1981 American National Standard (1981). *Guide for Loading Mineral-Oil-Immersed Overhead and Pad-Mounted Distribution Transformers Rated 500 kVA and Less with 65°C or 55°C Average Winding Rise*. New York: IEEE.

2 ANSI/IEEE C57.92.1981 American National Standard (1981). *Guide for Loading Mineral-Oil-Immersed Power Transformers Up to and Including 100 MVA with 55°C or 65°C Average Winding Rise*. New York: IEEE.

3 Hobson, J.E., Witzke, R.L., and Williams, J.S. (1950). Power transformers and reactors. In: *Electrical Transmission and Distribution Reference Book* (ed. A.C. Monteith and C.F. Wagner). East Pittsburgh, PA: Westinghouse Electric Corporation.

4 Dillard, J.K. (1959). *Distribution Systems: Electric Utility Engineering Reference Book*. East Pittsburgh, PA: Westinghouse Electric Corporation.

5 General Electric (1958). *General Electric Distribution Transformer Manual GET-2485*. Schenectady, NY: General Electric.

6 Bergseth, F.R., Venkata, S.S. (1986). Introduction to Electric Energy Devices, Englewood, New Jersey.

3

Distribution Line Models

3.1 Overview

In a distribution system, power between a substation and loads flows through three-, two-, or single-phase lines. These lines can be overhead or underground. In this chapter, we develop models to represent these lines in the analysis of distribution systems.

3.2 Conductor Types and Sizes

3.2.1 Sizes

The size of the conductors is measured in circular mils. A mil is one-thousandth of an inch, and circular mil is the area of a circle with diameter of one mil. Typically, larger conductors have their designation given in terms of their size in kcmil. Medium and small conductors are designated by a number according to the American Wire Gauge (AWG) in which larger number means smaller size. These numbers start at 0000 (four zeros or four-) to higher numbers. The conductors from 0000 to 0 are designated by the number followed by slash and a zero. For example, 0000 is called 4/0, and 0 is called 1/0 in utility jargon. Conductors smaller than these go from #1 to #6. Conductors smaller than #6 are too small for distribution systems. AWG sizes are also used for wires used in homes. Typically, house wiring uses #12 copper conductors. Extension cords use #14 to #18 copper conductors.

3.2.2 Overhead Feeders

Overhead feeders for the primary side of the distribution system most often use aluminum conductor steel-reinforced (ACSR) conductors. These conductors have

Electric Power and Energy Distribution Systems: Models, Methods, and Applications, First Edition.
Subrahmanyam S. Venkata and Anil Pahwa.

Companion website: www.wiley.com/go/Pahwa/ElectricPowerDistributionSystems

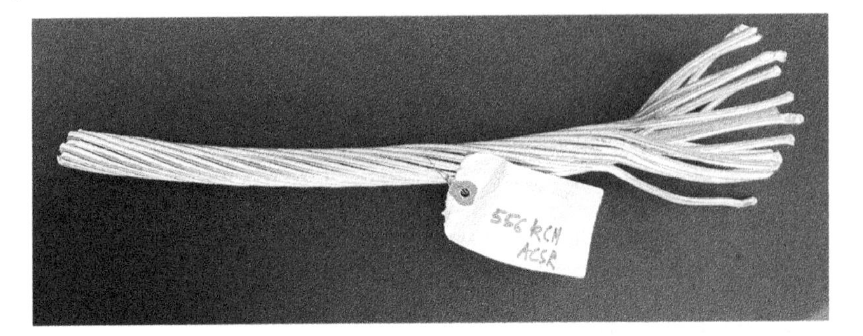

Figure 3.1 556-kcmil ACSR overhead conductor.

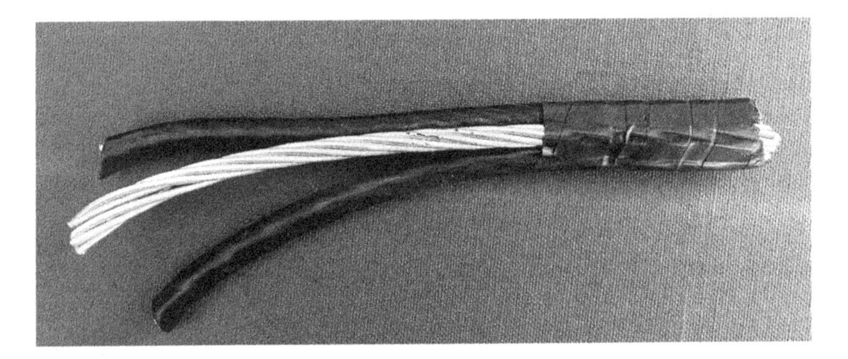

Figure 3.2 #2 triplex cable used for overhead service drops.

a steel core surrounded by aluminum strands. Figures 3.1 to 3.3 show images of different conductors of the available samples shot by authors. Readers must refer to literature and data from conductor manufacturers for additional information. Figure 3.1 shows a 556-kcmil ACSR conductor. Steel core is used for strength, and the aluminum strands are wound spirally to provide flexibility to the conductor. Overhead cables used for the secondary service drop are all aluminum. The live conductors are covered with a jacket to provide insulation, but the neutral is bare. Figure 3.2 shows #2 Triplex cable used for single-phase residential service. For three-phase service, quadraplex conductors are used.

3.2.3 Underground Feeders

Underground feeders are built with either copper or aluminum cables. The cables used for the primary side of the distribution system have a layer of insulation and an outer jacket. Figure 3.3 shows an example of an insulated 15-kV class copper

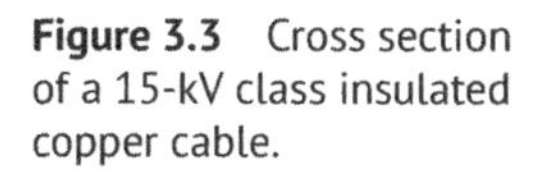

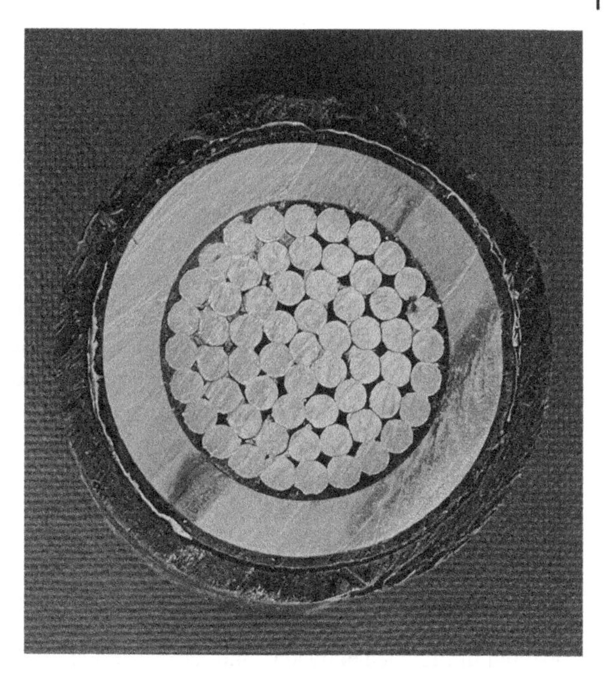

Figure 3.3 Cross section of a 15-kV class insulated copper cable.

cable. They are laid in ducts or directly buried. Triplex and quadraplex cables suitable for direct burial in ground are used for service drops.

3.2.4 Conductor Data

Manufacturers of conductors provide detailed data based on the construction geometry. In addition, electrical properties, such as resistance, are provided. Also provided is a quantity called geometric mean radius (GMR), which is used in the computation of line parameters. Let us consider a conductor made with N strands twisted together. The GMR for such a conductor is given by

$$\mathrm{GMR} = \sqrt[N^2]{\prod_i \prod_k d_{ij}};\ i, k = 1 \text{ to } N \tag{3.1}$$

where d_{ij} is the distance of each strand from every other strand in the ensemble, and d_{ii} is the radius of the individual strand. However, the radius is multiplied by a factor of $e^{-1/4}$ to account for internal magnetic flux while computing the inductance of conductors. These details are beyond the scope of this chapter, but the readers are referred to books on power system analysis for additional information on this topic. Since GMR is provided in the conductor data tables, we need not compute them.

3.3 Generalized Carson's Models

Carson [1] did seminal work in the 1920s on modeling of overhead lines. All the models used today for transmission and distribution feeders are based on his work. To gain an understanding of his work, consider an overhead line shown in Figure 3.4. It is assumed that the lines are long enough, which allows neglecting the end effects. It is also assumed that the earth has uniform conductivity and is bounded by a flat plane of infinite extent, and the conductors are parallel to that plane. However, in practice, the conductors attached to two poles sag. Therefore, the average height above ground is considered by adding height at midspan and one-third of sag. The figure also shows the image conductors, which are considered to be located at depths below the earth equal to the heights of the conductors above the earth.

The self-inductance of conductor i and the mutual impedance between the conductors for the given configuration are computed as follows:

Self-inductance of conductor i:

$$Z_{iig} = (R_{ii} + \Delta R_{iig}) + j\left(2\omega \cdot 10^{-4} \ln \frac{2h_i}{\mathrm{GMR}_i} + \Delta X_{iig}\right) \ \Omega/\mathrm{km} \tag{3.2}$$

Mutual inductance between conductors i and k:

$$Z_{ikg} = Z_{kig} = \Delta R_{ikg} + j\left(2\omega \cdot 10^{-4} \ln \frac{D_{ik}}{d_{ik}} + \Delta X_{ikg}\right) \ \Omega/\mathrm{km} \tag{3.3}$$

R_{ii}, resistance of conductor i in Ω/km;
h_i, height above ground of conductor i;
D_{ik}, distance between conductor i and image of conductor k;
d_{ik}, distance between conductors i and k;
GMR_i, geometric mean radius of conductor i;
$\omega = (2\pi f)$ with f = frequency in Hertz.

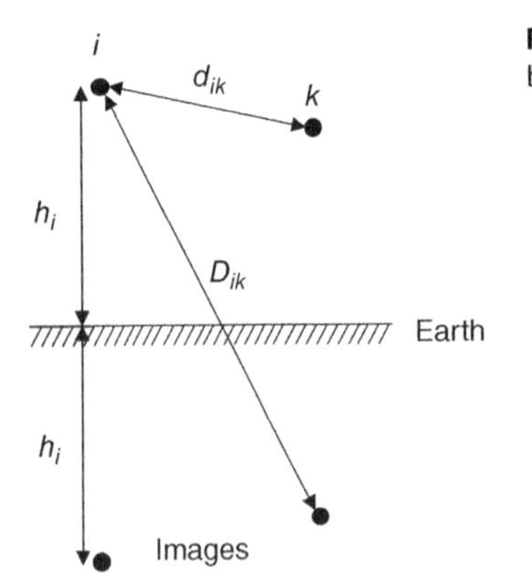

Figure 3.4 Overhead lines and their images below earth.

ΔR_{ii}, ΔX_{ii}, ΔR_{ik}, and ΔX_{ik} are corrections to the respective terms for the earth return effect. Carson described these corrections with integral equations, which do not have closed-form solutions. Various authors have provided different interpretations to his work [2–6]. The paper by Horak [6] is an excellent reference on this subject. While all aspects of his work have not been fully uncovered yet, the results are the same as interpreted by different researchers. Carson also provided infinite series equations as part of the solution. While it is hard to get the complete proof of Carson's work, follow-on work by other researchers shows that various approximations yield results that match practical values. Hence, there is widespread acceptance of his work. Some authors have shown that only the first few terms of the infinite series are significant for frequencies relevant to power systems. In fact, only including the first term itself gives good results. Without giving detailed proof, we provide the final equations that are widely accepted and prevalent for computing the self and mutual impedances of overhead conductors including the ground effect.

Self-inductance of conductor i:

$$\boldsymbol{Z_{iig}} = (R_{ii} + 0.00159f) + j\left(0.004657f\ \log\frac{2160\sqrt{\frac{\rho}{f}}}{\text{GMR}_i}\right)\ \Omega/\text{mile} \tag{3.4}$$

Mutual inductance between conductors i and k:

$$\boldsymbol{Z_{ikg}} = \boldsymbol{Z_{kig}} = 0.00159f + j\left(0.004657f\ \log\frac{2160\sqrt{\frac{\rho}{f}}}{d_{ik}}\right)\ \Omega/\text{mile} \tag{3.5}$$

R_{ii}, resistance of conductor i in Ω/mile;
d_{ik}, distance between conductors i and k in feet;
GMR_i, geometric mean radius of conductor i in feet;
f, frequency in Hertz;
ρ, resistivity of earth in ohm-meter (usually 100 Ω-m is used as the typical value).

Note the mixed nature of units in these equations. While feet and miles are used for all the quantities, resistivity of earth is given in ohm-meter. This is a leftover from the legacy work on this subject and continues to be used with mixed units.

3.4 Series Impedance Models of Overhead Lines

3.4.1 Three-phase Line

Figure 3.5 shows a typical configuration of a three-phase distribution feeder. We use the formulas presented in the previous section to compute the series

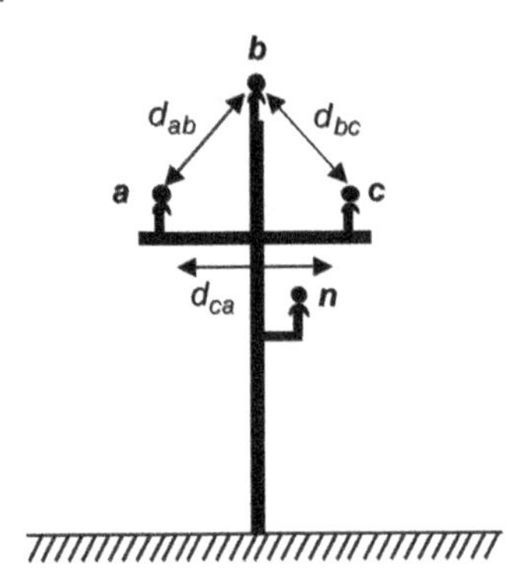

Figure 3.5 Typical three-phase overhead line configuration.

impedance matrix of this line. We assume that the neutral is grounded at multiple points including each pole. Also, we assume that the conductors are not transposed. Transposition is the practice of interchanging the position of conductors after a certain distance. Transposition is implemented in transmission systems, and it allows balancing the impedances of the three phases for a three-phase line. The shunt capacitance of most normal overhead distribution circuits is very low and makes no appreciable effect on the outcome of their steady-state performance.

Any feeder can be broken into sections, which are usually the portion of the feeder between two interconnection points or nodes. Each section has associated with it a matrix to describe the impedance including self and mutual impedances of conductors in that section. Using Ohm's law, the impedance matrix is postmultiplied by the current phasor vector of currents flowing on the conductors to get the voltage drop per phase in the section connected between buses x and y as given below:

$$\begin{bmatrix} \boldsymbol{V}_{xy}^{a} \\ \boldsymbol{V}_{xy}^{b} \\ \boldsymbol{V}_{xy}^{c} \\ \boldsymbol{V}_{xy}^{n} \end{bmatrix} = \begin{bmatrix} \boldsymbol{Z}_{aag} & \boldsymbol{Z}_{abg} & \boldsymbol{Z}_{acg} & \boldsymbol{Z}_{ang} \\ \boldsymbol{Z}_{bag} & \boldsymbol{Z}_{bbg} & \boldsymbol{Z}_{bcg} & \boldsymbol{Z}_{bng} \\ \boldsymbol{Z}_{cag} & \boldsymbol{Z}_{cbg} & \boldsymbol{Z}_{ccg} & \boldsymbol{Z}_{cng} \\ \boldsymbol{Z}_{nag} & \boldsymbol{Z}_{nbg} & \boldsymbol{Z}_{ncg} & \boldsymbol{Z}_{nng} \end{bmatrix} \begin{bmatrix} \boldsymbol{I}_{xy}^{a} \\ \boldsymbol{I}_{xy}^{b} \\ \boldsymbol{I}_{xy}^{c} \\ \boldsymbol{I}_{xy}^{n} \end{bmatrix} \tag{3.6}$$

$\boldsymbol{V}_{xy}^{i}$, phasor voltage drop across buses x and y on conductor i, $i = a, b, c, n$; $\boldsymbol{I}_{xy}^{i}$, phasor current flowing in conductor i between buses x and y, $i = a, b, c, n$; $\boldsymbol{Z}_{iig}$, self-impedance of conductor i including the ground effect, $i = a, b, c, n$; $\boldsymbol{Z}_{ikg}$, mutual impedance between conductors i and k including the ground effect; i, a, b, c, n; $k = a, b, c, n$; $i \neq k$.

Equations (3.4) and (3.5) are used to compute elements of the impedance matrix. For example,

$$\boldsymbol{Z}_{aag} = (R_{aa} + 0.00159f) + j\left(0.004657f \,\log \frac{2160\sqrt{\frac{\rho}{f}}}{\mathrm{GMR}_a}\right) \ \Omega/\text{mile} \tag{3.7}$$

and

$$Z_{abg} = Z_{bag} = 0.00159f + j\left(0.004657f \log \frac{2160\sqrt{\frac{\rho}{f}}}{d_{ab}}\right) \Omega/\text{mile} \tag{3.8}$$

Similarly, all the other elements of the matrix can be computed. Since the system analysis is usually focused on the three phases, we can remove the neutral from the equations using Kron's reduction as explained below. This results in 3×3 matrix from a 4×4 matrix. The assumption is that the voltage drop across the neutral conductor is zero because the neutral is grounded at multiple locations (details on grounding are discussed in Chapter 10). Also, some current will flow through the ground, but some of it will flow through the neutral. This gives

$$V_{xy}^{n} = 0 = Z_{nag}I_{xy}^{a} + Z_{nbg}I_{xy}^{b} + Z_{nbg}I_{xy}^{c} + Z_{nng}I_{xy}^{n} \tag{3.9}$$

Solving Eq. (3.9) gives

$$I_{xy}^{n} = \frac{Z_{nag}I_{xy}^{a} + Z_{nbg}I_{xy}^{b} + Z_{nbg}I_{xy}^{c}}{Z_{nng}} \tag{3.10}$$

Substituting Eq. (3.10) into (3.9) to compute V_{xy}^{a} yields

$$V_{xy}^{a} = Z_{aag}I_{xy}^{a} + Z_{abg}I_{xy}^{b} + Z_{acg}I_{xy}^{c} + Z_{ang} \cdot \left(\frac{Z_{nag}I_{xy}^{a} + Z_{nbg}I_{xy}^{b} + Z_{nbg}I_{xy}^{c}}{Z_{nng}}\right) \tag{3.11}$$

Grouping the proper coefficients for the currents in Eq. (3.12) gives

$$V_{xy}^{a} = \left(Z_{aag} - \frac{Z_{ang}Z_{nag}}{Z_{nng}}\right) I_{xy}^{a} + \left(Z_{abg} - \frac{Z_{ang}Z_{nbg}}{Z_{nng}}\right) I_{xy}^{b} + \left(Z_{acg} - \frac{Z_{ang}Z_{ncg}}{Z_{nng}}\right) I_{xy}^{c} \tag{3.12}$$

Similar equations can be obtained for V_{xy}^{b} and V_{xy}^{c}. Now, we can define the modified values of the self and mutual impedances for the untransposed lines with the following general equations:

$$Z'_{iig} = \left(Z_{iig} - \frac{(Z_{ing})^{2}}{Z_{nng}}\right); i = a, b, c \tag{3.13}$$

and

$$Z'_{ikg} = \left(Z_{ikg} - \frac{Z_{ing} Z_{nkg}}{Z_{nng}} \right); i, k = a, b, c \text{ and } i \neq k \tag{3.14}$$

where Z'_{iig}, self-impedance of conductor i including the ground and neutral current effects; Z'_{ikg}, mutual impedance between phases i and k including the ground and neutral current effects.

Equation (3.6) can be rewritten in the reduced form to get a 3×3 matrix to get the phase domain model for series line parameters.

$$\begin{bmatrix} V^a_{xy} \\ V^b_{xy} \\ V^c_{xy} \end{bmatrix} = \begin{bmatrix} Z'_{aag} & Z'_{abg} & Z'_{acg} \\ Z'_{bag} & Z'_{bbg} & Z'_{bcg} \\ Z'_{cag} & Z'_{cbg} & Z'_{ccg} \end{bmatrix} \begin{bmatrix} I^a_{xy} \\ I^b_{xy} \\ I^c_{xy} \end{bmatrix} \tag{3.15}$$

Note that in this matrix the mutual impedance terms between any two conductors are the same due to symmetry. For example, $Z'_{abg} = Z'_{bag}$ and so on for the other pairs of conductors.

3.4.2 Single- and Two-phase Line Modeling

Single- and two-phase line models can be deduced logically using the same approach as described for the three-phase models. The nonzero values of impedances are essentially the same as in the three-phase case, but the entries for the phases that do not exist are zero. The impedance matrix for the single-phase line for phase a is

$$\begin{bmatrix} Z'_{aag} & 0 & 0 \\ 0 & 0 & 0 \\ 0 & 0 & 0 \end{bmatrix} \tag{3.16}$$

Similarly, the impedance matrix for a two-phase line with phases a and b is

$$\begin{bmatrix} Z'_{aag} & Z'_{abg} & 0 \\ Z'_{bag} & Z'_{bbg} & 0 \\ 0 & 0 & 0 \end{bmatrix} \tag{3.17}$$

If the line has phases other than a and b, the matrix is developed accordingly.

3.4.3 Three-phase Line Example

Consider a three-phase line with the configuration given in Figure 3.6. Consider the phase conductors to be 636 kcmil 54/7 (54 strands of aluminum and 7 strands of steel) ACSR, and the neutral conductor to be 2/0 ACSR. Consider earth resistivity of 100 Ω-m.

Figure 3.6 Three-phase overhead line configuration.

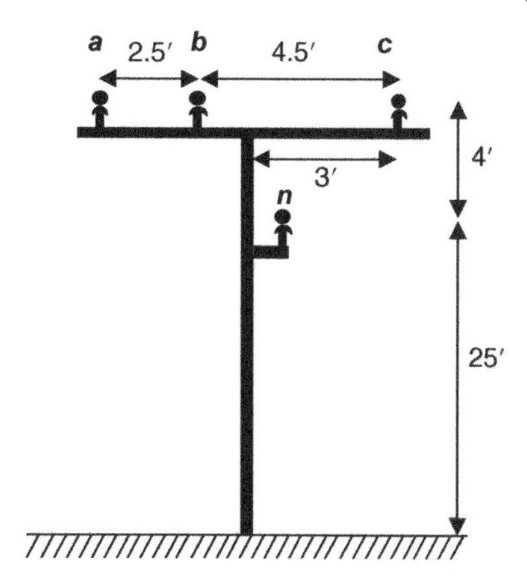

Solution We can compute the distances between the conductors for the configuration:

$$D_{ab} = 2.5\text{ ft}, D_{bc} = 4.5\text{ ft}, D_{ca} = 7\text{ ft}$$

and

$$D_{an} = \sqrt{4^2 + 4^2} = 5.657\text{ ft}$$

$$D_{bn} = \sqrt{(1.5)^2 + 4^2} = 4.272\text{ ft}$$

$$D_{cn} = \sqrt{3^2 + 4^2} = 5\text{ ft}$$

Now, we use the conductor data sheet to find the relevant information.

$$\text{GMR} = 0.0329\text{ ft}$$

Resistance has multiple values and different frequencies and temperatures. We select the value given for higher current at 60 Hz, or $R = 0.1688\ \Omega/\text{mile}$. Similarly, for the neutral conductor, GMR = 0.0051 ft and $R = 0.706\ \Omega/\text{mile}$. Note that we have selected the resistance at the lower current values for the neutral because in a three-phase line, neutral current is very small.

Next, we compute all the self and mutual impedances. For example,

$$\boldsymbol{Z_{aag}} = (R_{ii} + 0.00159f) + j\left(0.004657f \log \frac{2160\sqrt{\frac{\rho}{f}}}{\text{GMR}_i}\right)$$

$$= (0.1688 + 0.00159 \times 60) + j\left(0.004657 \times 60 \log \frac{2160\sqrt{\frac{100}{60}}}{0.0329}\right)$$

$$= 0.2572 + j1.3770\ \Omega/\text{mile}$$

$$Z_{abg} = Z_{bag} = 0.00159f + j\left(0.004657f \log \frac{2160\sqrt{\frac{\rho}{f}}}{d_{ab}}\right)$$

$$= 0.00159 \times 60 + j\left(0.004657 \times 60 \log \frac{2160\sqrt{\frac{100}{60}}}{2.5}\right)$$

$$= 0.0954 + j0.8515\ \Omega/\text{mile}$$

Repeating the same procedure gives the following impedance matrix for the line:

$$\begin{bmatrix} 0.2572 + j1.3770 & 0.0954 + j0.8515 & 0.0954 + j0.7265 & 0.0954 + j0.7523 \\ 0.0954 + j0.8515 & 0.2572 + j1.3770 & 0.0954 + j0.7801 & 0.0954 + j0.7865 \\ 0.0954 + j0.7265 & 0.0954 + j0.7801 & 0.2572 + j1.3770 & 0.0954 + j0.7674 \\ 0.0954 + j0.7523 & 0.0954 + j0.7865 & 0.0954 + j0.7674 & 0.8013 + j1.6032 \end{bmatrix} \Omega/\text{mile}$$

Now, we apply Kron's reduction to eliminate the neutral and obtain a 3×3 impedance matrix. For example,

$$Z'_{aag} = \left(Z_{aag} - \frac{(Z_{ang})^2}{Z_{nng}}\right)$$

$$= \left(0.2572 + j1.3770 - \frac{(0.0954 + j0.7523)^2}{0.8013 + j1.6032}\right) = 0.3244 + j1.0632$$

and

$$Z'_{abg} = \left(Z_{abg} - \frac{Z_{ang}Z_{nbg}}{Z_{nng}}\right)$$

$$= \left(0.0954 + j0.8515 - \frac{(0.0954 + j0.7523)(0.0954 + j0.7865)}{0.8013 + j1.6032}\right)$$

$$= 0.1674 + j0.5241$$

Similarly, we can obtain all the other elements to get the following 3×3 matrix:

$$\begin{bmatrix} 0.3244 + j1.0632 & 0.1674 + j0.5214 & 0.1674 + j0.4067 \\ 0.1674 + j0.5214 & 0.3343 + j1.0353 & 0.1697 + j0.4464 \\ 0.1674 + j0.4067 & 0.1697 + j0.4464 & 0.3244 + j1.0632 \end{bmatrix} \Omega/\text{mile}$$

3.5 Series Impedance Models of Underground Lines

3.5.1 Nonconcentric Neutral Cables

Among the underground cables used, one usually encounters either nonconcentric (separate) neutral conductors or concentric neutral conductors. Figure 3.7

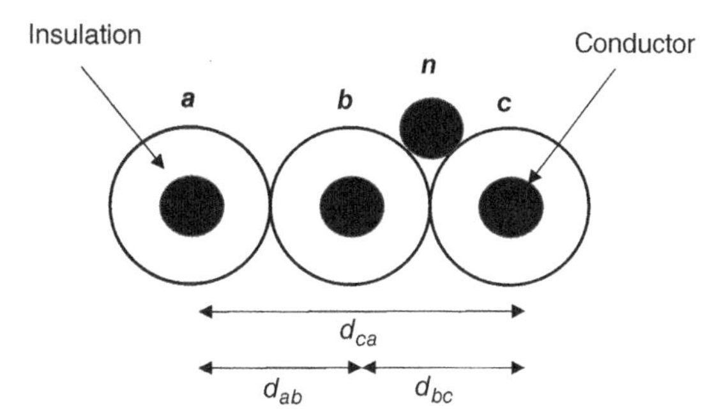

Figure 3.7 Three-phase underground cable with nonconcentric neutral. Distances from centers of the phase conductors and the neutral can be computed for the given cable and neutral sizes.

shows an example of a three-phase cable with separate neutral conductor. For such cables used in underground distribution systems, both the self and mutual impedance elements for phase conductors are evaluated using the same equations (3.4) and (3.5) that were used for the overhead lines.

3.5.2 Concentric Neutral Cables

Concentric neutral cables have several neutral strands at the periphery of the cable as shown in Figure 3.8. These neutral strands help in the distribution of electric field in the cable to reduce stress on the insulation.

3.5.2.1 Single-phase Cable

To compute the series impedance of this cable, consider the following quantities:

N number of neutral strands around the phase conductor
D_c diameter of the phase conductor
D outer diameter of the cable over the neutral strands
D_s diameter of the individual neutral stands
GMR_p geometric mean radius of the phase conductor
GMR_n geometric mean radius of each neutral strand
GMR_N geometric mean radius of ensemble of neutral strands of each phase
D_{an} distance from the center of each neutral strand to the center of the phase conductor
R_a resistance of the phase conductor in Ω/mile
R_n resistance of a single neutral strand in Ω/mile.

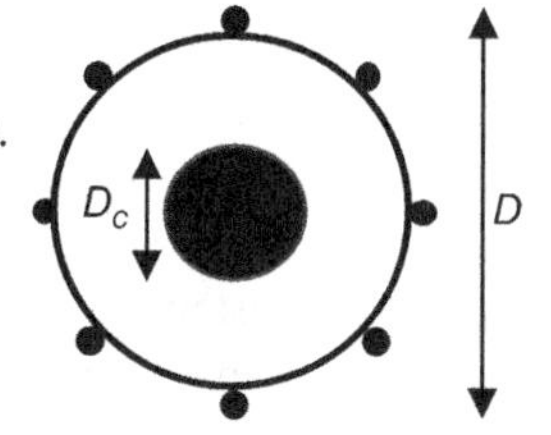

Figure 3.8 Schematic of a single-phase cable with concentric neutral conductor.

Note that all the distances associated with cable geometry and configuration must be converted to feet (ft) to use Eqs. (3.4) and (3.5). Also, we use single subscript for resistance of phase and neutral conductors.

Next, we compute all the information that is needed for computation of line impedance:

$$D_{an} = \frac{D - D_s}{2} \tag{3.18}$$

$$\text{GMR}_n = e^{-1/4}\frac{D_s}{2} = 0.7788\,\frac{D_s}{2} \tag{3.19}$$

Since the individual neutral stands are solid, the factor $e^{-1/4}$ is used to compute its GMR. This factor accounts for the magnetic field internal to the conductor. In the next step, we compute GMR_N, for which a formula has been suggested in the existing literature [5]. We will accept it without seeking a detailed proof.

$$\text{GMR}_N = \sqrt[N]{N.\text{GMR}_n(D_{an})^{(N-1)}} \tag{3.20}$$

Now, we can compute self-impedance of the phase and equivalent neutral conductor and mutual impedance using the general Eqs. (3.4) and (3.5):

$$\boldsymbol{Z}_{aag} = (R_a + 0.00159f) + j\left(0.004657f\,\log\frac{2160\sqrt{\frac{\rho}{f}}}{\text{GMR}_p}\right)\,\Omega/\text{mile} \tag{3.21}$$

$$\boldsymbol{Z}_{nng} = (R_n + 0.00159f) + j\left(0.004657f\,\log\frac{2160\sqrt{\frac{\rho}{f}}}{\text{GMR}_N}\right)\,\Omega/\text{mile} \tag{3.22}$$

$$\boldsymbol{Z}_{ang} = \boldsymbol{Z}_{nag} = 0.00159f + j\left(0.004657f\,\log\frac{2160\sqrt{\frac{\rho}{f}}}{D_{an}}\right)\,\Omega/\text{mile} \tag{3.23}$$

These impedances give us the 2×2 matrix of the cable, which is

$$\begin{bmatrix} \boldsymbol{Z}_{aag} & \boldsymbol{Z}_{ang} \\ \boldsymbol{Z}_{nag} & \boldsymbol{Z}_{nng} \end{bmatrix} \tag{3.24}$$

Finally, we use Kron's reduction to find the series impedance of the cable with an assumption that the neutrals are grounded.

$$\boldsymbol{Z}'_{aag} = \left(\boldsymbol{Z}_{aag} - \frac{(\boldsymbol{Z}_{ang})^2}{\boldsymbol{Z}_{nng}}\right) \tag{3.25}$$

3.5.2.2 Three-phase Cable

Figure 3.9 shows an example of a three-phase cable with concentric neutrals. We can compute the required quantities in a manner similar to the single-phase cable. Distances between the phase conductors are already specified. However, equivalent distances of neutral strands of one conductor to the other phase conductors can be approximated by distances from the center of the conductor to other conductors. For example, the equivalent distance of phase a neutral strands to phase b conductor is $D_{n_a b} = d_{ab}$.

For computing the impedance matrix of this cable arrangement, we consider three separate neutrals associated with each of the phase cables. The

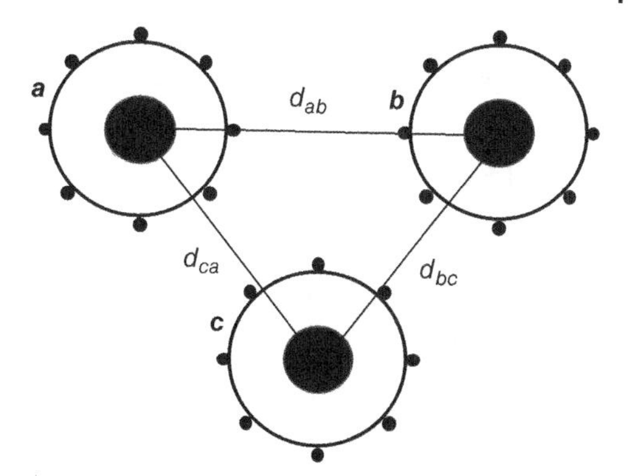

Figure 3.9 Schematic of a three-phase cable with concentric neutral conductor.

self-impedance of each phase will be the same and is given by Eq. (3.21) similarly, the self-impedance of each of the neutral ensemble will be the same and is given by Eq. (3.22). Following that, we compute mutual impedances between the phase conductors, between phase conductors and neutrals, and between neutrals of separate phases.

Mutual impedances between phase conductors:

$$Z_{ikg} = Z_{kig} = 0.00159f + j\left(0.004657f \log \frac{2160\sqrt{\frac{\rho}{f}}}{d_{ik}}\right) \ \Omega/\text{mile}$$

$$i, k = a, b, c \text{ and } i \neq k \tag{3.26}$$

Mutual impedances between phase conductors and neutral strands of the same conductor are given by Eq. (3.23).

Mutual impedances between phase conductors and neutral strands of another conductor:

$$Z_{in_kg} = Z_{n_kig} = 0.00159f + j\left(0.004657f \log \frac{2160\sqrt{\frac{\rho}{f}}}{d_{ik}}\right) \ \Omega/\text{mile}$$

$$i, k = a, b, c \text{ and } i \neq k \tag{3.27}$$

Mutual impedances between neutral strands of one phase and neutral strands of another phase:

$$Z_{n_in_kg} = Z_{n_kn_ig} = 0.00159f + j\left(0.004657f \log \frac{2160\sqrt{\frac{\rho}{f}}}{d_{ik}}\right) \ \Omega/\text{mile}$$

$$i, k = a, b, c \text{ and } i \neq k \tag{3.28}$$

With three phase conductors and three equivalent neutral conductors, we get a 6×6 impedance matrix. We can partition this matrix into four partitions as shown below:

$$\begin{bmatrix} \mathbf{Z}_{pp} & | & \mathbf{Z}_{pn} \\ -- & | & -- \\ \mathbf{Z}_{np} & | & \mathbf{Z}_{nn} \end{bmatrix} \tag{3.29}$$

where $\mathbf{Z}_{pp}$ represents the self and mutual impedances of phase conductors, $\mathbf{Z}_{pn}$ and $\mathbf{Z}_{np}$ represent the mutual impedances between phase and neutral conductors, and $\mathbf{Z}_{nn}$ represents self and mutual impedances of neutral conductors. We can apply Kron's reduction to remove the neutrals and obtain the impedance matrix for the phase conductors.

$$\left[\mathbf{Z}'_{pp}\right] = \left[\mathbf{Z}_{pp}\right] - \left[\mathbf{Z}_{pn}\right] \cdot \left[\mathbf{Z}_{nn}\right]^{-1} \cdot \left[\mathbf{Z}_{np}\right] \tag{3.30}$$

Example

Consider a single-phase 15-kV cable with concentric neutral conductors, as shown in Figure 3.8. The main conductor is 350-kcmil copper with a diameter of 0.661 inch, the outer diameter of the cable is 1.486 inches, and there are 16 concentric neutral conductors of #9 AWG. The GMR of the main conductor is 0.0214 ft, the alternating current (AC) resistance of the cable at 90 °C is 0.064 Ω/1000 ft, and the direct current (DC) resistance at 25 °C of each neutral strand is 0.832 Ω/1000 ft. Find the 2×2 impedance matrix of the cable for the phase and neutral conductors with impedances represented in Ω/mile. Remove the neutral using Kron's reduction to determine the impedance of the cable with ground and neutral effects included.

Solution

First, we compute the distance from the center of the neutral conductors to the center of the main conductor. The outer diameter of the cable is given, but the diameter of #9 AWG neutral wire is needed. A search for data on wires gives 0.1144 inch as the diameter of this wire. Therefore:

$$D_{an} = \frac{D - D_s}{2} = \frac{1.486 - 0.1144}{2} = 0.6858 \text{ inch or } \frac{0.6858}{12} = 0.05715 \text{ ft}$$

$$\text{GMR}_n = 0.7788 \frac{D_s}{2} = 0.7788 \times \frac{0.1144}{2 \times 12} = 0.003712 \text{ ft}$$

$$\text{GMR}_N = \sqrt[N]{N \cdot \text{GMR}_n (D_{an})^{(N-1)}} = \sqrt[16]{16 \cdot 0.003712\,(0.05715)^{(16-1)}}$$

$$= 0.5728 \text{ ft}$$

$$R_a = 0.064 \times 5.28 = 0.33792\ \Omega/\text{mile}$$

$$R_n = \frac{0.832 \times 5.28}{16} = 0.27456\ \Omega/\text{mile}$$

Using Eqs. (3.21)–(3.23)

$$Z_{aag} = (0.33792 + 0.00159 \times 60) + j\left(0.004657 \times 60 \log \frac{2160\sqrt{\frac{100}{60}}}{0.0214}\right)$$

$$= 0.4333 + j1.4292\ \Omega/\text{mile}$$

$$Z_{nng} = (0.27456 + 0.00159 \times 60) + j\left(0.004657 \times 60 \log \frac{2160\sqrt{\frac{100}{60}}}{0.5728}\right)$$

$$= 0.3699 + j1.0303\ \Omega/\text{mile}$$

$$Z_{ang} = Z_{nag} = 0.00159 \times 60 + j\left(0.004657 \times 60 \log \frac{2160\sqrt{\frac{100}{60}}}{0.5715}\right)$$

$$= 0.0954 + j1.0305\ \Omega/\text{mile}$$

Therefore, the 2×2 impedance matrix of the cable is

$$\begin{bmatrix} 0.4333 + j\,1.4292 & 0.0954 + j\,1.0305 \\ 0.0954 + j\,1.0305 & 0.3699 + j\,1.0303 \end{bmatrix} \Omega/\text{mile}$$

Further,

$$Z'_{aag} = \left(0.4333 + j\,1.4292 - \frac{(0.0954 + j\,1.0305)^2}{0.3699 + j\,1.0303}\right)$$
$$= 1.9892 + j0.46341\ \Omega/\text{mile}$$

Problems

3.1 Figure 3.10 shows a single-phase 7.2-kV overhead line. The phase conductor is #1 ACSR, and the neutral conductor is #2 ACSR. Compute the 2×2 series impedance matrix for the phase and neutral in Ω/mile. Use Kron's reduction to find Z'_{bbg}.

3.2 Figure 3.11 shows a three-phase 12.47-kV overhead line. The phase conductors are 556-kcmil ACSR, and the neutral conductor is 1/0 ACSR. Compute the 4×4 series impedance matrix for the line in Ω/mile. Use Kron's reduction to find the 3×3 $[Z'_{abc}]$ matrix. Note: Ignore the height of the insulators computing distances between conductors. For example, D_{na} is 3 ft.

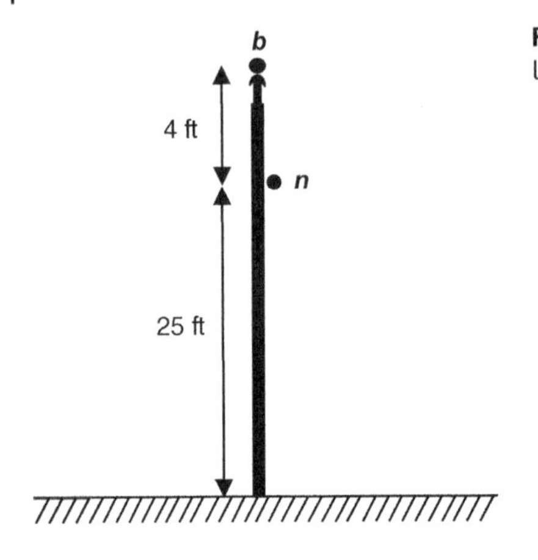

Figure 3.10 A single-phase 7.2-kV overhead line.

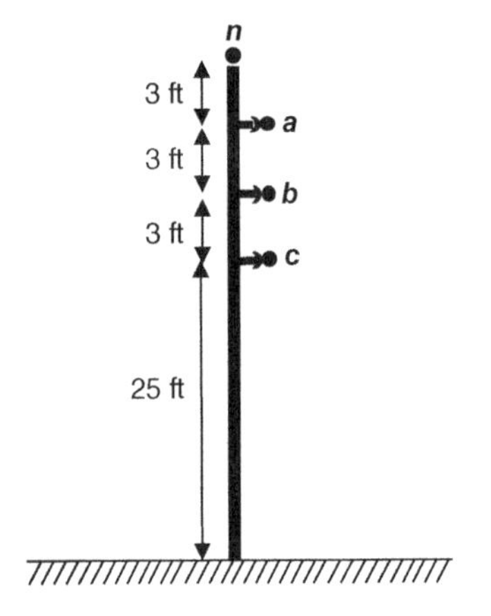

Figure 3.11 A three-phase 12.47-kV overhead line.

3.3 Consider the three-phase cable arrangement shown in Figure (3.9). The phase conductors are 4/0 copper with a diameter of 0.498 inch, the outer diameter of the individual cable is 1.269 inches, and there are 13 concentric neutral conductors of #10AWG. The GMR of the main conductor is 0.01668 ft, the AC resistance of the cable is 0.303 Ω/mile, and the resistance of each neutral strand is 5.9026 Ω/mile. The diameter of each neutral strand is 0.1019 inch. The distance from the centers of the cables to the other cable in the ensemble is 6 inches. Find the 4×4 impedance matrix of the cable for the phase and neutral conductors with impedances represented in Ω/mile. Remove the neutral using Kron's reduction to determine 3×3 $\left[Z'_{abc}\right]$ matrix.

3.4 Figure 3.12 shows the cross section of a 600-V all aluminum conductor (AAC) quadraplex cable, which is used for three-phase secondary overhead

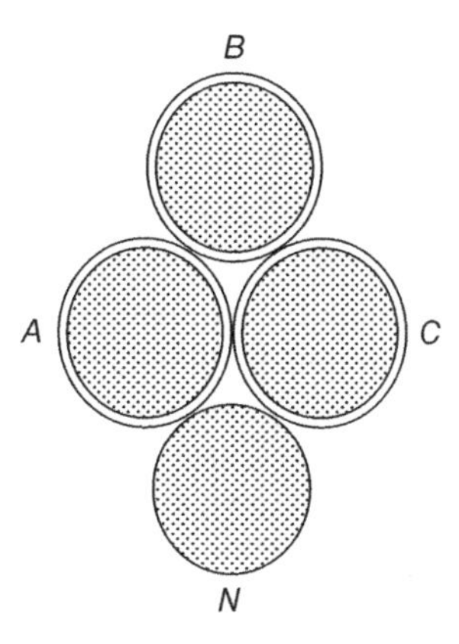

Figure 3.12 Cross section of a 600-V all aluminum conductor (AAC) quadraplex cable.

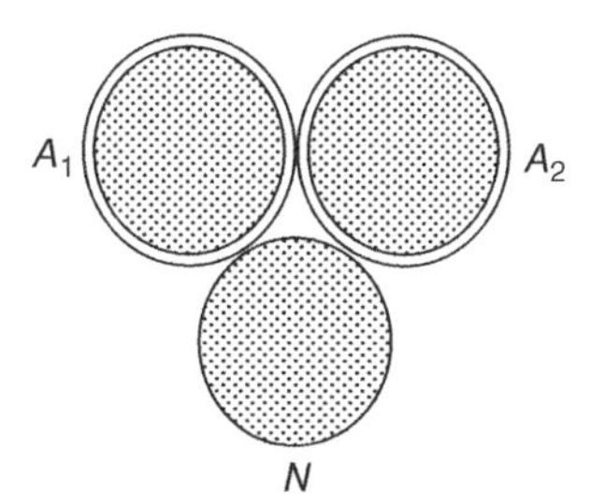

Figure 3.13 Cross section of a 600-V all aluminum conductor (AAC) triplex cable.

service drops to customers. The phase and the neutral conductors are #2 aluminum. The outside radius of all the conductors is 0.1415 inch, and the radius of the phase conductors with jacket is 0.1865 inch. The GMR of conductors is 0.0086 ft, and the resistance of conductors is 1.7809 Ω/mile. Find the 6×6 impedance matrix of the cable for the phase and neutral conductors with impedances represented in Ω/mile. Remove the entries for the neutrals using Kron's reduction to determine the 3×3 $\left[Z'_{abc}\right]$ matrix.

3.5 Figure 3.13 shows the cross section of a 600-V AAC triplex cable, which is used for single-phase 120/240 V secondary overhead service drops to customers. The phase and the neutral conductors are 1/0 aluminum. The outside diameter of all the conductors is 0.357 inch, and the diameter of the phase conductors with jacket is 0.477 inch. The GMR of the conductors is 0.0111 ft, and the resistance of the conductors is 0.1837 Ω/1000 ft. Find the 3×3 impedance matrix of a 100-ft-long cable for the phase and neutral conductors. Remove the neutral using Kron's reduction to determine the 2×2 matrix.

References

1 Carson, J.R. (1926). Wave propagation in overhead wires with ground return. *Bell System Technical Journal* 5: 539–554.
2 Monteith, A.C. and Wagner, C.F. (1964). *Electrical Transmission and Distribution Reference Book*. ABB/Westinghouse.
3 Dommel, H. (1985). Overhead line parameters from handbook formulas and computer programs. *IEEE Transactions on Power Apparatus and Systems* PAS-104 (2): 366–372.
4 Anderson, P.M. (1995). *Analysis of Faulted Power Systems (Text)*. IEEE/Wiley Press.
5 Kersting, W.H. (2002). *Distribution Modeling and Analysis*. CRC Press.
6 Horak, J. (2006). Zero sequence impedance of overhead transmission lines. Annual Conference for Protective Relay Engineers, College Station, TX (April 2006).

4

Distribution System Analysis

4.1 Introduction

The objective of distribution system analysis is to determine the state of the system including voltages, real and reactive power flow on lines, and losses in the system. This requires modeling all the components in the system such as lines and transformers and their interconnections based on the topology. In addition, models for loads and sources connected to the system are needed. Since distribution system is connected to transmission systems, which are connected to large generators, we model the whole upper level system at the point of coupling as an equivalent source. However, distributed energy resources (DERs) directly connected to the distribution system are modeled based on their characteristics.

4.2 Modeling of Source Impedance

A method that is valid for both radial and loop systems is to derive the sequence impedance values from the results of a fault study. See Appendix B for definitions of symmetrical components including positive, negative, and zero sequences. The procedure is to consider different faults at the bus of interest and use the equivalent circuit for that fault to determine the source impedance for positive, negative, and zero sequences as given below for bus m. Subscripts 1, 2, and 0 are used to represent positive-, negative-, and zero-sequence quantities, respectively. Since this approach requires three-phase, line-to-line (LL), and single-line-to-ground (SLG) faults, it can be applied only to a bus that has all three phases.

$$\boldsymbol{Z}_1^{sm} = \frac{\boldsymbol{V}}{\boldsymbol{I}_{f3\emptyset}^m} - Z_f \tag{4.1}$$

$$\boldsymbol{Z}_2^{sm} = \frac{j\sqrt{3}\boldsymbol{V}}{\boldsymbol{I}_{fLL}^m} - \boldsymbol{Z}_{s1}^m - Z_f \tag{4.2}$$

Electric Power and Energy Distribution Systems: Models, Methods, and Applications, First Edition.
Subrahmanyam S. Venkata and Anil Pahwa.

Companion website: www.wiley.com/go/Pahwa/ElectricPowerDistributionSystems

$$Z_0^{sm} = \frac{3V}{I_{fLG}^m} - Z_{s1}^m - Z_{s2}^m - 3Z_f \tag{4.3}$$

where V, system nominal voltage (line neutral); $I_{f3\emptyset}^m$, three-phase fault current at bus m; I_{fLL}^m, line-to-line fault current at bus m; I_{fLG}^m, SLG fault current at bus m; Z_f, fault impedance.

Knowing the sequence values, the corresponding values in the phase domain can be obtained by the proper symmetrical component similarity transformation. However, the obtained result will be an approximation because all the lines must be fully transposed to balance the three phases for decoupling of the sequence impedances. This assumption works well in transmission systems but not for distribution systems. So, if we have to represent the entire transmission system at the substation, this approach will work well.

For loop systems, if $[Z_{BUS}]$ matrix of the entire system in phase domain is known, the diagonal submatrix $[Z_{abc}^{sm}]$ corresponding to bus m is the source impedance (3 × 3) matrix. This is also called the driving point impedance matrix. That is

$$[Z_{abc}^m] = \begin{bmatrix} Z_{aa}^{sm} & Z_{ab}^{sm} & Z_{ac}^{sm} \\ Z_{ab}^{sm} & Z_{bb}^{sm} & Z_{bc}^{sm} \\ Z_{ac}^{sm} & Z_{bc}^{sm} & Z_{cc}^{sm} \end{bmatrix} \tag{4.4}$$

Note that the driving point impedance obtained based on the positive-sequence network topology of the system should not be used for fault calculations because it does not account for transformer connections that are important for the zero-sequence network.

4.3 Load Models

Under steady state, complex power S at any location in a distribution system varies with voltage and can be described as a function of voltage V at that point, that is

$$S = VI^* = P + jQ = f(V) \tag{4.5}$$

There are different modeling approaches used to describe this relationship. We present these approaches in this section.

4.3.1 Load Model I

It is usual to represent both P and Q as a general polynomial function of V. That is

$$P = a_0 + a_1 V + a_2 V^2 + a_{-1} V^{-1} + a_{-2} V^{-2} + \dots \tag{4.6}$$

$$Q = b_0 + b_1 V + b_2 V^2 + b_{-1} V^{-1} + b_{-2} V^{-2} + \dots \tag{4.7}$$

Such a representation is valid for individual type of loads or aggregate (composite) type of loads. Different conditions for the coefficients give different models. For example, if only a_0 and b_0 are nonzero, and all other coefficients are zero, we get

$$P = a_0 \tag{4.8}$$

and

$$Q = b_0 \tag{4.9}$$

or both P and Q are constant and (4.8) and (4.9) give equations for constant power load. If only a_1 and b_1 are nonzero, we get

$$P = a_1 V \tag{4.10}$$

and

$$Q = b_1 V \tag{4.11}$$

Equations (4.10) and (4.11) represent constant current models. Further, if only a_2 and b_2 are nonzero, we get

$$P = a_2 V^2 \tag{4.12}$$

and

$$Q = b_2 V^2 \tag{4.13}$$

Equations (4.12) and (4.13) provide a model for constant impedance loads. If a load is a combination of the three above mentioned types, we can combine them to find a composite expression. For examples, tests and subsequent regression analysis on the data show the following models for common load types [1]:

(a) Air-conditioning load demand (per-unit values)

$$P = 2.97 - 4.00V + 2.02V^2 \tag{4.14}$$

$$Q = 12.90 - 26.8V + 14.90V^2 \tag{4.15}$$

(b) Fluorescent lighting

$$P = 2.18 + 0.286V - 1.45V^{-1} \tag{4.16}$$

$$Q = 6.31 - 15.60V + 10.30V^2 \tag{4.17}$$

(c) Induction motor

$$P = 0.720 + 0.109V + 0.172V^{-1} \tag{4.18}$$

$$Q = 2.80 + 1.63V - 7.60V^2 + 4.89V^3 \tag{4.19}$$

4.3.2 Load Model II

Composite loads, which are assumed to be mixtures of the types discussed in the previous section, can be represented as

$$P = P_n\left(\frac{V}{V_n}\right)^k \tag{4.20}$$

$$Q = Q_n\left(\frac{V}{V_n}\right)^l \tag{4.21}$$

where both k and l vary between 0 and 3, V_n is the initial or base value of voltage, P_n is the initial or base value of real power, and Q_n is the initial or base value of reactive power. Note that in all these models, one should be aware of the range of V for which Eqs. (4.20) and (4.21) are valid. Some examples of using this model are given below [2].

(a) If $k = 1$ and $l = 0$, it implies that the load is a constant current type with unity power factor.
(b) If $k = l = 2$, the load is a constant impedance type.
(c) If $k = 2.5$ and $l = 2.7$, it represents an aluminum reduction plant. This is a simple model and can be determined empirically from measurements.

4.3.3 Load Model III

Loads are seldom modeled to include frequency effects. However, if loads (or demands) are sensitive to frequency, these effects should be included. In that case, voltage and frequency dependence are described by the following relationships:

$$P(f, V) = P_n\left(\frac{\omega}{\omega_n}\right)^\alpha\left(\frac{V}{V_n}\right)^k \tag{4.22}$$

$$Q(f, V) = Q_n\left(\frac{\omega}{\omega_n}\right)^\beta\left(\frac{V}{V_n}\right)^l \tag{4.23}$$

where $\omega = 2\pi f$, $\omega_n = 2\pi f_n$, f_n is the base frequency, and α and β are constant exponents.

Instead of determining the exponent for this model, it is often a practice to determine the four sensitivity coefficients, which are $\frac{\partial P}{\partial f}$, $\frac{\partial P}{\partial V}$, $\frac{\partial Q}{\partial f}$, and $\frac{\partial Q}{\partial V}$. Knowing these coefficients, the new values for ΔP and ΔQ can be determined from Δf and ΔV. All the changes are assumed to be small; thus

$$\begin{bmatrix} \Delta P \\ \Delta Q \end{bmatrix} = \begin{bmatrix} \frac{\partial P}{\partial f} & \frac{\partial P}{\partial V} \\ \frac{\partial Q}{\partial f} & \frac{\partial Q}{\partial V} \end{bmatrix} \begin{bmatrix} \Delta f \\ \Delta V \end{bmatrix} \tag{4.24}$$

Table 4.1 Suggested values of sensitivities of real power and reactive power to voltage and frequency changes based on load survey results.

Load type	$\frac{\Delta P/S}{\Delta V}$	$\frac{\Delta Q/S}{\Delta V}$	$\frac{\Delta P/S}{\Delta f}$	$\frac{\Delta Q/S}{\Delta f}$
Composite load	0.7–1.2	1.0–2.0	0.6–1.5	0 to −0.6
Residential	1.0–1.5	1.0–1.4	0.5	−0.7
Commercial	1.2	1.17	−0.185	−0.488
Industrial	0.7–1.5	1.0–2.0		

Then, $P = P_n + \Delta P$ and

$$Q = Q_n + \Delta Q \tag{4.25}$$

for $f = f_n + \Delta f$ and

$$V = V_n + \Delta V \tag{4.26}$$

Table 4.1 shows these coefficients based on the results of a data survey conducted by Electric Power Research Institute (EPRI) [3]. The values have been normalized to apparent power, S.

4.3.4 Load Model IV

This model is particularly suitable for modeling uncertainties in aggregate loads at a node knowing the demand profiles for a day, a season, or a year [4]. Assuming Gaussian distribution for load demand

$$P_p = P_n + k_p\sigma \tag{4.27}$$

and

$$Q_p = Q_n + k_p\sigma \tag{4.28}$$

where P_p is the power value at which the probability of load exceeding that value is p percent, k_p is the coefficient related to p, and σ is the standard deviation of the load.

Typically, p of 10% is used in voltage-drop calculations, and p of 50% or mean values of load are used for loss calculations. Smaller values of p are used for overload and determination of emergency conditions.

4.4 Distributed Energy Resources (DERs)

Traditional distribution systems were not designed to accommodate active generation and storage. However, with decreasing cost and advances in technology, such devices are being deployed in distribution systems. By definition, DERs are sources of electric power that are not directly connected to the bulk power system but are connected to the distribution system [5]. They are limited in size to 10 MVA or less. DER includes generators of different types and energy storage devices with the ability to inject power into the system. Generators include rotating induction or synchronous rotating machines driven by burning diesel, natural gas, bio gas, propane, or by wind or water flow. The static types of DERs include solar photovoltaic (PV) and batteries. While a synchronous machine producing alternating current (ac) power at the operating frequency can be directly connected to a system, generators that produce ac power at variable or a different frequency require a power electronics interface for connection to the system. For example, microturbine runs at very high speed and generates ac voltage at 1600 Hz. Wind-based generator produces ac power with a singly- or doubly-fed induction generator. Both solar PV and batteries provide dc power. So, the majority of DERs are connected to the system through a power electronic interface. For the resources that generate ac power, the converter changes it to dc power, and an inverter changes it back to ac power at the system frequency. For the resources that produce dc power, the inverter changes it to ac power at the system frequency. The DERs that use a power electronics controller with embedded inverters are called inverter-based resources (IBRs). In this chapter, we will not address the details of the electricity generation process of different DERs but instead focus on their modeling for various distribution system analyses.

Addition of DERs in distribution systems, especially radial systems, can impact the voltage at the point of coupling. However, in the early stages of DER deployment, DERs were not permitted to actively regulate voltage. They operated at fixed power factor. To mitigate adverse effects on voltages caused by increased deployment of DERs, the IEEE 1547 standard was revised in 2018 [5] to permit DERs to regulate voltage within the operating range by injecting reactive power. According to this standard, DERs are divided into two categories, Category A for systems with lower penetration of DERs and Category B for systems with higher penetration of DERs or systems with frequent large variations in power output. The required reactive power capabilities of the DERs for Category A and Category B are shown in Figure 4.1. These requirements apply to DERs for continuous operation when the voltage is 0.88 and 1.1 times the nominal voltage.

Although constant power factor mode with unity power factor setting is the default mode for DER operation unless otherwise specified by the distribution system operator, the distribution system operator has to approve active participation of DERs in voltage regulation. To be able to do so, the 1547 standard requires DERs to have the ability to control voltage, reactive power, and real power within the operating region specified in Figure 4.1. These control function requirements are specified in Table 4.2, and the control modes are illustrated in Figures 4.2–4.4.

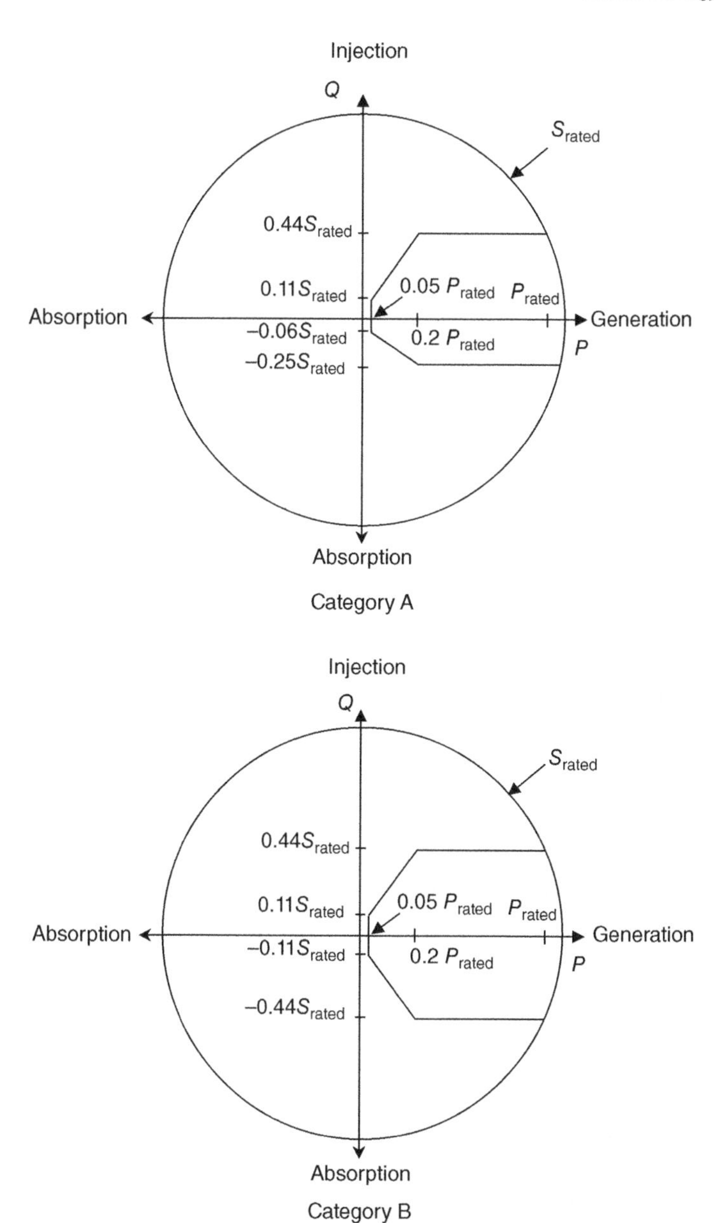

Figure 4.1 Minimum reactive power capability of Category A and B DER. Source: IEEE 1547-2018 [5]/with permission of IEEE.

Table 4.2 Voltage, reactive power, and real power control function requirements for DER [5].

Control mode	Category A	Category B
Constant power factor	Mandatory	Mandatory
Volt-var	Mandatory	Mandatory
Watt-var	Not required	Mandatory
Constant reactive power	Mandatory	Mandatory
Volt–watt	Not required	Mandatory

Source: IEEE 1547-2018 [5]/with permission of IEEE.

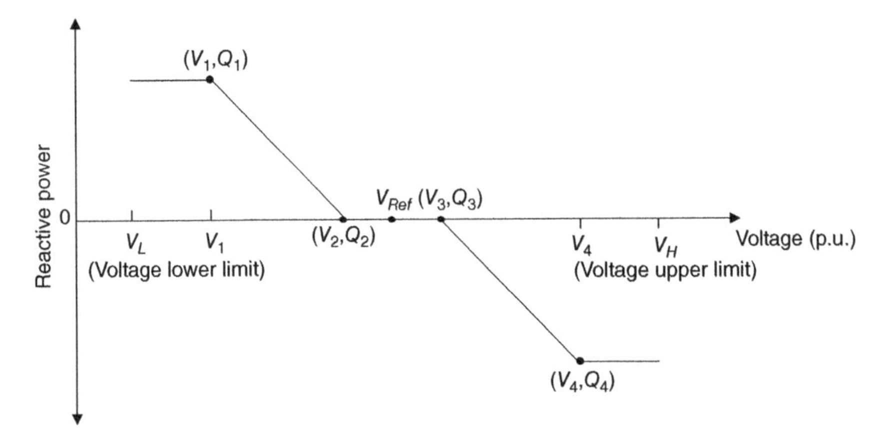

Figure 4.2 Volt-var characteristic for DER control.

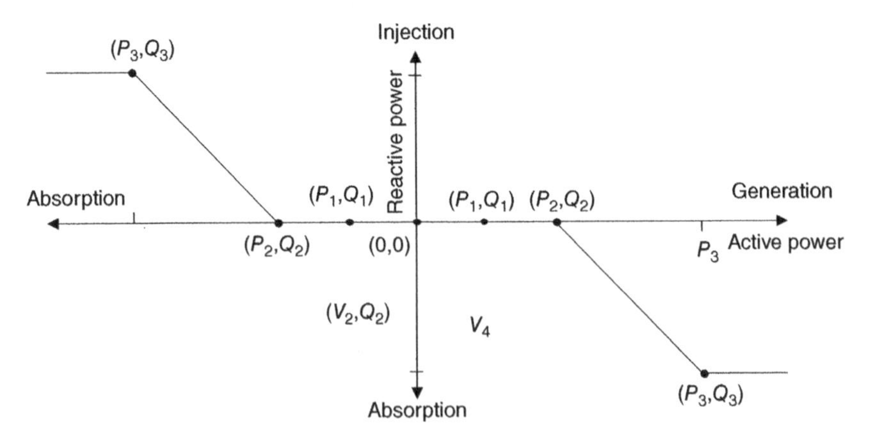

Figure 4.3 Watt-var characteristic for DER control.

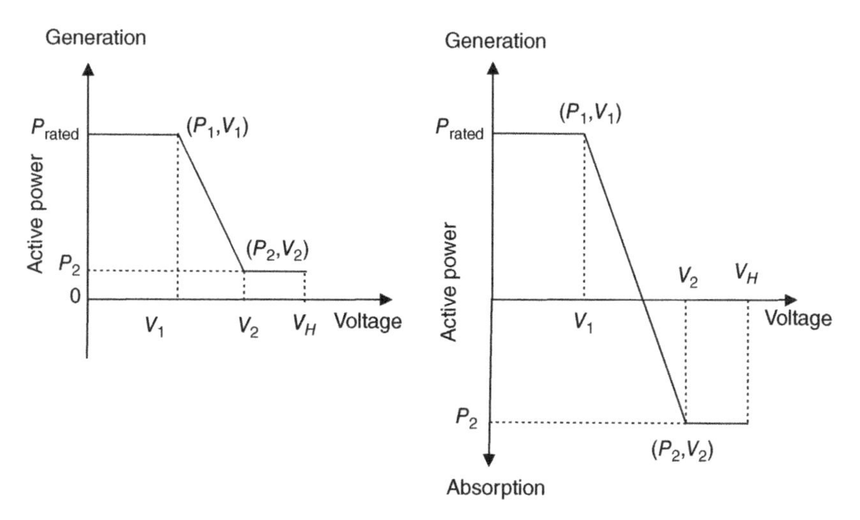

Figure 4.4 Volt–watt characteristic for DER control.

Standard 1547 [5] provides suggested values for various set points shown in these figures. The DERs will provide the capabilities of mutually exclusive reactive power control modes listed in this table and will be capable of initiating any of these modes one at a time. DER operator is responsible for implementing changes to settings and mode selections upon request by the system operator within a specified time. Other control modes mutually agreeable to the DER operator and the system operator can be implemented.

Irrespective of the type of DER and the selected control mode, DERs supply real power and either supply or absorb reactive power. Hence, at any given time, they can be represented by a model similar to that of load with known P and Q values. Thus, while considering DER as a load, one must consider its real power and reactive power as negative if delivering it, and positive if absorbing it.

4.5 Power Flow Studies

Since a typical distribution system is unbalanced due to both design and operation, it is imperative that the system's steady-state performance evaluation or analysis be conducted in three-phase domain, or to consider each phase separately, in the analysis. This is unlike transmission systems, where usually single-phase power flow studies are conducted while considering the system to be balanced. To facilitate distribution system analysis, various models shown in Table 4.3 are used for the components of the system.

Table 4.3 Power flow models.

Component	Mathematical model
Substation	Infinite source as a reference bus, where voltage magnitude can be controlled using regulators and/or taps on the transformers
Feeders and laterals	Three-phase series impedance and shunt admittance matrices for each line section. Shunt admittance matrix is included for long underground cable with significant charge currents
Load	Complex power $\left(P_i^{abc} + jQ_i^{abc}\right)$ for each bus i
DER	Complex power $\left(P_i^{abc} + jQ_i^{abc}\right)$ for each bus i

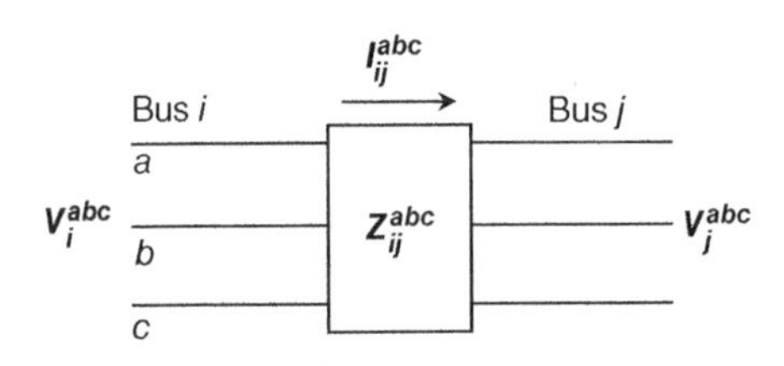

Figure 4.5 Schematic of a three-phase line connected between buses i and j.

4.5.1 Line Model

Figure 4.5 shows the schematic of a three-phase line connected between buses i and j. For simplicity, only the series impedance is shown in this figure. $\boldsymbol{Z}_{ij}^{abc}$ is the impedance matrix of the line, $\boldsymbol{V}_i^{abc}$ is the vector of phase a, b, and c voltages at bus i, $\boldsymbol{V}_j^{abc}$ is the vector of phase a, b, and c voltages at bus j, and $\boldsymbol{I}_{ij}^{abc}$ is the vector of currents flowing in phases a, b, and c from bus i to j of the line. An expression for the vector of voltage differences between buses i and j can be determined as follows:

$$\boldsymbol{V}_{ij}^{abc} = \boldsymbol{Z}_{ij}^{abc}.\boldsymbol{I}_{ij}^{abc} \tag{4.29}$$

Also,

$$\boldsymbol{V}_{ij}^{abc} = \boldsymbol{V}_i^{abc} - \boldsymbol{V}_j^{abc} \tag{4.30}$$

$\boldsymbol{Z}_{ij}^{abc}$ is a 3×3 matrix, which is realized after reducing a fourth-order model of the line by Kron's reduction. For a two-phase line, this matrix has nonzero entries in the 2×2 submatrix corresponding to the phases of the line, and for a single-phase line, it has nonzero entry in the diagonal entry corresponding to the phase of the line.

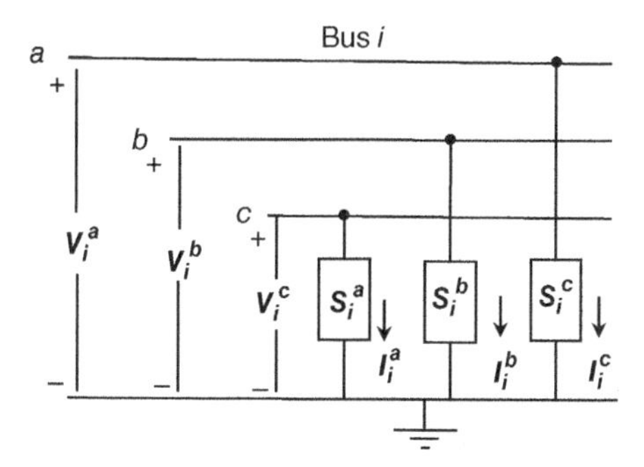

Figure 4.6 Representation of a three-phase Y-connected load on bus i.

4.5.2 Load and DER Model

For modeling loads and DER (negative load), we consider that the real and reactive power for each of the three phases are known separately. For a three-phase Y-connected load, the complex power drawn is shown in Figure 4.6. For this load connected at bus i, the vector $\boldsymbol{S}_i^{abc}$ of complex power is given by

$$\boldsymbol{S}_i^{abc} = \begin{bmatrix} \boldsymbol{S}_i^a \\ \boldsymbol{S}_i^b \\ \boldsymbol{S}_i^c \end{bmatrix} = \begin{bmatrix} P_i^a + jQ_i^a \\ P_i^b + jQ_i^b \\ P_i^c + jQ_i^c \end{bmatrix} \tag{4.31}$$

Also, the vectors of voltages and currents at bus i are given by

$$\boldsymbol{V}_i^{abc} = \begin{bmatrix} \boldsymbol{V}_i^a \\ \boldsymbol{V}_i^b \\ \boldsymbol{V}_i^c \end{bmatrix} \tag{4.32}$$

and

$$\boldsymbol{I}_i^{abc} = \begin{bmatrix} \boldsymbol{I}_i^a \\ \boldsymbol{I}_i^b \\ \boldsymbol{I}_i^c \end{bmatrix} \tag{4.33}$$

Now, consider a matrix operation $U\boldsymbol{V}_i^{abc}$, which gives the following matrix:

$$U\boldsymbol{V}_i^{abc} = \begin{bmatrix} \boldsymbol{V}_i^a & 0 & 0 \\ 0 & \boldsymbol{V}_i^b & 0 \\ 0 & 0 & \boldsymbol{V}_i^c \end{bmatrix} \tag{4.34}$$

Hence,

$$\begin{bmatrix} \boldsymbol{S}_i^a \\ \boldsymbol{S}_i^b \\ \boldsymbol{S}_i^c \end{bmatrix} = \begin{bmatrix} \boldsymbol{V}_i^a & 0 & 0 \\ 0 & \boldsymbol{V}_i^b & 0 \\ 0 & 0 & \boldsymbol{V}_i^c \end{bmatrix} \begin{bmatrix} \boldsymbol{I}_i^a \\ \boldsymbol{I}_i^b \\ \boldsymbol{I}_i^c \end{bmatrix}^* \tag{4.35}$$

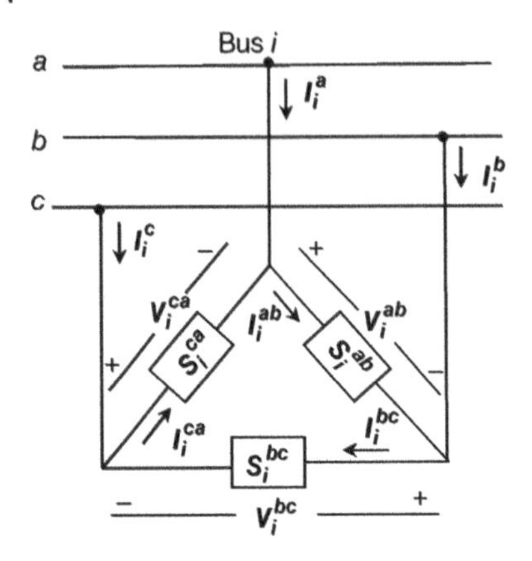

Figure 4.7 Representation of a three-phase Δ-connected load on bus *i*.

and

$$\begin{bmatrix} I_i^a \\ I_i^b \\ I_i^c \end{bmatrix}^* = \begin{bmatrix} V_i^a & 0 & 0 \\ 0 & V_i^b & 0 \\ 0 & 0 & V_i^c \end{bmatrix}^{-1} \begin{bmatrix} S_i^a \\ S_i^b \\ S_i^c \end{bmatrix} \tag{4.36}$$

Note that for two- or single-phase loads, S_i^{abc}, V_i^{abc}, and I_i^{abc} must be truncated by removing the entries for phases that do not exist in the load.

Now, consider a Δ-connected load as shown in Figure 4.7.

A relationship between complex powers, voltages, and currents is given below:

$$\begin{bmatrix} S_i^{ab} \\ S_i^{bc} \\ S_i^{ca} \end{bmatrix} = \begin{bmatrix} V_i^{ab} & 0 & 0 \\ 0 & V_i^{bc} & 0 \\ 0 & 0 & V_i^{ca} \end{bmatrix} \begin{bmatrix} I_i^{ab} \\ I_i^{bc} \\ I_i^{ca} \end{bmatrix}^* \tag{4.37}$$

and

$$\begin{bmatrix} I_i^{ab} \\ I_i^{bc} \\ I_i^{ca} \end{bmatrix}^* = \begin{bmatrix} V_i^{ab} & 0 & 0 \\ 0 & V_i^{bc} & 0 \\ 0 & 0 & V_i^{ca} \end{bmatrix}^{-1} \begin{bmatrix} S_i^{ab} \\ S_i^{bc} \\ S_i^{ca} \end{bmatrix} \tag{4.38}$$

Following that, we can compute I_{ij}^{abc} using Kirchhoff's current law (KCL), or

$$I_i^a = I_i^{ab} - I_i^{ca} \tag{4.39}$$

$$I_i^b = I_i^{bc} - I_i^{ab} \tag{4.40}$$

$$I_i^c = I_i^{ca} - I_i^{bc} \tag{4.41}$$

If there is a load connected between two phases as shown in Figure 4.8, we can use a similar procedure to compute currents.

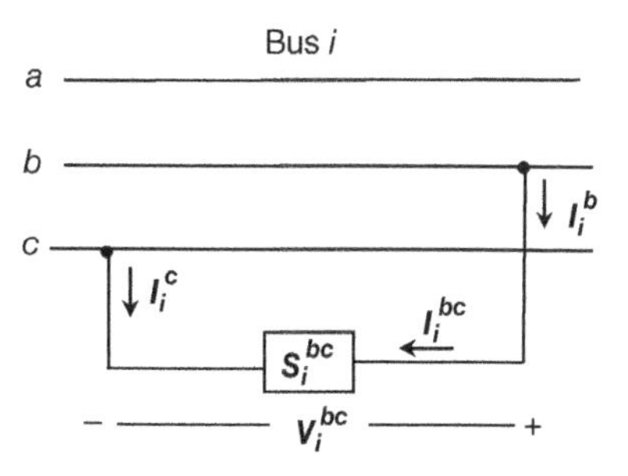

Figure 4.8 A two-phase load connected between phases *b* and *c*.

For this load

$$S_i^{bc} = V_i^{bc} \cdot I_i^{bc^*} \tag{4.42}$$

or

$$I_i^{bc} = \left({}^{S_i^{bc}}/_{V_i^{bc}} \right)^* \tag{4.43}$$

Therefore,

$$I_i^b = I_i^{bc}, I_i^c = -I_i^{bc}, \text{ and } I_i^a = 0 \tag{4.44}$$

Similar approach can be used to compute currents for a two-phase load connected between phases *a* and *b* or *c* and *a*.

4.5.3 Computing Currents

Knowing the load or demand vectors S^{abc} at each bus, the corresponding current vectors I^{abc} can be found using Eq. (4.36), given the voltage vectors V^{abc} at the buses. Figure 4.9 shows two three-phase feeders connected between buses *i*, *j*, and *k* with respective loads at these buses. I_i^{abc}, I_j^{abc}, and I_k^{abc} are load current vectors at buses *i*, *j*, and *k*, respectively. Also, I_{ij}^{abc} and I_{jk}^{abc} are current vectors of currents

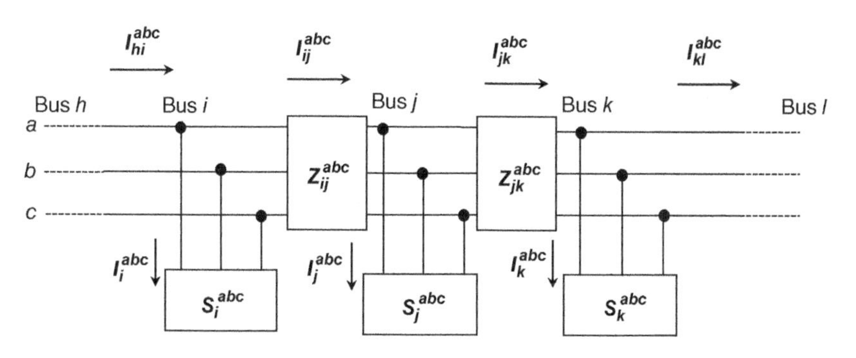

Figure 4.9 Schematic showing line sections and connected loads.

flowing on lines between buses i and j, and j and k. Similarly, $\boldsymbol{I}_{hi}^{abc}$ is vector of currents entering bus i from bus h, and $\boldsymbol{I}_{kl}^{abc}$ is vector of currents leaving bus k toward bus l.

Now, applying KCL at buses, i, j, and k, we get

$$\boldsymbol{I}_{hi}^{abc} = \boldsymbol{I}_{ij}^{abc} + \boldsymbol{I}_{i}^{abc} \tag{4.45}$$

$$\boldsymbol{I}_{ij}^{abc} = \boldsymbol{I}_{jk}^{abc} + \boldsymbol{I}_{j}^{abc} \tag{4.46}$$

and

$$\boldsymbol{I}_{jk}^{abc} = \boldsymbol{I}_{kl}^{abc} + \boldsymbol{I}_{k}^{abc} \tag{4.47}$$

Similar equations can be written for additional sections in the system. If there are additional feeders splitting off of the main feeder, the KCL can be expanded. For example, consider a case where two feeders are splitting at bus k as shown in Figure 4.10.

Applying the KCL at bus k gives

$$\boldsymbol{I}_{jk}^{abc} = \boldsymbol{I}_{kl}^{abc} + \boldsymbol{I}_{km}^{abc} + \boldsymbol{I}_{k}^{abc} \tag{4.48}$$

Again, if some of the feeders are two or single phase, the equations are modified to include only the phases that exist for a given line section.

4.5.4 Power Flow Algorithm

Source-Load-Iteration Method is an efficient method for power flow analysis of radial distribution system. The steps to implement it are described below. For ease of presentation, we are considering computation in per-unit (pu) representation, which implies that all the impedances and loads are converted to pu values using the appropriate base values.

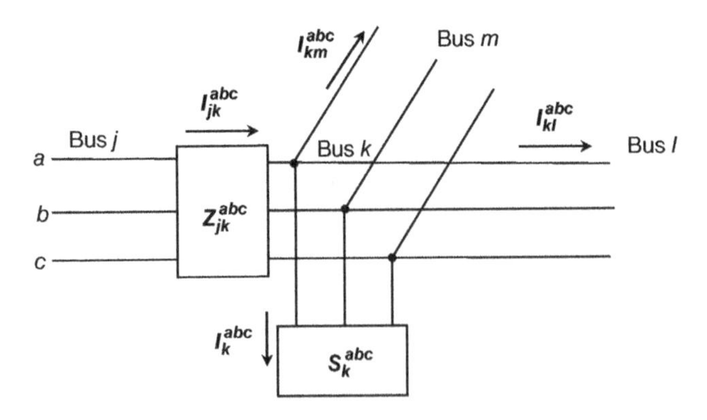

Figure 4.10 Feeders splitting at bus k.

Step 1: The voltage at the substation is fixed or regulated; therefore, it is known. With substation declared as bus 1 of the system, we get

$$\begin{bmatrix} V_1^a \\ V_1^b \\ V_1^c \end{bmatrix} = \begin{bmatrix} 1.0\angle 0^\circ \\ 1.0\angle -120^\circ \\ 1.0\angle 120^\circ \end{bmatrix} \tag{4.49}$$

Step 2: Assume a flat profile for all the other bus voltages and set initial values of voltage vectors at all the buses equal to the substation bus, that is

$$V_2^{abc(0)} = V_3^{abc(0)} = \ldots = V_n^{abc(0)} = V_1^{abc} \tag{4.50}$$

Note that (0) in the superscript is the iteration count, and n is the total number of buses in the system.

Step 3: Find the current vectors for all the buses that have loads connected to them using Eq. (4.36), or

$$I_i^{abc(m)} = \begin{bmatrix} I_i^{a(m)} \\ I_i^{b(m)} \\ I_i^{c(m)} \end{bmatrix}^* = \begin{bmatrix} V_i^{a(m)} & 0 & 0 \\ 0 & V_i^{b(m)} & 0 \\ 0 & 0 & V_i^{c(m)} \end{bmatrix}^{-1} \begin{bmatrix} S_i^a \\ S_i^b \\ S_i^c \end{bmatrix}; i = 2, 3, \ldots, n \tag{4.51}$$

Note that m in the superscript is the iteration count, and it should not be confused with bus number m used in the previous section. For DERs, determine P_i^{abc} and Q_i^{abc} corresponding to the voltage at bus i and the selected control mode while considering the reactive power capability specified in Figure 4.1 for the specified DER category.

Step 4: Find $I_{jk}^{abc(m)}$ for each feeder section j–k (section between bus j and bus k) using Eq. (4.40) starting from the bus at the edge of the system and sequentially moving toward the source. If the system has multiple branches emanating from the main feeder, this must be done for all branches starting from the bus at the edge.

Step 5: Determine the voltage drop in each feeder section by proceeding from source to load using Eq. (4.39), or

$$V_{ij}^{abc(m)} = Z_{ij}^{abc} . I_{ij}^{abc(m)} \tag{4.52}$$

To determine $V_i^{abc(m+1)}$, we start from bus 1 (or source) with known voltages and move toward the loads to compute the bus voltages sequentially starting with bus 2,

$$V_2^{abc(m+1)} = V_1^{abc} - V_{12}^{abc(m)} \tag{4.53}$$

And for all other buses,

$$V_j^{abc(m+1)} = V_i^{abc(m+1)} - V_{ij}^{abc(m)}, j = 3 \text{ to } n, i = 2 \text{ to } n-1, \text{ and } j < i. \tag{4.54}$$

Step 6: Check for the tolerance level, which is the difference between voltages in two successive iterations for every bus and for all three phases at each bus.

$$\Delta V_i^{abc(m+1)} = \left| V_i^{abc(m+1)} - V_i^{abc(m)} \right|, i = 2 \text{ to } n \tag{4.55}$$

If all the elements of $\boldsymbol{\Delta V}_i^{abc(m+1)}$ are less than ε, which is an arbitrarily selected small number for tolerance, for all the buses, the power flow converges. Typically, a value of 0.001 for ε gives good results. If convergence is not achieved, repeat steps 3–6.

4.6 Voltage Regulation

Maintaining proper voltage in the distribution system is very important. There are numerous methods to improve and control the voltage of primary distribution systems:

1. Use of load tap changing (LTC) transformers.
2. Application of voltage regulators in the distribution substation as well as on the feeders.
3. Application of shunt (or series) capacitors on the feeders or at the distribution substation.
4. Balancing of loads on primary feeders.
5. Increasing feeder conductor size.
6. Increasing primary voltage level.
7. Changing feeder sections from single phase to three phase.
8. Transferring of loads from existing feeders to new feeders.
9. Installation of new substations and primary feeders.

While several of these options are usually considered during the planning stages, LTCs, regulators, and capacitors provide the best means of achieving good voltage regulation during the operational stages on a continuous basis.

4.6.1 Voltage Regulation Definition

Voltage regulation is the voltage difference between the two ends of a line defined as percentage of the receiving end or downstream voltage. Mathematically, it can be expressed as

$$\%\ \text{Voltage Regulation} = \frac{(V_s - V_r)100}{V_r} \tag{4.56}$$

where V_s is the magnitude of the sending-end voltage, and V_r is the magnitude of the receiving-end voltage.

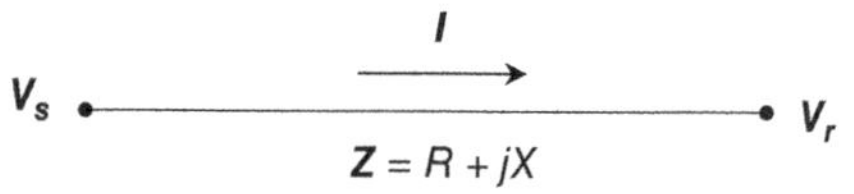

Figure 4.11 A feeder of resistance R and reactance X.

4.6.2 Approximate Method for Voltage Regulation

Consider a feeder of impedance $\boldsymbol{Z}$ with resistance R and reactance X as shown in Figure 4.11. $\boldsymbol{I}$ is the current flowing on the feeder corresponding to the specified load at the sending end.

From this figure

$$\boldsymbol{V}_s = \boldsymbol{V}_r + \boldsymbol{I}(R + jX) \tag{4.57}$$

And the voltage drop on the feeder is

$$\boldsymbol{V}_s - \boldsymbol{V}_r = \boldsymbol{I}(R + jX) \tag{4.58}$$

A phasor diagram for this equation with the condition that current lags voltage by an angle $ø$ is shown in Figure 4.12.

With $\boldsymbol{V}_r$ as the reference voltage or $\boldsymbol{V}_r = V_r \angle 0°$, $\boldsymbol{V}_s = V_s \angle \delta°$, and $\boldsymbol{I} = I \angle -Ø°$, we get

$$V_s \cos\delta + jV_s \sin\delta - V_r = I(R \cos Ø + X \sin Ø) + jI(X \cos Ø - R \sin Ø) \tag{4.59}$$

Typically, in distribution systems, R is approximately equal to X, and δ is very small. Hence, we can consider $\cos\delta \approx 1$ and $\sin\delta \approx 0$, which makes the imaginary part of (4.59) go to zero, and the real part giving voltage drop per mile becomes

$$V_s - V_r = I(R \cos Ø + X \sin Ø) \tag{4.60}$$

And Eq. (4.56) for voltage regulation becomes

$$\% \text{ Voltage Regulation} = \frac{I(R \cos Ø + X \sin Ø)}{V_r} \times 100 \tag{4.61}$$

Now, if we consider a load $\boldsymbol{S} = P + jQ$ connected at the receiving end, we can compute the corresponding current

$$\boldsymbol{I} = \left(\frac{P + jQ}{\boldsymbol{V}_r}\right)^* \tag{4.62}$$

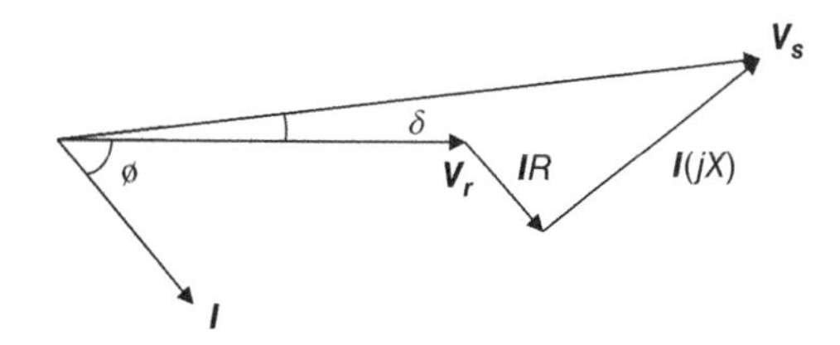

Figure 4.12 Phasor diagram for Eq. (4.58).

and

$$I = \frac{S}{V_r} \tag{4.63}$$

which gives

$$\%\ \text{Voltage Regulation} = \frac{S(R\cos\emptyset + X\sin\emptyset)}{V_r^2} \times 100 = \frac{(R\,P + X\,Q)}{V_r^2} \times 100 \tag{4.64}$$

In practice, the approximate voltage drop is an acceptable measure since the error between the exact and approximate values is negligible.

In distribution systems, usually $\boldsymbol{V_s}$, the voltage at the substation, is controlled and held constant for varying loading conditions, which implies that the voltage drop and hence $\boldsymbol{V_r}$ at other buses changes. Obtaining $\boldsymbol{V_r}$ at a given bus for the given impedance and load values requires iterative solution using power flow method as discussed in Section 4.5.

Example

Consider a 12.47-kV feeder with three point loads as shown in Figure 4.13. The loads are

$S_2 = 2.5$ MVA with 0.92 lagging power factor,
$S_3 = 3.0$ MVA with 0.90 lagging power factor, and
$S_4 = 2.0$ MVA with 0.95 lagging power factor.

The given impedance of the feeder, z, is $(0.258 + j\ 0.6644)\ \Omega/\text{mi}$ or $0.7127\angle 68.78°\ \Omega/\text{mi}$. Find the percent voltage drop at bus 4 of the primary feeder for the stipulated load conditions using the approximate method.

Solution

First, compute impedances, $\boldsymbol{Z}$, of the line sections by multiplying z by l, which is the length of the feeder sections, or

$$\begin{aligned} \boldsymbol{Z_{12}} &= z \cdot l_{12} = (0.7127\angle 68.78°\ \Omega/\text{mi})(1.5\ \text{mi}) = 1.0691\angle 68.78°\ \Omega \\ &= 0.3870 + j0.9966\ \Omega, \end{aligned}$$

$$\begin{aligned} \boldsymbol{Z_{23}} &= z \cdot l_{23} = (0.7127\angle 68.78°\ \Omega/\text{mi})(1\ \text{mi}) = 0.7127\angle 68.78°\ \Omega \\ &= 0.258 + j0.6644\ \Omega, \end{aligned}$$

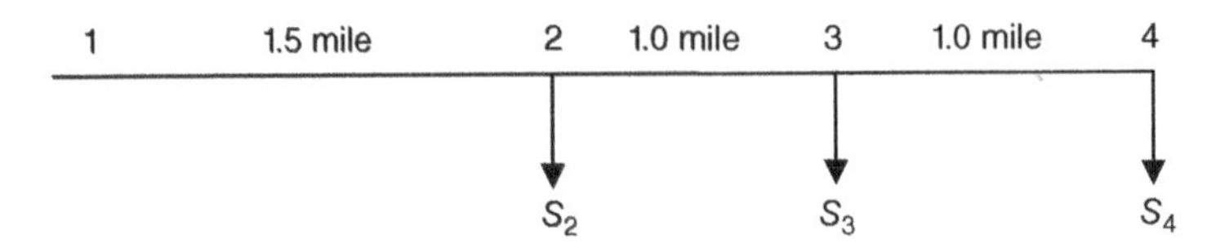

Figure 4.13 A single-phase feeder supplying three point loads.

$$\boldsymbol{Z}_{34} = \boldsymbol{z} \cdot l_{34} = (0.7127\angle 68.78°\ \Omega/\text{mi})(1\ \text{mi}) = 0.7127\angle 68.78°\ \Omega$$
$$= 0.258 + j0.6644.$$

Next, compute the line-to-neutral voltage at bus 1 for the given line-to-line voltage for the feeder, or

$$V_1 = \frac{12{,}470}{\sqrt{3}} = 7200\ \text{V}$$

Assume this voltage to be the reference voltage, or $\boldsymbol{V}_1 = 7200\angle 0°$. We also assume that the voltage at each bus remains at $7200\angle 0°$, and we compute the load currents for each bus, or

$$\boldsymbol{I}_4 = \left(\frac{\boldsymbol{S}_4}{\boldsymbol{V}_4}\right)^* = \frac{2.0 \times 10^6}{7200}\angle - \cos^{-1}(0.95)°\ \text{A}$$
$$= 92.60\angle - 18.19°\ \text{A}$$

$$\boldsymbol{I}_3 = \left(\frac{\boldsymbol{S}_3}{\boldsymbol{V}_3}\right)^* = \frac{3.0 \times 10^6}{7200}\angle - \cos^{-1}(0.92)°\ \text{A}$$
$$= 138.89\angle - 25.84°\text{A}$$

$$\boldsymbol{I}_2 = \left(\frac{\boldsymbol{S}_2}{\boldsymbol{V}_2}\right)^* = \frac{2.5 \times 10^6}{7200}\angle - \cos^{-1}(0.92)°$$
$$= 115.74\angle - 23.07°\ \text{A}$$

Now, compute the currents in the feeder sections using KCL.

$$\boldsymbol{I}_{34} = \boldsymbol{I}_4 = 92.60\angle - 18.19°\ \text{A}$$

$$\boldsymbol{I}_{23} = \boldsymbol{I}_{34} + \boldsymbol{I}_3 = 92.60\angle - 18.19°\text{A} + 138.89\angle - 25.84°\text{A}$$
$$= 230.9950\angle - 22.78°\text{A}$$

$$\boldsymbol{I}_{12} = \boldsymbol{I}_{23} + \boldsymbol{I}_2 = 230.9950\angle - 22.78°\text{A} + 115.74\angle - 23.07°\text{A}$$
$$= 346.73\angle - 22.78°\text{A}$$

Compute voltage drops in each feeder section using Eq. (4.60)

$$\text{VD}_{34} = I_{34}\ [R_{34}\cos(18.90°) + X_{34}\sin(18.19°)]$$
$$= 92.60[0.2580 \times 0.95 + 0.6644 \times 0.3122]$$
$$= 41.90\ \text{V}$$

Similarly,

$$\text{VD}_{23} = I_{23}\ [R_{23}\cos(22.78°) + X_{23}\sin(22.78°)]$$
$$= 230.9950[0.2580 \times 0.9219 + 0.6644 \times 0.3872]$$
$$= 114.37\ \text{V}$$

Finally,

$$\begin{aligned} \mathrm{VD}_{12} &= I_{12}\,[R_{12}\cos(22.88°) + X_{12}\sin(22.88°)] \\ &= 346.7340[0.3870 \times 0.9213 + 0.6644 \times 0.3888] \\ &= 257.97\ \mathrm{V} \end{aligned}$$

Thus, the total voltage drop in the entire feeder is the sum of the voltage drops in individual sections, or

$$\begin{aligned} \mathrm{VD}_{14} &= \mathrm{VD}_{34} + \mathrm{VD}_{23} + \mathrm{VD}_{12} \\ &= 41.90 + 114.37 + 257.97 \\ &= 414.24\ \ \mathrm{V} \end{aligned}$$

and

$$V_r = 7200 - 414.24 = 6815.76\ \mathrm{V}$$

Therefore, the percent voltage regulation is

$$\%\ \text{Voltage Regulation} = \frac{414.24}{6815.74} \times 100 = 6.07\%$$

Verify the accuracy on this result by finding the exact solution for this problem by determining voltages with power flow using the source-to-load iterative method. You will find that the error is minimal in this case due to the power factors of all three loads being close to each other.

4.6.3 Voltage Drop on Radial Feeders with Uniformly Distributed Load

In the previous section, we considered spot loads connected at the buses. In this section, we consider a case where the loads are distributed uniformly across the length of the feeder in a rectangular service area. Also, we generalize it for a three-phase feeder, assuming that all the three phases have identical values of resistance and reactance. Consider the current density to be k A/mile for the feeder shown in Figure 4.14. With l as the length of the feeder, the current at the sending end is $I_s = kl$, and the current at the receiving end is 0.

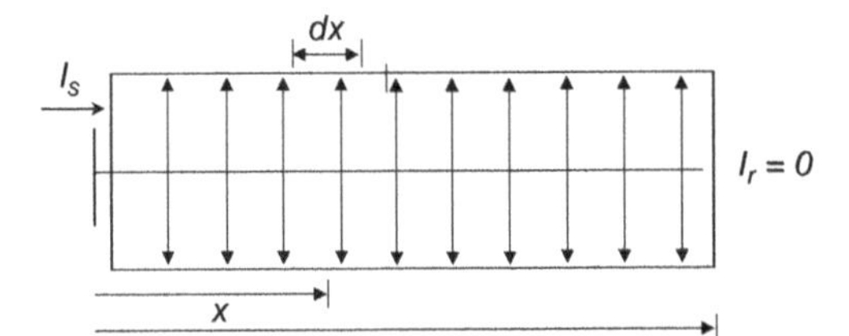

Figure 4.14 Feeder with load distributed uniformly in a rectangular service area.

Assuming that the current throughout the feeder has the same phase angle, we can draw a plot showing the magnitude of current as a function of distance as shown in Figure 4.15.

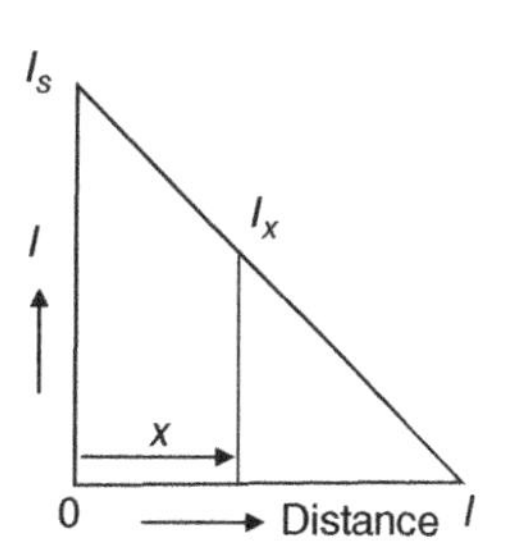

Figure 4.15 Current on the feeder.

From this figure, we get

$$I_x = I_s - kx \tag{4.65}$$

or

$$I_x = I_s - \frac{I_s}{l}x \tag{4.66}$$

$$I_x = I_s\left(1 - \frac{x}{l}\right) \tag{4.67}$$

Now, consider a differential element of the feeder of length dx at a distance x. With z ohms/mile as impedance of the feeder, the differential voltage across dx is

$$dV_D = I_x z dx \tag{4.68}$$

$$= I_s\left(1 - \frac{x}{l}\right) z dx \tag{4.69}$$

Therefore, the voltage drop on the feeder is

$$V_{Dl} = \int_0^l I_s\left(1 - \frac{x}{l}\right) z dx \tag{4.70}$$

$$= \frac{1}{2} I_s z l \text{ V} \tag{4.71}$$

We can also compute losses on the feeder by considering differential power loss with r ohms/mile as the resistance of the feeder.

$$dP_{LSx} = I_x^2 r dx \tag{4.72}$$

$$= I_s^2\left(1 - \frac{x}{l}\right)^2 r dx \tag{4.73}$$

Then, the power loss per phase is

$$P_{LS} = \int_0^l I_s^2\left(1 - \frac{x}{l}\right)^2 r dx \tag{4.74}$$

$$= \frac{1}{3} I_s^2 r l \tag{4.75}$$

and the total power loss for the three phases is

$$P_{LS3\emptyset} = I_s^2 r l \tag{4.76}$$

The results shown above are useful for planning studies because while planning we do not have the system topology and actual load values available. All we have are projected values. Also, these approximations are useful in optimizing operating

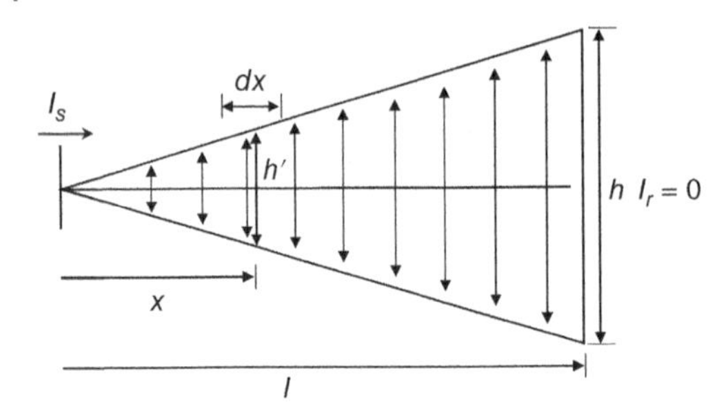

Figure 4.16 Feeder serving a triangular area with fixed load density.

scenarios where precise calculations may slow down the process of obtaining the final solution. In those cases, approximation can be used to narrow the solution space, and precise calculations can be implemented on a smaller solution space, thus, speeding the overall computation in search for the optimal solution.

4.6.4 Voltage Drop on a Radial Feeder Serving a Triangular Area

Now, consider a case in which the service area of a feeder is triangular with load distributed uniformly in the service area as shown in Figure 4.16. Such cases are common in distribution systems.

Let a be the current density in amperes per square mile. In terms of the dimensions of the service area, we get an expression for a, or

$$a = \frac{2I_s}{lh} \tag{4.77}$$

Next, we get an expression for current I_x at distance x.

$$I_x = I_s - a\frac{h'x}{2} \tag{4.78}$$

Substituting the value of a gives

$$I_s - I_s\frac{2}{lh}\frac{h'x}{2} \tag{4.79}$$

Note that $\frac{h'}{h} = \frac{x}{l}$. Substituting this in (4.81) gives

$$I_x = I_s - I_s\frac{x^2}{l^2} \tag{4.80}$$

Again, consider a differential element dx and write an expression for the differential voltage drop

$$dV_D = I_x z dx \tag{4.81}$$

Further, the voltage drop across the feeder is

$$V_{Dl} = \int_0^l I_s\left(1 - \frac{x^2}{l^2}\right) z dx \tag{4.82}$$

or

$$V_{Dl} = \frac{2}{3} I_s z l \tag{4.83}$$

Similarly, we can find power loss, which is

$$dP_{LSx} = I_x^2 r dx \tag{4.84}$$

$$= I_S^2 \left(1 - \frac{x^2}{l^2}\right)^2 r dx \tag{4.85}$$

Then, the power loss per phase is

$$P_{LS} = \int_0^l I_S^2 \left(1 - \frac{x^2}{l^2}\right)^2 r dx$$

$$= \frac{8}{15} I_S^2 r l \tag{4.86}$$

and

$$P_{LS3\emptyset} = \frac{8}{5} I_S^2 r l \tag{4.87}$$

4.7 Fault Calculations

The standard procedure for fault calculation in power systems requires determining Thevenin's equivalent for positive-, negative-, and zero-sequence networks at the point of fault. The sequence networks are connected to each other based on the type of fault. However, decoupling of circuits in the sequence domain works only under the conditions of symmetry. For example, consider a three-phase feeder with impedance matrix in the phase domain as shown below

$$\begin{bmatrix} \mathbf{Z}_{aa} & \mathbf{Z}_{ab} & \mathbf{Z}_{ac} \\ \mathbf{Z}_{ba} & \mathbf{Z}_{bb} & \mathbf{Z}_{bc} \\ \mathbf{Z}_{ca} & \mathbf{Z}_{cb} & \mathbf{Z}_{cc} \end{bmatrix} \tag{4.88}$$

The conditions of symmetry require that

$$\mathbf{Z}_{aa} = \mathbf{Z}_{bb} = \mathbf{Z}_{cc}$$
$$\text{and} \tag{4.89}$$
$$\mathbf{Z}_{ab} = \mathbf{Z}_{ac} = \mathbf{Z}_{bc}$$

While for transmission lines these are satisfied with transposition, they are not feasible for distribution feeders. Distribution feeders are rarely transposed and also can be of two or one phase in addition to three phases. Since single- and two-phase feeders do not have models in the sequence domain, symmetrical component-based analysis is not applicable to them. Also, for three-phase part of the system, symmetrical components do not provide any advantage because the sequence impedance matrix is not diagonal. Hence, positive-, negative-, and zero-sequence impedances cannot be decoupled from one another. However, we can implement a solution technique in phase domain to determine fault currents in distribution systems. We initially consider a radial distribution system connected to the bulk power system with no additional sources in the system. However, later we expand the method to include additional sources.

4.7.1 Prefault System

Consider a general distribution feeder connected to the substation bus with multiple feeder sections as shown in Figure 4.17. While the main feeder is a three-phase feeder, the laterals can be three, two, or single phase. Each feeder section is modeled by its impedance matrix. For illustration, we are considering delta-wye-grounded connection for the transformer, but the method will work for other configurations too.

The system shown in the oval consists of the entire power system upstream of the transformer. This system can be represented by a Thevenin equivalent impedance matrix and a voltage source vector for the three phases. The equivalent voltage source $\boldsymbol{V}_s$ is considered balanced with 1 pu magnitude, but other values can also be used. Therefore,

$$\begin{bmatrix} V_s^a \\ V_s^b \\ V_s^c \end{bmatrix} = \begin{bmatrix} 1.0\angle 0^\circ \\ 1.0\angle -120^\circ \\ 1.0\angle 120^\circ \end{bmatrix} \tag{4.90}$$

where subscript s is the source. All the prefault load currents are considered to be 0 with the assumption that load currents are much smaller compared to the fault current.

Since directly determining the Thevenin equivalent matrix of the system in the phase domain is difficult, we consider an indirect approach. Usually, the

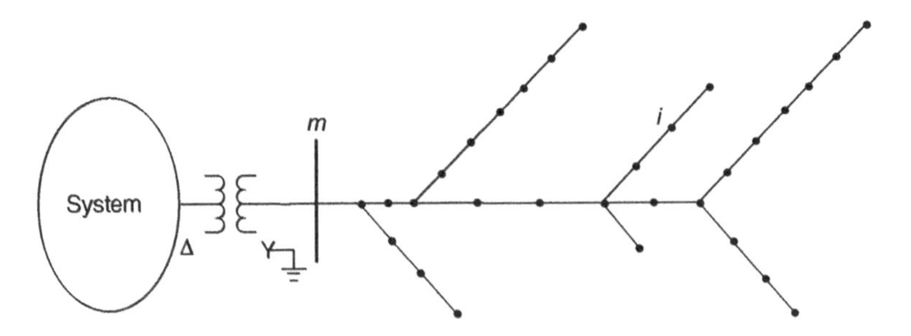

Figure 4.17 An example distribution system.

utilities have information on equivalent positive-, negative-, and zero-sequence impedance values at different buses in the transmission system. They can also be computed using the approach discussed in Section 4.2. Let their values be $\boldsymbol{Z}_1^{sm}$, $\boldsymbol{Z}_2^{sm}$, and $\boldsymbol{Z}_0^{sm}$ at bus m. Note that the zero-sequence impedance value will have only the impedance of the transformer due to the selected transformer connection. Delta/wye-grounded transformers create an open circuit in zero-sequence circuit, where the current is able to flow through the transformer only on the wye-grounded side. We assume that the transmission system is fully balanced, and as a result of that, the sequence networks are decoupled. Hence, we can write the impedance matrix of the system in the sequence domain as follows:

$$\begin{bmatrix} \boldsymbol{Z}_0^{sm} & 0 & 0 \\ 0 & \boldsymbol{Z}_1^{sm} & 0 \\ 0 & 0 & \boldsymbol{Z}_2^{sm} \end{bmatrix} \tag{4.91}$$

Now, we use transformation to convert this matrix to phase domain

$$\begin{bmatrix} \boldsymbol{Z}_{aa}^{sm} & \boldsymbol{Z}_{ab}^{sm} & \boldsymbol{Z}_{ac}^{sm} \\ \boldsymbol{Z}_{ba}^{sm} & \boldsymbol{Z}_{bb}^{sm} & \boldsymbol{Z}_{bc}^{sm} \\ \boldsymbol{Z}_{ca}^{sm} & \boldsymbol{Z}_{cb}^{sm} & \boldsymbol{Z}_{cc}^{sm} \end{bmatrix} = \begin{bmatrix} 1 & 1 & 1 \\ 1 & \mathbf{a}^2 & \mathbf{a} \\ 1 & \mathbf{a} & \mathbf{a}^2 \end{bmatrix} \begin{bmatrix} \boldsymbol{Z}_0^{sm} & 0 & 0 \\ 0 & \boldsymbol{Z}_1^{sm} & 0 \\ 0 & 0 & \boldsymbol{Z}_2^{sm} \end{bmatrix} \frac{1}{3} \begin{bmatrix} 1 & 1 & 1 \\ 1 & \mathbf{a} & \mathbf{a}^2 \\ 1 & \mathbf{a}^2 & \mathbf{a} \end{bmatrix} \tag{4.92}$$

where $\boldsymbol{a}$ is a complex number equal to $1\angle 120°$ as discussed in Appendix B. Note that we have used the same terminology as Eq. (4.4) to represent the matrix on the left, but the two matrices are not the same.

Let bus i be the candidate bus for faults. Since all the feeder sections between the substation bus (bus m) and the faulted bus are in series, we can add their impedance matrices to compute the equivalent impedance matrix. All the feeder sections not in the path from the substation to the faulted bus do not influence the fault current and thus are not included in the calculations. Consider this matrix to be

$$\begin{bmatrix} \boldsymbol{Z}_{aa}^{mi} & \boldsymbol{Z}_{ab}^{mi} & \boldsymbol{Z}_{ac}^{mi} \\ \boldsymbol{Z}_{ba}^{mi} & \boldsymbol{Z}_{bb}^{mi} & \boldsymbol{Z}_{bc}^{mi} \\ \boldsymbol{Z}_{ca}^{mi} & \boldsymbol{Z}_{cb}^{mi} & \boldsymbol{Z}_{cc}^{mi} \end{bmatrix} \tag{4.93}$$

Now, this matrix is added to the system impedance matrix to get the impedance matrix from the equivalent source to the faulted bus, or

$$\begin{bmatrix} \boldsymbol{Z}_{aa}^{si} & \boldsymbol{Z}_{ab}^{si} & \boldsymbol{Z}_{ac}^{si} \\ \boldsymbol{Z}_{ba}^{si} & \boldsymbol{Z}_{bb}^{si} & \boldsymbol{Z}_{bc}^{si} \\ \boldsymbol{Z}_{ca}^{si} & \boldsymbol{Z}_{cb}^{si} & \boldsymbol{Z}_{cc}^{si} \end{bmatrix} = \begin{bmatrix} \boldsymbol{Z}_{aa}^{sm} & \boldsymbol{Z}_{ab}^{sm} & \boldsymbol{Z}_{ac}^{sm} \\ \boldsymbol{Z}_{ba}^{sm} & \boldsymbol{Z}_{bb}^{sm} & \boldsymbol{Z}_{bc}^{sm} \\ \boldsymbol{Z}_{ca}^{sm} & \boldsymbol{Z}_{cb}^{sm} & \boldsymbol{Z}_{cc}^{sm} \end{bmatrix} + \begin{bmatrix} \boldsymbol{Z}_{aa}^{mi} & \boldsymbol{Z}_{ab}^{mi} & \boldsymbol{Z}_{ac}^{mi} \\ \boldsymbol{Z}_{ba}^{mi} & \boldsymbol{Z}_{bb}^{mi} & \boldsymbol{Z}_{bc}^{mi} \\ \boldsymbol{Z}_{ca}^{mi} & \boldsymbol{Z}_{cb}^{mi} & \boldsymbol{Z}_{cc}^{mi} \end{bmatrix} \tag{4.94}$$

Note that this summation can only be done if both matrices on the right-hand side are in per unit, or impedances are referred to the low-voltage side of the

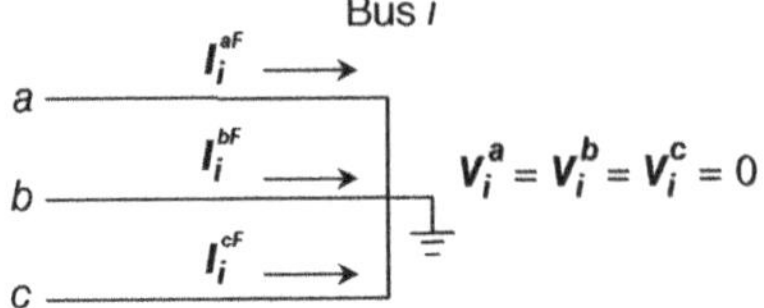

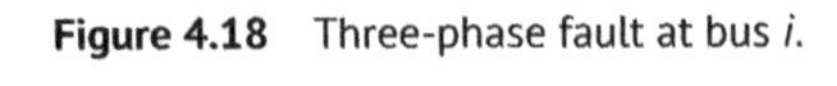
Figure 4.18 Three-phase fault at bus *i*.

transformer due to different voltage levels of the distribution system and the bulk power system.

4.7.2 Three-phase Fault

For a three-phase fault at bus *i*, current will flow from the substation to bus *i*. Since there is no additional source, no other currents will flow. Also, we assume that the fault impedance is 0. If the fault is a high impedance fault, these assumptions will not be fully valid. Figure 4.18 shows the conditions for a three-phase fault.

The fault currents are

$$I_i^{abcF} = \begin{bmatrix} I_i^{aF} \\ I_i^{bF} \\ I_i^{cF} \end{bmatrix} \tag{4.95}$$

Now, we can write an equation for voltage drop from the source to bus *i*.

$$\begin{bmatrix} V_s^a - V_i^a \\ V_s^b - V_i^b \\ V_s^c - V_i^c \end{bmatrix} = \begin{bmatrix} Z_{aa}^{si} & Z_{ab}^{si} & Z_{ac}^{si} \\ Z_{ba}^{si} & Z_{bb}^{si} & Z_{bc}^{si} \\ Z_{ca}^{si} & Z_{cb}^{si} & Z_{cc}^{si} \end{bmatrix} \begin{bmatrix} I_i^{aF} \\ I_i^{bF} \\ I_i^{cF} \end{bmatrix} \tag{4.96}$$

Note that for a solidly grounded fault, all the voltages at bus *i* will be 0. Applying this condition and substituting values from Eq. (4.90) gives

$$\begin{bmatrix} 1.0\angle 0^\circ - 0 \\ 1.0\angle -120^\circ - 0 \\ 1.0\angle 120^\circ - 0 \end{bmatrix} = \begin{bmatrix} Z_{aa}^{si} & Z_{ab}^{si} & Z_{ac}^{si} \\ Z_{ba}^{si} & Z_{bb}^{si} & Z_{bc}^{si} \\ Z_{ca}^{si} & Z_{cb}^{si} & Z_{cc}^{si} \end{bmatrix} \begin{bmatrix} I_i^{aF} \\ I_i^{bF} \\ I_i^{cF} \end{bmatrix} \tag{4.97}$$

In the next step, we get

$$\begin{bmatrix} I_i^{aF} \\ I_i^{bF} \\ I_i^{cF} \end{bmatrix} = \begin{bmatrix} Z_{aa}^{si} & Z_{ab}^{si} & Z_{ac}^{si} \\ Z_{ba}^{si} & Z_{bb}^{si} & Z_{bc}^{si} \\ Z_{ca}^{si} & Z_{cb}^{si} & Z_{cc}^{si} \end{bmatrix}^{-1} \begin{bmatrix} 1.0\angle 0^\circ \\ 1.0\angle -120^\circ \\ 1.0\angle 120^\circ \end{bmatrix} \tag{4.98}$$

If the system has additional sources, we can apply them one at a time while removing the equivalent source at the substation. However, we have to know the source impedance matrix and determine the equivalent impedance matrix from each source to bus *i* and apply Eq. (4.85) to compute fault currents. Additionally, the voltages at the source terminal will be needed for the calculations. If the sources are inverter based, we need to know their characteristics under faults and the operating rules. Some inverter-based sources are automatically disconnected during fault, and most are adjusted to produce a smaller current by adjusting the impedance of inverter, which makes fault calculations challenging. However, if we can compute fault currents for all the sources one at a time, all the fault currents can be added using superposition to compute the cumulative current. Since the circuit is linear, superposition can be applied without affecting the results.

4.7.3 Double-Line-to-Ground (DLG) Fault

For a DLG fault at bus i, current will flow from the substation to bus i on the two faulted phases and no current on the third phase. For illustration, we consider a fault between phases b and c as shown in Figure 4.19.

The currents for this fault are

$$\boldsymbol{I}_i^{abcF} = \begin{bmatrix} 0 \\ \boldsymbol{I}_i^{bF} \\ \boldsymbol{I}_i^{cF} \end{bmatrix} \tag{4.99}$$

Now we can write an equation for voltage drop from the source to bus i.

$$\begin{bmatrix} 1.0\angle 0^\circ - \boldsymbol{V}_i^a \\ 1.0\angle -120^\circ - 0 \\ 1.0\angle 120^\circ - 0 \end{bmatrix} = \begin{bmatrix} \boldsymbol{Z}_{aa}^{si} & \boldsymbol{Z}_{ab}^{si} & \boldsymbol{Z}_{ac}^{si} \\ \boldsymbol{Z}_{ba}^{si} & \boldsymbol{Z}_{bb}^{si} & \boldsymbol{Z}_{bc}^{si} \\ \boldsymbol{Z}_{ca}^{si} & \boldsymbol{Z}_{cb}^{si} & \boldsymbol{Z}_{cc}^{si} \end{bmatrix} \begin{bmatrix} 0 \\ \boldsymbol{I}_i^{bF} \\ \boldsymbol{I}_i^{cF} \end{bmatrix} \tag{4.100}$$

Discard the first row and the first column and write the equations in a reduced form, or

$$\begin{bmatrix} 1.0\angle -120^\circ \\ 1.0\angle 120^\circ \end{bmatrix} = \begin{bmatrix} \boldsymbol{Z}_{bb}^{si} & \boldsymbol{Z}_{bc}^{si} \\ \boldsymbol{Z}_{cb}^{si} & \boldsymbol{Z}_{cc}^{si} \end{bmatrix} \begin{bmatrix} \boldsymbol{I}_i^{bF} \\ \boldsymbol{I}_i^{cF} \end{bmatrix} \tag{4.101}$$

Further,

$$\begin{bmatrix} \boldsymbol{I}_i^{bF} \\ \boldsymbol{I}_i^{cF} \end{bmatrix} = \begin{bmatrix} \boldsymbol{Z}_{bb}^{si} & \boldsymbol{Z}_{bc}^{si} \\ \boldsymbol{Z}_{cb}^{si} & \boldsymbol{Z}_{cc}^{si} \end{bmatrix}^{-1} \begin{bmatrix} 1.0\angle -120^\circ \\ 1.0\angle 120^\circ \end{bmatrix} \tag{4.102}$$

The same approach can be used for faults on any other two phases.

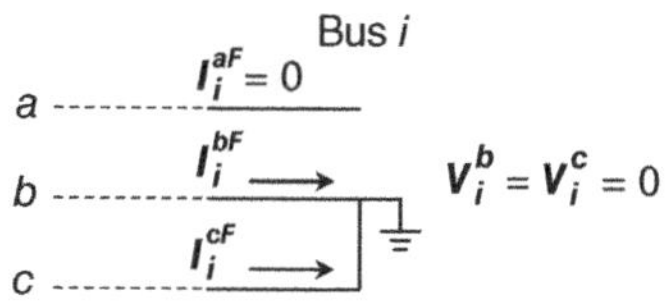

Figure 4.19 DLG fault between phases b and c at bus i.

4.7.4 Single-Line-to-Ground (SLG) Fault

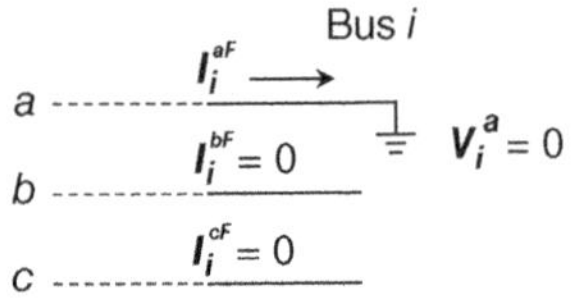

For a SLG fault at bus i, current will flow from the substation to bus i on the faulted phase and no current on the other two phases. For illustration, we consider a fault on phase a as shown in Figure 4.20.

Figure 4.20 SLG fault on phase a at bus i.

The currents for this fault are

$$I_i^{abcF} = \begin{bmatrix} I_i^{aF} \\ 0 \\ 0 \end{bmatrix} \tag{4.103}$$

We can write an equation for voltage drop from the equivalent source to bus i.

$$\begin{bmatrix} 1.0\angle 0^\circ - 0 \\ 1.0\angle -120^\circ - V_i^b \\ 1.0\angle 120^\circ - V_i^c \end{bmatrix} = \begin{bmatrix} Z_{aa}^{si} & Z_{ab}^{si} & Z_{ac}^{si} \\ Z_{ba}^{si} & Z_{bb}^{si} & Z_{bc}^{si} \\ Z_{ca}^{si} & Z_{cb}^{si} & Z_{cc}^{si} \end{bmatrix} \begin{bmatrix} I_i^{aF} \\ 0 \\ 0 \end{bmatrix} \tag{4.104}$$

Discard the last two equations and keep only the first one, or

$$1.0\angle 0^\circ = Z_{aa}^{si} I_i^{aF} \tag{4.105}$$

Therefore,

$$I_i^{aF} = \frac{1.0\angle 0^\circ}{Z_{aa}^{1i}} \tag{4.106}$$

Again, the same approach can be used for faults on other phases.

4.7.5 Line-to-Line (LL) Fault

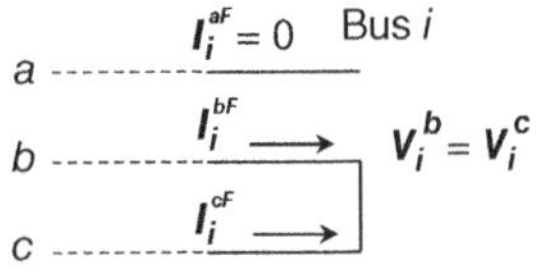

For an LL fault at bus i, current will flow from the substation to bus i on the faulted phases and no current on the third phases. For illustration, we consider a fault between phase b and c as shown in Figure 4.21.

Figure 4.21 LL fault between phases b and c at bus i.

For this fault, the currents will not flow to the ground but return through the second phase, or

$$I_i^{cF} = -I_i^{bF} \tag{4.107}$$

Hence, the currents for this fault are

$$I_i^{abcF} = \begin{bmatrix} 0 \\ I_i^{bF} \\ -I_i^{bF} \end{bmatrix} \tag{4.108}$$

Also, the voltages at bus i for the faulted phases are equal, or

$$V_i^b = V_i^c \tag{4.109}$$

Therefore, the equation for voltage drop from the equivalent source to bus i is

$$\begin{bmatrix} 1.0\angle 0^\circ - V_i^a \\ 1.0\angle -120^\circ - V_i^b \\ 1.0\angle 120^\circ - V_i^b \end{bmatrix} = \begin{bmatrix} Z_{aa}^{si} & Z_{ab}^{si} & Z_{ac}^{si} \\ Z_{ba}^{si} & Z_{bb}^{si} & Z_{bc}^{si} \\ Z_{ca}^{si} & Z_{cb}^{si} & Z_{cc}^{si} \end{bmatrix} \begin{bmatrix} 0 \\ I_i^{bF} \\ -I_i^{bF} \end{bmatrix} \tag{4.110}$$

Expanding the last two equations gives

$$1.0\angle -120^\circ - V_i^b = Z_{bb}^{si} I_i^{bF} - Z_{bc}^{si} I_i^{bF} = \left(Z_{bb}^{si} - Z_{bc}^{si}\right) I_i^{bF} \tag{4.111}$$

and

$$1.0\angle 120^\circ - V_i^b = Z_{cb}^{si} I_i^{bF} - Z_{cc}^{si} I_i^{bF} = \left(Z_{cb}^{si} - Z_{cc}^{si}\right) I_i^{bF} \tag{4.112}$$

Rearranging and expressing in the matrix form gives

$$\begin{bmatrix} I_i^{bF} \\ V_i^b \end{bmatrix} = \begin{bmatrix} \left(Z_{bb}^{si} - Z_{bc}^{si}\right) & 1 \\ \left(Z_{cb}^{si} - Z_{cc}^{si}\right) & 1 \end{bmatrix}^{-1} \begin{bmatrix} 1.0\angle -120^\circ \\ 1.0\angle 120^\circ \end{bmatrix} \tag{4.113}$$

Solving it gives both I_i^{bF} and V_i^b as well as I_i^{cF} and V_i^c.

4.7.6 Symmetrical Component-based Fault Analysis

Although symmetrical component-based fault analysis does not work well for distribution systems, it can be used to get approximate results only for the part of the system that has all the three phases. The first step is to determine the three-phase impedance matrix from the substation bus (bus m) to the faulted bus (bus i), which in general is given by Eq. (4.91). The next step is to transform this matrix to sequence domain, which gives

$$\begin{bmatrix} Z_{00}^{mi} & Z_{01}^{mi} & Z_{02}^{mi} \\ Z_{10}^{mi} & Z_{11}^{mi} & Z_{12}^{mi} \\ Z_{20}^{mi} & Z_{21}^{mi} & Z_{22}^{mi} \end{bmatrix} = \frac{1}{3} \begin{bmatrix} 1 & 1 & 1 \\ 1 & a & a^2 \\ 1 & a^2 & a \end{bmatrix} \begin{bmatrix} Z_{aa}^{mi} & Z_{ab}^{mi} & Z_{ac}^{mi} \\ Z_{ba}^{mi} & Z_{bb}^{mi} & Z_{bc}^{mi} \\ Z_{ca}^{mi} & Z_{cb}^{mi} & Z_{cc}^{mi} \end{bmatrix} \begin{bmatrix} 1 & 1 & 1 \\ 1 & a^2 & a \\ 1 & a & a^2 \end{bmatrix} \tag{4.114}$$

As discussed previously, the resulting sequence impedance matrix will not be diagonal, but for approximation we can discard the off-diagonal terms and consider the diagonal entries as the zero-, positive-, and negative-sequence impedances of the distribution system from the substation bus to the point of fault. To determine the prefault circuits in the sequence domain, we add the respective impedances from the equivalent source to the substation bus and the impedances from the substation bus to the faulted bus. Again, all the impedances

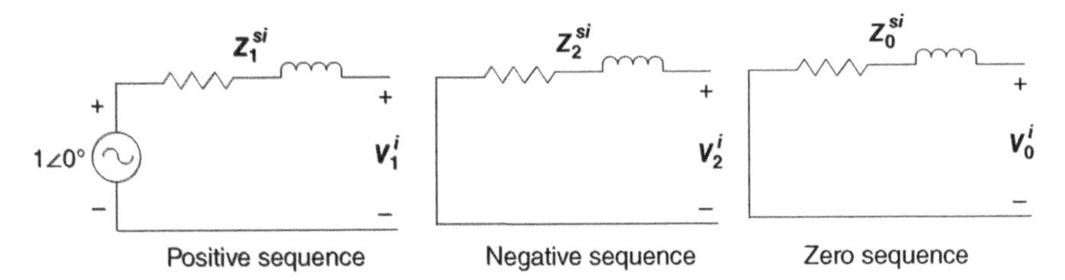

Figure 4.22 Prefault positive-, negative-, and zero-sequence equivalent circuits with respect to bus i.

must be converted to per unit before we can add them. Also, note that since the voltage source is balanced in the phase domain, it gives a source only for the positive sequence. Hence, we can determine the three prefault sequence domain circuits as shown in Figure 4.22.

Note that in the figure

$$\begin{aligned} Z_1^{si} &= Z_1^{sm} + Z_{11}^{mi} \\ Z_2^{si} &= Z_2^{sm} + Z_{22}^{mi} \\ Z_0^{si} &= Z_0^{sm} + Z_{00}^{mi} \end{aligned} \tag{4.115}$$

4.7.6.1 Three-phase Fault

For a three-phase fault, the system stays balanced, and the negative- and the zero-sequence currents are 0. To compute the positive-sequence current, we create a short circuit across the positive-sequence circuit as shown in Figure 4.23.

From this circuit

$$I_1^F = \frac{1.0\angle 0°}{Z_1^{si}} \tag{4.116}$$

Now, we convert them to the phase domain

$$\begin{bmatrix} I_i^{aF} \\ I_i^{bF} \\ I_i^{cF} \end{bmatrix} = \begin{bmatrix} 1 & 1 & 1 \\ 1 & a^2 & a \\ 1 & a & a^2 \end{bmatrix} \begin{bmatrix} 0 \\ I_1^F \\ 0 \end{bmatrix} = \begin{bmatrix} I_1^F \\ I_1^F\angle -120° \\ I_1^F\angle 120° \end{bmatrix} \tag{4.117}$$

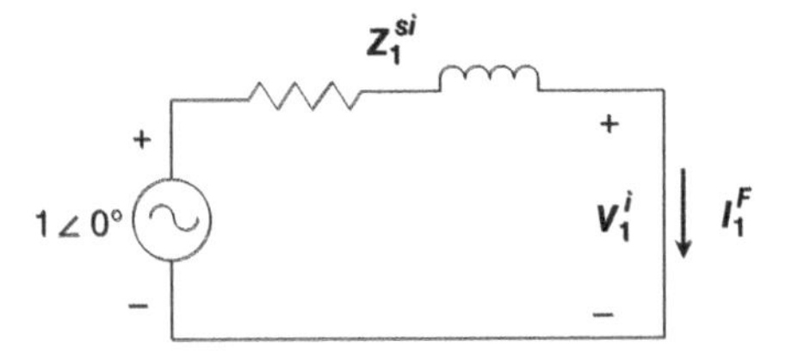

Figure 4.23 Three-phase fault in sequence domain.

4.7.6.2 DLG Fault

The conditions for this fault in phase domain are $\boldsymbol{I}_i^{aF} = 0$ and $\boldsymbol{V}_i^b = \boldsymbol{V}_i^c$. Transforming these conditions to sequence domain gives

$$\boldsymbol{I}_0^F + \boldsymbol{I}_1^F + \boldsymbol{I}_2^F = 0 \tag{4.118}$$

and

$$\boldsymbol{V}_0^F + \boldsymbol{a}\boldsymbol{V}_1^F + \boldsymbol{a}^2\boldsymbol{V}_2^F = \boldsymbol{V}_0^F + \boldsymbol{a}^2\boldsymbol{V}_1^F + \boldsymbol{a}\boldsymbol{V}_2^F \tag{4.119}$$

Simplifying gives

$$(\boldsymbol{a} - \boldsymbol{a}^2)\boldsymbol{V}_1^F = (\boldsymbol{a} - \boldsymbol{a}^2)\boldsymbol{V}_2^F \tag{4.120}$$

or

$$\boldsymbol{V}_1^F = \boldsymbol{V}_2^F \tag{4.121}$$

Also,

$$\boldsymbol{V}_i^b = \boldsymbol{V}_0^F + \boldsymbol{a}\boldsymbol{V}_1^F + \boldsymbol{a}^2\boldsymbol{V}_2^F = 0 \tag{4.122}$$

or

$$\boldsymbol{V}_0^F + (\boldsymbol{a} + \boldsymbol{a}^2)\boldsymbol{V}_1^F = \boldsymbol{V}_0^F - \boldsymbol{V}_1^F = 0 \tag{4.123}$$

Therefore,

$$\boldsymbol{V}_0^F = \boldsymbol{V}_1^F = \boldsymbol{V}_2^F \tag{4.124}$$

We use conditions given by (4.115) and (4.124) to create the circuit in sequence domain shown in Figure 4.24.

From this circuit, we get

$$\boldsymbol{I}_1^F = \frac{1.0\angle 0^\circ}{\boldsymbol{Z}_1^{si} + (\boldsymbol{Z}_2^{si} \| \boldsymbol{Z}_0^{si})} = \frac{1.0\angle 0^\circ}{\boldsymbol{Z}_1^{si} + \left(\frac{\boldsymbol{Z}_2^{si} + \boldsymbol{Z}_0^{si}}{\boldsymbol{Z}_2^{si}\,\boldsymbol{Z}_0^{si}}\right)} \tag{4.125}$$

$$\boldsymbol{I}_2^F = -\boldsymbol{I}_1^F\left(\frac{\boldsymbol{Z}_0^{si}}{\boldsymbol{Z}_2^{si} + \boldsymbol{Z}_0^{si}}\right) \tag{4.126}$$

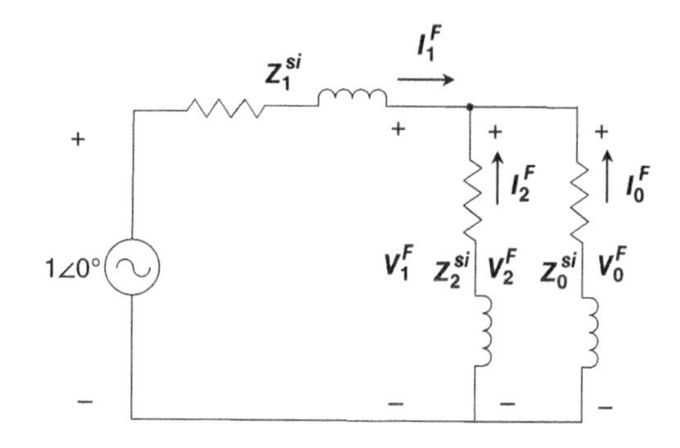

Figure 4.24 DLG fault in sequence domain.

and

$$I_0^F = -I_1^F \left(\frac{Z_2^{si}}{Z_2^{si} + Z_0^{si}} \right) \tag{4.127}$$

The sequence currents and voltages can be transformed to the phase domain using the relevant equations.

4.7.6.3 SLG Fault

The conditions for this fault in phase domain are $I_i^{bF} = I_i^{cF} = 0$ and $V_i^a = 0$. Transforming these conditions to sequence domain gives

$$\begin{bmatrix} I_0^F \\ I_1^F \\ I_2^F \end{bmatrix} = \frac{1}{3} \begin{bmatrix} 1 & 1 & 1 \\ 1 & a & a^2 \\ 1 & a^2 & a \end{bmatrix} \begin{bmatrix} I_i^{aF} \\ 0 \\ 0 \end{bmatrix} = \frac{1}{3} \begin{bmatrix} I_i^{aF} \\ I_i^{aF} \\ I_i^{aF} \end{bmatrix} \tag{4.128}$$

or

$$I_0^F = I_1^F = I_2^F = \frac{1}{3} I_i^{aF} \tag{4.129}$$

Also,

$$V_i^a = V_0^F + V_1^F + V_2^F = 0 \tag{4.130}$$

Based on (4.129) and (4.130), we can create a circuit shown in Figure 4.25.

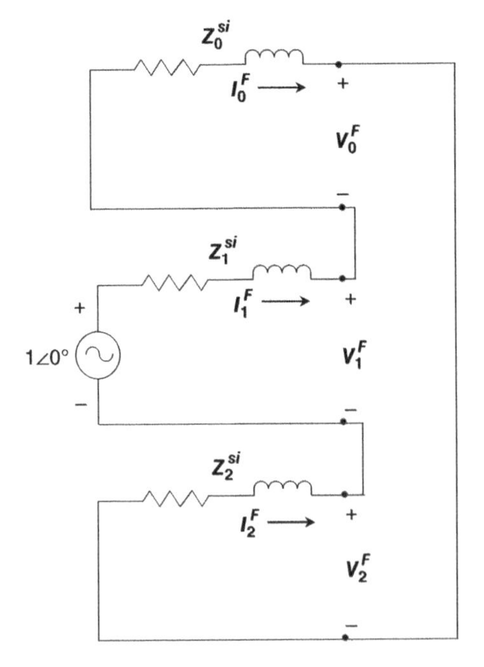

Figure 4.25 SLG fault in sequence domain.

From this circuit, we get

$$I_0^F = I_1^F = I_2^F = \frac{1.0\angle 0^\circ}{Z_0^{si} + Z_1^{si} + Z_2^{si}} \tag{4.131}$$

and

$$I_i^{aF} = 3I_1^F \tag{4.132}$$

4.7.6.4 LL Fault

The conditions for this fault in the phase domain are $I_i^{aF} = 0$, $I_i^{cF} = -I_i^{bF}$, and $V_i^b = V_i^c$. Transforming these conditions to sequence domain gives

$$\begin{bmatrix} I_0^F \\ I_1^F \\ I_2^F \end{bmatrix} = \frac{1}{3}\begin{bmatrix} 1 & 1 & 1 \\ 1 & a & a^2 \\ 1 & a^2 & a \end{bmatrix}\begin{bmatrix} 0 \\ I_i^{bF} \\ -I_i^{bF} \end{bmatrix} = \frac{1}{3}\begin{bmatrix} 0 \\ (a-a^2)I_i^{bF} \\ (a^2-a)I_i^{bF} \end{bmatrix} \tag{4.133}$$

and

$$V_0^F + a^2V_1^F + aV_2^F = V_0^F + aV_1^F + a^2V_2^F \tag{4.134}$$

Therefore,

$$I_2^F = -I_1^F = \frac{1}{3}(a-a^2)I_i^{bF} \tag{4.135}$$

and

$$(a^2-a)V_1^F = (a^2-a)V_2^F \tag{4.136}$$

or

$$V_1^F = V_2^F \tag{4.137}$$

Based on (4.135) and (4.137), we can create a circuit shown in Figure 4.26.

$$I_1^F = -I_2^F = \frac{1.0\angle 0^\circ}{Z_1^{si} + Z_2^{si}} \tag{4.138}$$

and

$$\begin{aligned} I_i^{bF} &= I_0^F + a^2I_1^F + aI_2^F = (a^2-a)I_1^F \\ &= -j\sqrt{3}I_1^F \end{aligned} \tag{4.139}$$

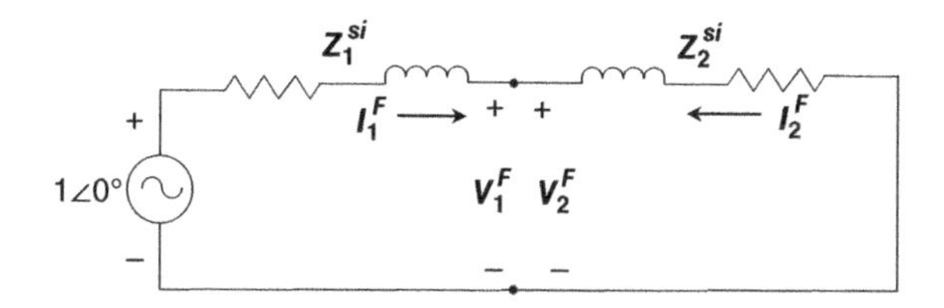

Figure 4.26 LL fault in sequence domain.

Problems

4.1 Consider a single-phase load with $P = 800$ kW and $Q = 600$ kvar connected to a 7.2-kV voltage source. Find current drawn by the load and the impedance of the load.

4.2 Consider that the load of Problem 4.1 is connected to the 7.2-kV source through a line of impedance of $j5\ \Omega$. Compute the voltage at the load terminals, current, and real and reactive power drawn by the load while considering the load to be (i) constant power, (ii) constant current, and (iii) constant impedance as determined in Problem 4.1.

4.3 In the 12.47-kV system shown in Figure 4.27, Line 1 is a 1.5-mile-long three-phase line, Line 2 is a 0.5-mile-long two-phase line with phases A and B, and Line 3 is a 0.3-mile-long single-phase line with only phase C. Their impedance matrices are given below. Consider the following loads at the end of the feeders:

Phase A: 400 kVA at 0.95 lagging power factor
Phase B: 600 kVA at 0.85 leading power factor
Phase C: 500 kVA at 0.90 lagging power factor

Using the Source-Load-Iteration Method, manually compute voltages in all phases A, B, and C at buses 2, 3, and 4 at the end of first iteration.

Write a MATLAB program to solve voltages in the given system based on the Source-Load-Iteration Method. Run forward–backward iterations until voltages at all the buses in subsequent iterations are within 0.1%.

Line 1:

$$[Z^{abc}] = \begin{bmatrix} 0.34 + j1.05 & 0.16 + j0.50 & 0.15 + j0.39 \\ 0.16 + j0.50 & 0.35 + j1.02 & 0.16 + j0.42 \\ 0.15 + j0.39 & 0.16 + j0.42 & 0.34 + j1.03 \end{bmatrix} \Omega/\text{mile}$$

Line 2:

$$[Z^{abc}] = \begin{bmatrix} 0.46 + j1.08 & 0.16 + j0.50 \\ 0.16 + j0.50 & 0.35 + j1.02 \end{bmatrix} \Omega/\text{mile}$$

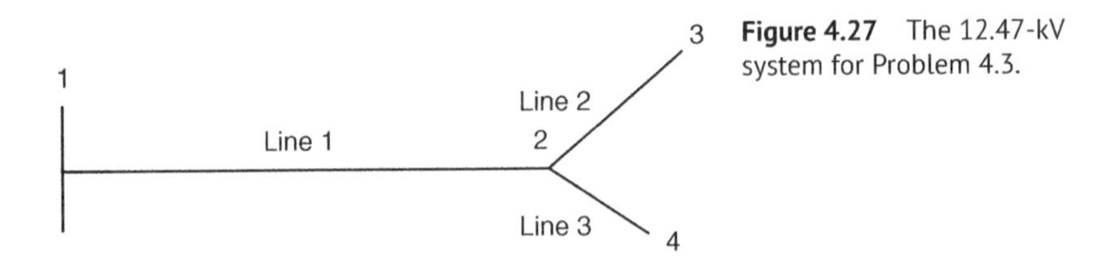

Figure 4.27 The 12.47-kV system for Problem 4.3.

Line 3:

$$[Z^{abc}] = \begin{bmatrix} 0 & 0 & 0 \\ 0 & 0 & 0 \\ 0 & 0 & 0.60 + j1.20 \end{bmatrix} \Omega/\text{mile}$$

4.4 Use approximate method for voltage drop to determine voltage drops on phases a, b, and c of Line 1, Line 2, and Line 3. Determine voltage regulation for all the respective phases at buses 1, 2, and 3. What are the voltage regulations based on the results of power flow in Problem 4.3? (Note: Use only the self-impedance of the lines for these calculations.)

4.5 Consider that a feeder serves an area shown in Figure 4.28. Then consider the load density to be *a* A/square mile. Find an expression for voltage drop on the feeder.

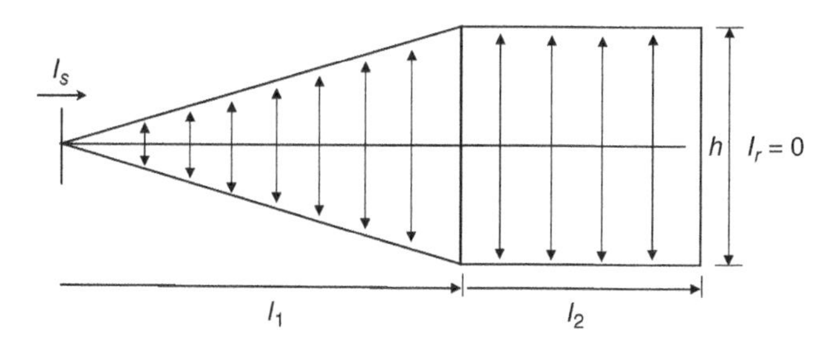

Figure 4.28 Service area for Problem 4.5.

4.6 Consider that the system given in Problem 4.3 is connected to a system with positive- and negative-sequence impedances of $0.255 + j\ 2.291$ ohms and zero-sequence impedance of $0 + j1.089$ ohms. Use the transformation given in Eq. (4.92) to determine the equivalent source impedance in the phase domain. Consider the source voltages to be balanced with 1 pu magnitude. Using the phase domain method

(a) Find currents for an SLG fault on bus 4.
(b) Find currents for DLG and LL faults on bus 3.

4.7 Using the phase domain analysis find currents for three phase, DLG, SLG, and LL faults on bus 2 of the system of Problem 4.3.

4.8 Determine the sequence impedances of Line 1 using the relevant transformation. Using the symmetrical component method, find current for three phases, DLG, SLG, and LL faults on bus 2 of the system of Problem 4.3. Do these results match with those found in Problem 4.7? If not, what are the errors in percentage based on the values obtained in Problem 4.7?

References

1 Adler, R.B. and Mosher, C.C. (1974). (This book has a series of papers on modeling.). Steady-state power characteristics for power loads. In: *Stability of Large Electric Power Systems* (ed. R.T. Byerly and E.W. Kimbark), 147–153. IEEE Press.
2 Kent, M.H., Schmus, W.R., and McCrackin, F.A. Dynamic modeling of loads in the stability studies. In: *Stability of Large Electric Power Systems*, 139–146. IEEE Press.
3 EPRI Project Report (1974). Long Term Power System Dynamics, Vol. I, EPRI RP. 90-7-0.
4 Lakervi, E. and Holmes, E.J. (1989). *Electricity Distribution Network Design*. Peter Peregrinus Ltd.
5 IEEE 1547-2018 (2018). IEEE Standard for Interconnection and Interoperability of Distributed Energy Resources with Associated Electric Power Systems Interfaces. IEEE Standards Coordinating Committee 2, IEEE, New York.

5

Distribution System Planning

5.1 Introduction

In a simple sense, distribution system planning means being prepared to meet the electricity needs of the customers in the future. Since the loads change continuously due to changes in population, technology, and in habits of people, planning for electricity needs becomes a very complex and capital-intensive process. It requires the consideration of several objectives simultaneously while also meeting the technical constraints. The overall goal is to minimize the total cost while ensuring that the system has adequate capacity to supply the load for the future with adequate reliability and acceptable voltage quality. Good plans should be able to address the short-term needs of one to two years as well as the long-term needs ranging from 5 to 10 years.

The first stage in planning is to forecast the load of the future. In addition to knowing the load growth, it is important to know where the load will be growing. After the areas of load growth and future loads are known, the next step is the location and capacity of substations. The number and size of transformers need to be included. The planning results may also include reinforcing the existing substations as well as building new substations. The final step is the feeder design to deliver power from the substations to the customers. Feeder design includes both the primary and the secondary systems. The decisions include the number of primary feeders, the size of conductors and their routing.

At the secondary level, the decisions include the location and size of the distribution transformers. Most of the secondary construction in the United States has been overhead so far. However, now there is a trend toward making new secondary construction underground. Although the underground secondary system can be 5–10 times more expensive than the overhead secondary system, the underground secondary system is preferred by many localities for greater reliability and aesthetics.

Electric Power and Energy Distribution Systems: Models, Methods, and Applications, First Edition.
Subrahmanyam S. Venkata and Anil Pahwa.

Companion website: www.wiley.com/go/Pahwa/ElectricPowerDistributionSystems

5.2 Traditional vs. Modern Approaches to Planning

Legacy distribution systems have very little metering information available from the system. Usually, only loads on the substation transformers and the feeders are measured. Thus, utilities have very little information beyond the feeder. Utilities typically rely on customer billing data and the total installed distribution transformer capacity to get an estimate of the existing load. As part of the load research activity, utilities also installed recording devices at selected customer locations to record loads at a predetermined interval (5 minutes, 15 minutes, 30 minutes, or 1 hour) to get an idea of the daily load profiles for different classes of customers. These load profiles are also useful for planning purposes. However, with limited information available from the system, distribution planning was ad hoc, which would often result in overdesigned systems.

Modern distribution systems have significantly more metering capabilities, both at the system level due to distribution automation and at the customer level due to advanced metering infrastructure. Although many advances in automation and metering have been made over the years, by and large the existing distribution systems around the world do not have much metering capabilities. As more systems become automated, more accurate data from different parts of the system become available to the distribution planners. Such availability of data removes the need to make assumptions about various factors about which no data of information were available.

Traditionally, the distribution systems are designed to meet peak demand. The modern approach, based on risk analysis, requires the number of hours spent at different load levels. These data allow the planners to determine the risk of insufficient capacity to meet the demand over a period of time. Specifically, the number of hours under peak conditions is very important for risk assessment. Modern distribution planning should consider several other factors, such as system reliability, resiliency, aging assets, equipment loading close to margin, and regulatory environment in the planning process. Including details of all these issues is beyond the scope of this chapter. We will give details on only a few of the important factors related to distribution system planning.

5.3 Long-term Load Forecasting

Load forecasting entails determining load in a given area at a future time, which can be from seconds to years. Short-term load forecasts, such as hour ahead or day ahead, are needed for system operation. In contrast, long-term forecasts of loads one year or more into the future are needed for system planning. Since this chapter deals with planning, we only focus on long-term forecasting. Knowledge of future loads is the most important thing for a planner to know. Based on this knowledge the planner can design facilities in the distribution system to meet these loads with a certain level of reliability. Specifically, the planners need to know the peak load and the time of the peak load. In addition to temporal forecasts, spatial forecasts are important because they tell the locations of load growth in addition to the quantity of load growth. Spatial forecasts allow utilities to focus on areas within their territory where there is a need for new facilities.

Load forecasts depend on the local as well as national trends. For example, every city has its forecast of population growth for the future years. It also has knowledge of new industrial or commercial facilities planned for their localities. At the national level, there may be a push to encourage sales of high-efficiency appliances for energy conservation or a promotion to use light-emitting diode (LED) lamps for lighting. Appliance manufacturers and consumer organizations conduct regular surveys to keep track of consumer behaviors to determine these trends. A planners' job is to use information from such surveys and from local sources to create load forecasts. Past trends in actual load measurements within the distribution system are good indicators of future loads. Another approach to forecasting is based on end-use modeling in which loads of different end uses, such as lighting, water heater, and air conditioner, for different customers are added in a hierarchical manner to obtain the composite load.

If we examine the composite load over the years for a utility's service area, we will most likely find a gradual increase in load. However, for small areas, this model is not very appropriate. As discussed in Ref. [1], load growth in a small area follows an S curve as shown in Figure 5.1. Three parts of this curve are the dormant period, the growth period, and the saturation period. The dormant period typically has very little or no load growth. The growth period is when the load rises rapidly, and the saturation period is when the area is fully built, with very little new construction. The load may continue to grow slightly in the saturation

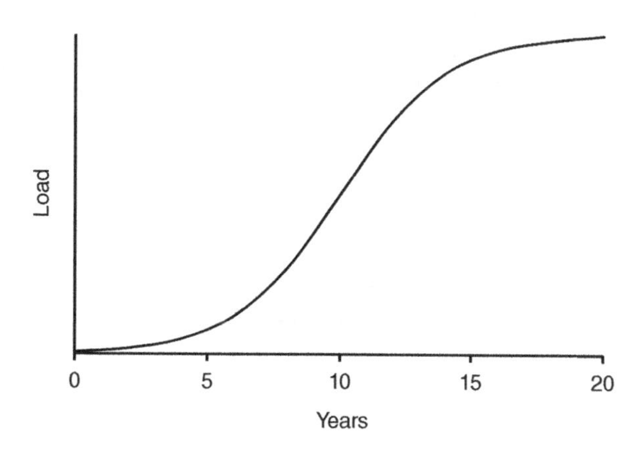

Figure 5.1 S-curve showing load growth in a small area.

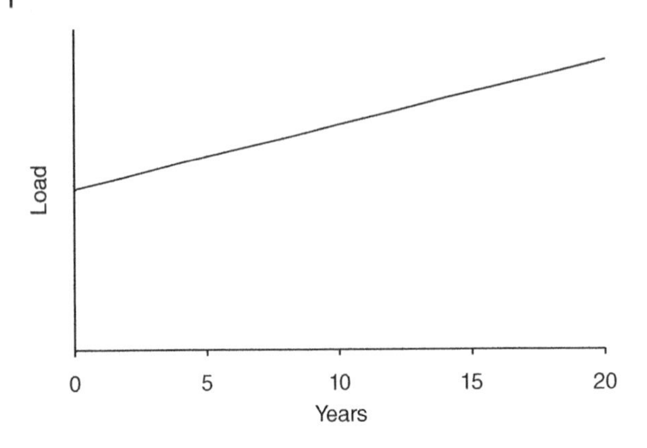

Figure 5.2 Load growth of a service area over the years.

period due to lifestyle changes of the customers in the area. Since all small areas have different S curves, both in terms of load and timing of load, when we combine all the small areas within a utility, the following happens: the composite load shows the linear characteristics with a certain slope showing load growth for a service area as shown in Figure 5.2. Breaking up utility service area into smaller areas is particularly useful for spatial load forecasting.

5.4 Load Characteristics

5.4.1 Customer Classes

Utilities serve customers of different classes, which include residential, commercial, industrial, and agricultural customers. While load demands of customers in different classes differ from one another significantly due to different electricity usage patterns, load demands of customers within a certain class are different, but they follow a similar pattern. For example, load demands of residential customers are low in the night and start to increase in the morning when people start waking up. The load demand continues to rise during the day reaching a peak and then starts to decline in the evening hours. Similarly, commercial and industrial load demands are high during the day but low during the night. Variation in the commercial load demand from day to night is larger than that of industrial loads because several industries tend to operate throughout the 24 hours of the day. Agricultural loads have very different demand patterns, and they depend on the irrigation needs of different crops. The load demands in each class also fluctuate with season. Utilities maintain a database of load demands in each class for planning purposes. Examples of typical load demand characteristics of customers of different classes for a utility are shown in Figure 5.3.

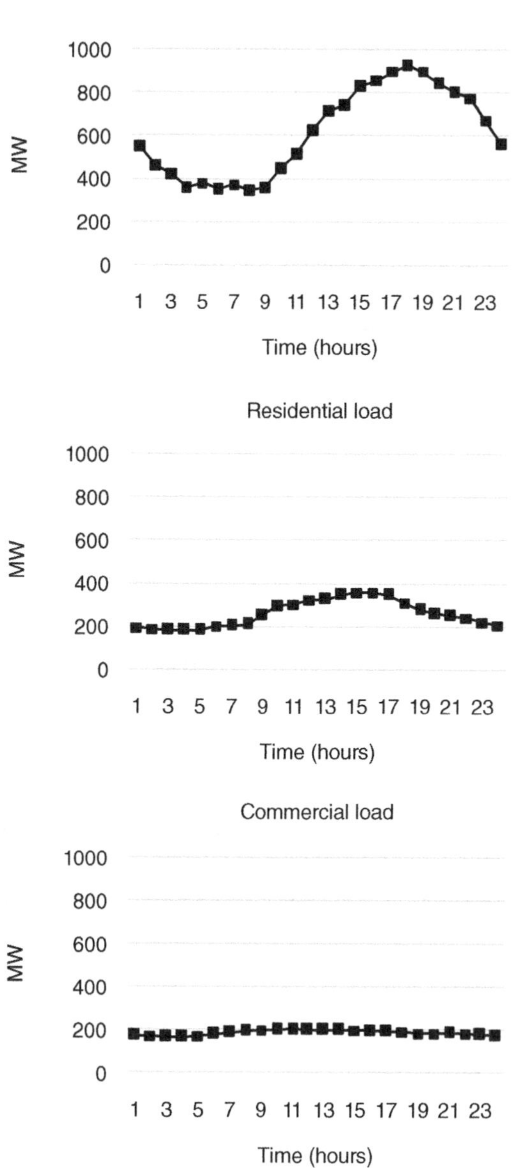

Figure 5.3 Typical aggregate load demands of residential, commercial, and industrial classes at a utility on a summer day.

5.4.2 Loads in a Modern House

Loads in houses today are very different from those 25 years ago. In addition to more efficient appliances, modern homes have new types of loads. Principal among them are computer and entertainment devices. Lighting loads for houses went through a transformation due to replacement of incandescent lamps with compact fluorescent lamps (CFLs) and now LED lamps, which consume substantially lower energy than the replaced ones.

According to a survey done by Energy Information Administration (EIA) for 2015, the average monthly electric energy consumption of 129,811,718 residential customers in the United States was 901 kWh, with an average monthly bill of $114.03 and an average price of $0.1265/kWh. Residents of the state of Washington enjoyed the lowest price for electricity at $0.0909/kWh, while the residents of the state of Hawaii paid the highest price for electricity at $0.296/kWh. In the lower 48 states, the state of Connecticut had the highest price of electricity at $0.2094/kWh. The average monthly electricity consumption was the highest in Louisiana at 1286 kWh, and it was the lowest in Maine at 556 kWh. The hot and humid climate of Louisiana required more electricity for air conditioning. Maine has cool summers but brutal winters. Since heating in winter is mainly based on gas, it did not impact the electricity consumption of Maine residents.

The EIA survey also reports that in 2015 electricity contributed to 47%, and natural gas contributed to 44% of energy consumed in homes. Heating, cooling, and ventilation accounted for about half of the home energy consumption. Table 5.1 shows the energy consumed by different usages. Air conditioning used the largest amount of energy, which was followed by space cooling and water

Table 5.1 Energy usage in the US homes in 2015 [2].

Usage	Percentage of energy used
Air conditioning	17
Space heating	15
Water heating	14
Lighting	10
Refrigerators	7
Television and related equipment	7
Clothes dryer	5
Miscellaneous	12
Other	13

Source: Data from U.S. Energy Information Administration [2].

heating. For homes with natural gas-based heating and water heating, space cooling or air conditioning is the largest electrical load with ratings ranging from 1 to 6 kW for a window unit to a large central air conditioner. Lighting consumed 10%, refrigerators 7%, televisions (TVs) and associated equipment 7%, and clothes dryers consumed 5%. Freezers, dishwashers, clothes washers, cooking ranges, microwaves, air handlers, fans, pool/hot tub pumps, hot tub heaters, humidifiers, dehumidifiers, and evaporative coolers consumed 12% of electricity (listed as miscellaneous in the table). The remaining 13% was consumed by computers, home audio, coffee makers, and miscellaneous appliances (listed as other in the table). Of these loads, the refrigerator is rated about 600 W, and it cycles on and off throughout the day to maintain the specified temperature inside it. Some homes have an additional stand-alone freezer, which has a typical rating of 500 W. The clothes washer is rated about 500 W, the dishwasher is about 1.2 kW, and the clothes dryer is about 4 kW. Several appliances are used a few times in a day or once in a few days, depending on the size of the family. Over the years, the electronics portion has gradually increased due to the popularity of large flat screen TVs and computers. On the other hand, the portion for lighting has been decreasing due to the proliferation of LED lighting.

Energy efficiency is a major focus of the government, the industry, and, now, the consumers. In the future, we should expect the electrical rating and energy consumption of different household loads to go down as the appliances, as well as the houses, become more efficient.

5.4.3 Time Aggregation

Instantaneous power, which is obtained by multiplying voltage and current waveforms, shows large fluctuations and is not very useful for planning. Instead, demand (in kW), which is the average power over a selected time period, is used. Typical time periods used by the utility are 1 minute, 5 minutes, 15 minutes, 30 minutes, and 1 hour. Thus, to determine demand for a time period, energy consumed (kWh) over that time period is divided by that time period in hours. As an example, consider that a load consumes 2 kWh of energy in a given 15-minute (0.25 h) period. Therefore, demand for this period is $2/0.25 = 8$ kW. Similarly, if the same load consumes 7.5 kWh of energy over a one-hour period, the demand for this period will be $7.5/1 = 7.5$ kW. Although the demand is over a period of time, the demand reported at a given time is the average load for the time period that ended at the time reported. So, if we are looking at demand at 2:00 a.m. reported on one-hour basis, the reported value at 2:00 a.m. is the average load between 1:00 a.m. and 2:00 a.m. Similarly, if the demand reported duration is 15 minutes, the reported value at 2:00 a.m. is the average load between 1:45 a.m. and 2:00 a.m. Similar logic can be used for other time durations for recording demands.

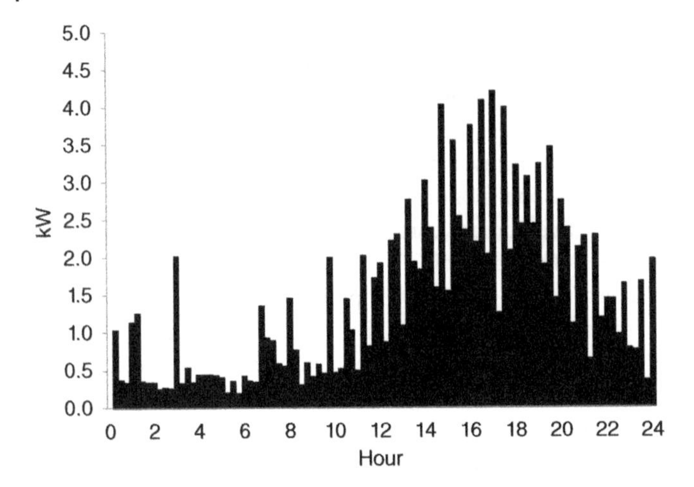

Figure 5.4 Demand of a house on 15-minute basis.

The demand characteristics of a single house based on a small time period have large fluctuations as shown in Figure 5.4. The main reason for these fluctuations is random switching on and off of loads within a house either automatically or by its occupants. Air conditioners, central heating, refrigerators, and water heaters are some loads that turn on and off automatically. On the other hand, lighting, cooking appliances, washer, dryer, computers, and entertainment loads are turned on and off by people. Even while on, some loads may require different amount of power at different times.

The load demand characteristics start to become smoother as we increase the time period. Figures 5.5 and 5.6 show the load demand characteristics of the same house on 30-minutes and 1-hour basis. Usually, time interval larger than one hour is not used to determine the load demand characteristics for planning purposes. Notice that the time aggregation reduces the peak value of the load demand.

5.4.4 Diversity and Coincidence

We noticed in the previous section that temporal aggregation reduces the peak demand of a house by averaging the demand over the selected time duration. Similarly, if we add demands of several houses, we find that the load demand characteristics become smoother as shown in Figures 5.7–5.10. This happens because loads in one house are going on and off at different times than at another house. In utility terminology, this phenomenon is called *load diversity* (LD). Another thing to keep in mind is that all houses have their peak demand at different times. Therefore, the average demand of a group of houses at any time, when any of the houses has its own peak, is lower than the peak demand of that house. For planning purposes,

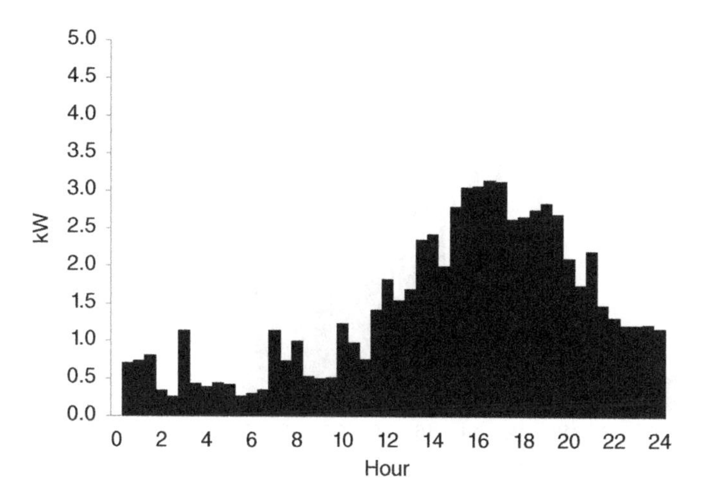

Figure 5.5 Demand of a house on 30-minute basis.

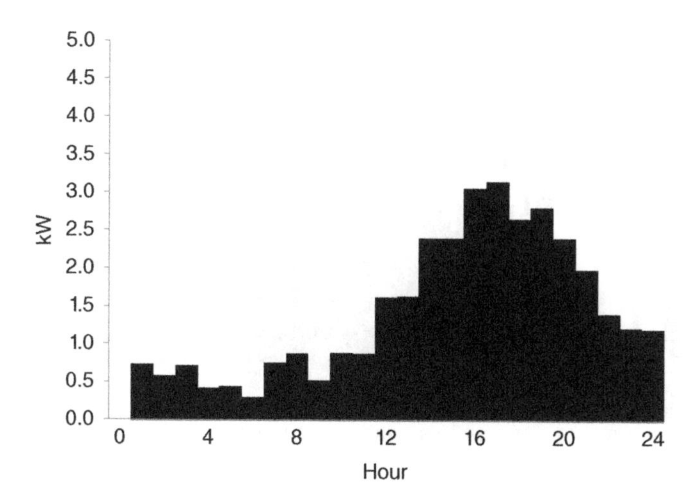

Figure 5.6 Demand of a house on one-hour basis.

the maximum value of the combined demand or *coincident peak demand* over a period of time for a group of loads is of interest. So when we decide the size of the conductor for service to a house, we use the peak value of the combined demand of the devices in the house on the hottest day (or the coldest day in winter-peaking regions). However, when the utility decides on the size of the transformer for supplying power to a house, they look at how many more houses are being served by the same transformer. Since the actual demands of the houses cannot be known in advance, the typical load demand characteristics of houses of a given type for

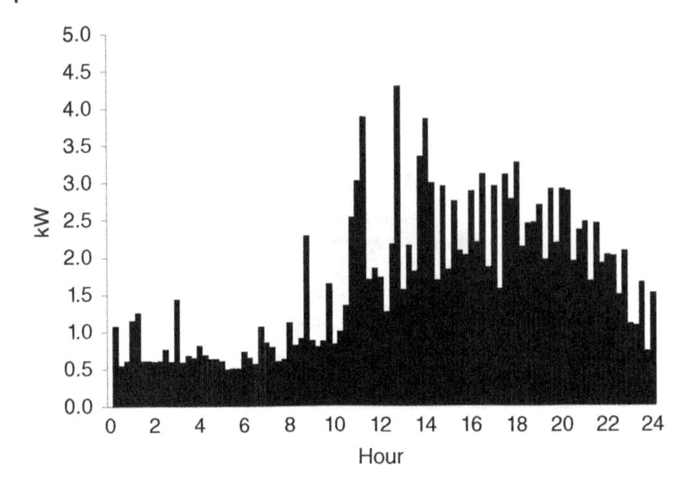

Figure 5.7 Fifteen-minute average load of two houses.

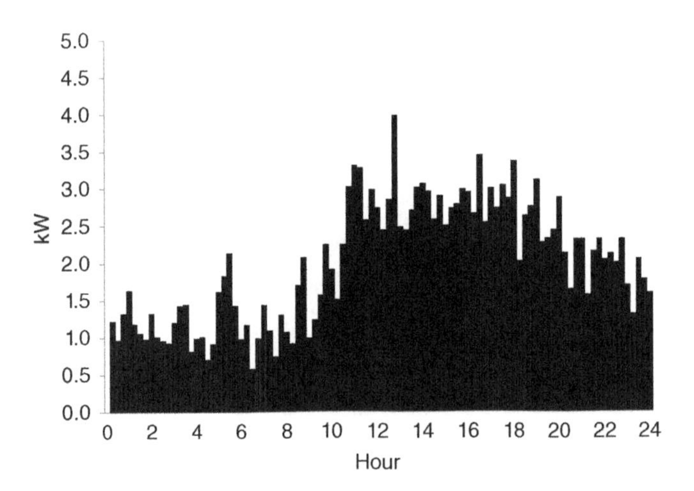

Figure 5.8 Fifteen-minute average load of five houses.

the selected housing subdivision are used in the design. The maximum loads of houses in the group are aggregated and adjusted by a factor called *coincidence factor* to get the coincident peak demand of the group. Let D_{mi} be the peak demand of the ith load in a group of n loads, and D_{mG} be the coincident peak demand of the group. Then, we define the *coincidence factor* (C) as follows:

$$C = \frac{D_{mG}}{\sum_{i=1}^{n} D_{mi}} \tag{5.1}$$

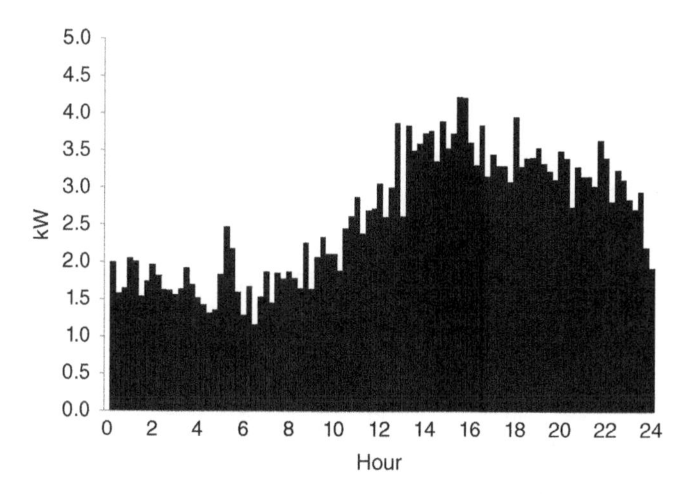

Figure 5.9 Fifteen-minute average load of 10 houses.

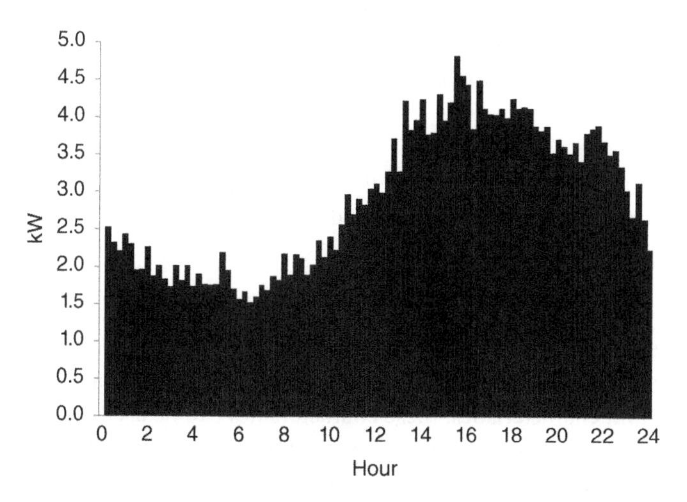

Figure 5.10 Fifteen-minute average load of 20 houses.

The value of the coincidence factor ranges from 0 to 1, and it is a function of the number of houses in a group with its value decreasing with increase in the number of houses. It is not unusual to find the coincidence factor in 0.3–0.6 range for a large group of houses. A graph similar to the one shown in Figure 5.11 can be obtained showing variation in the coincidence factor as a function of n, which is the number of houses in the group. The value of the coincidence factor decreases rapidly for lower values of n but saturates after a certain value. Typically, saturation takes place with n in 15–20 range.

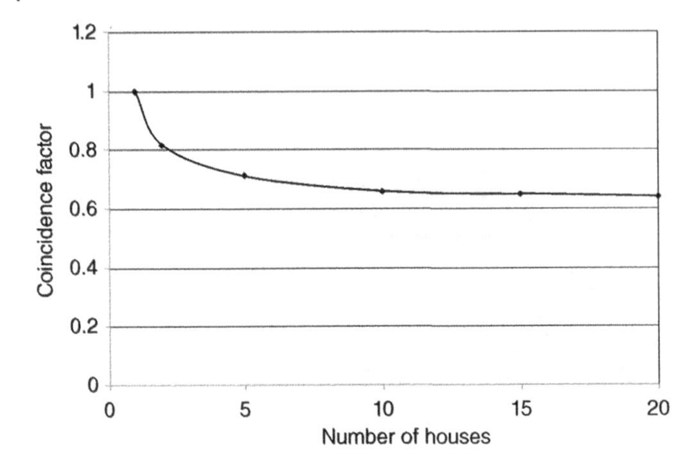

Figure 5.11 Coincidence factor as a function of number of houses.

The inverse of the coincidence factor is defined as the *diversity factor (D)*, and it gives an indication of the separation in individual house's peak demands in comparison to the peak demand of the group. Therefore,

$$D = \frac{1}{C} \tag{5.2}$$

Further, the difference between the sum of the individual peak demands and the coincident peak demand of the group is defined as LD, or

$$\text{LD} = \sum_{i=1}^{n} D_{mi} - D_{mG} \tag{5.3}$$

The concept of coincidence can also be applied to loads of different types. For example, if a feeder has to serve residential, commercial, and industrial loads with given load demand characteristics, these characteristics can be used to find the coincident peak demand. This and other factors are then used to find the size of the feeder that would be adequate to serve the combined load.

Example

Find the transformer size needed to serve a group of four houses with average peak load of 5 kW at 0.85 power factor for each house.

Solution

We use Figure 5.10 to find C for $n = 4$, which is 0.73.

Therefore, the coincident peak load of the group = $0.73 \times 5 \times 4 = 14.6$ kW

$$\begin{aligned}\text{kVA at peak load} &= \text{kW at peak/power factor}\\ &= 14.6/0.85 = 17.18\ \text{kVA}\end{aligned}$$

Since this is not a standard size, we look for the next higher standard size, which is 20 kVA.

5.4.5 Demand Factor

The ratio of the maximum load demand over a period to the connected load for any load or class of load is defined as the *demand factor (DF).*

$$\text{DF} = \frac{\text{Maximum Load Demand}}{\text{Connected Load}} \tag{5.4}$$

For example, a house has many loads, but they are not used simultaneously. Only a small portion of them is used simultaneously. Connected load is the total sum of the rated values of all the loads. If we count the ratings of all the light bulbs in a house, we might get a value close to 2 kW. However, if we count the rated values of the lights that are on simultaneously at a time when the maximum number of lights are on, we might get a value of 600 W. Hence, the DF of the lighting load in the given house will be $0.6/2 = 0.3$. Similarly, if we have a 6-kW air conditioner in a house and on the hottest day in summer, it runs for a total of 30 min, the average demand for the hour will be 3 kW. Therefore, the air conditioner's DF is $3/6 = 0.5$. The concept of the DF can be used for individual loads or a group of loads.

5.4.6 Load Duration Curve

So far we have seen that the load in a distribution system changes with time. For short durations such as a day, examination of hourly loads gives useful information related to the highest and the lowest loads with their time of occurrence for the day and pattern of fluctuation of the load during the day. However, if we have to examine the load for a longer duration, such as a year, examination of hourly load for each day of the year becomes a very tedious task. To simplify this task, we condense the load information into *load duration curve.* Load duration curve is a graph of load vs. time for which the load is higher than the specified load. We can easily obtain the two extreme ends of the load duration curve by looking at the highest and the lowest values of the load over a period of time. Since load is never higher than the peak load (L_p), it is higher than L_p for 0% of the time. However, it is always higher than or equal to the lowest load (L_l), or it is higher than or equal to L_l for 100% of the time. Note that we have made an exception for the lowest load for which we are considering greater than equal to instead of greater than the specified load. In between points can be obtained by quantizing the load in equal steps. For example, if the peak load is 17 MW, we could use steps of 2 MW. Too large steps will not give enough data points, and too small steps will give too many data points. Judgment of the user is important in determining the step size.

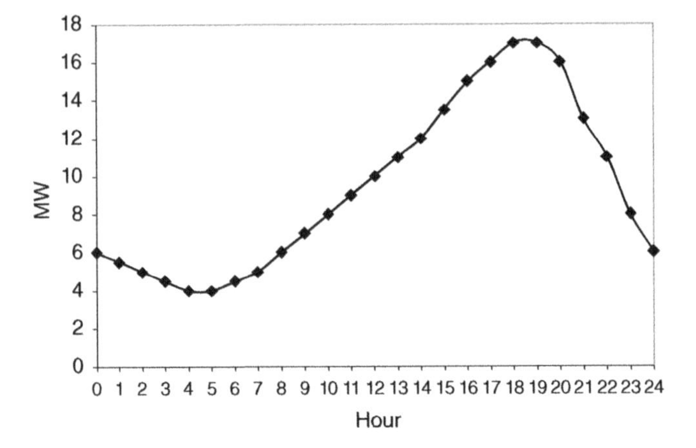

Figure 5.12 Hourly load at a substation. The load shown at a given time is the average load for the hour ending at that time.

To illustrate the idea, let us consider the load characteristics of a load on a substation transformer for a given day, as shown in Figure 5.12. Note that this figure is not based on real data, but this example is created to illustrate the concepts. From this figure, we see that $L_p = 17$ MW and $L_l = 4$ MW. For illustration, let us consider steps of 2 MW, and from the figure count the number of hours for which the load is higher than the selected load. For example, the load is higher than 10 MW for 10 h. Similar counting will give us the data shown in Table 5.2, and the corresponding load duration curve is shown in Figure 5.13. Note that the load at hour zero is not included in this calculation. Load duration curves, especially representing a year, are very important for planning studies. They are used to plan for equipment needed in the system and to estimate losses.

Table 5.2 Data for load duration curve obtained from Figure 5.12.

Load (MW)	Number of hours for which load is higher than this value	% of time for which load is higher than this value
17	0	0
16	2	8.33
14	5	20.83
12	7	29.17
10	10	41.67
8	12	50.00
6	15	62.50
4	24	100.00

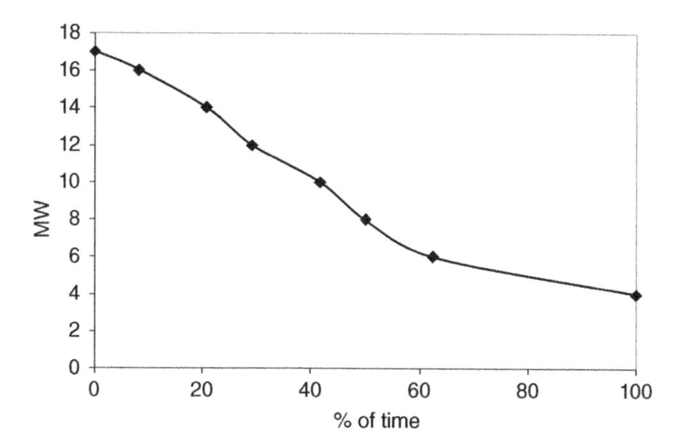

Figure 5.13 Load duration curve corresponding to the load characteristics of Figure 5.12.

5.4.7 Load Factor

Load factor is a term that relates to changes in load over a period of time with respect to the peak load. In other words, it describes the extent to which the load stays close to the peak load. A high load factor implies that the load stays close to the peak load for a large part of the time duration. Mathematically, it is the average load over a period of time divided by the peak load within that time duration, or

$$\mathrm{LF} = \frac{L_a}{L_p} = \frac{\left[\frac{\sum_{i=0}^{n} L_i}{N}\right]}{L_p} \tag{5.5}$$

where L_i is the hourly load at hour i, and N is the total number of hours in the time period under consideration. Notice that the numerator of Eq. (5.5) is equal to the total energy consumed during N hours. For the load characteristics shown in Figure 5.12, with $L_p = 17$ MW and $L_a = 9.5$ MW, we get a load factor of 0.527.

5.4.8 Loss Factor

Similar to the load factor, loss factor (LsF) is the ratio of the average hourly losses to peak hourly losses in the system. LsF is very useful for computing the total losses for a period of time if losses at the peak load are known. For load characteristics in which peak load exists for a very short duration, an approximate value for the LsF can be obtained by taking the square of the load factor.

Consider a discrete version of a load duration curve as shown in Figure 5.14. In this figure, L_p is the peak load, and L_1, L_2, and L_3 are the other load levels, which exist for certain time durations as shown in the figure. The total time duration is

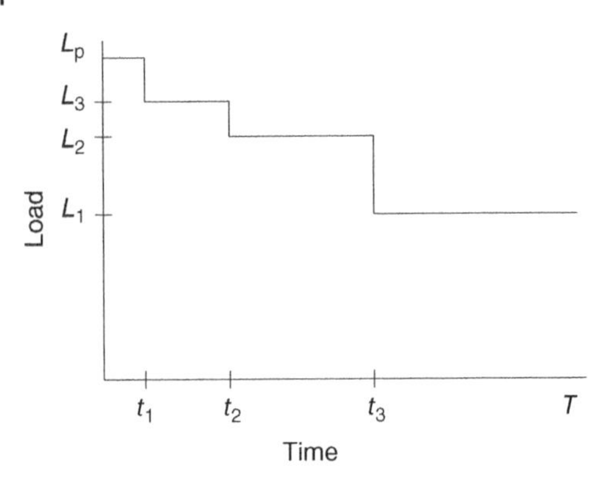

Figure 5.14 Discrete load duration curve.

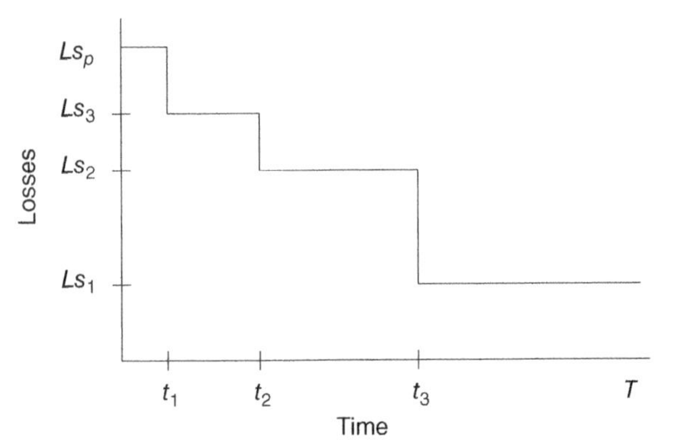

Figure 5.15 Discrete loss duration curve.

T hours. Since resistive losses (which are the predominant losses in distribution systems) at any load are proportional to the square of current, they are proportional to the square of load if we consider voltage to be fixed. Hence, we can obtain loss characteristics for the same duration as shown in Figure 5.15. Ls_p are the average hourly losses at peak load, and Ls_1, Ls_2, and Ls_3 are the losses at other load levels.

From Figure 5.14, we get

$$\mathrm{LF} = \frac{L_p t_1 + L_3(t_2 - t_1) + L_2(t_3 - t_2) + L_1(T - t_3)}{TL_p} \tag{5.6}$$

Similarly,

$$\mathrm{LsF} = \frac{Ls_p t_1 + Ls_3(t_2 - t_1) + Ls_2(t_3 - t_2) + Ls_1(T - t_3)}{TLs_p} \tag{5.7}$$

For a special case where L_1, L_2, and L_3 are equal, we get

$$\mathrm{LF} = \frac{t_1}{T} + \frac{L_1(T - t_1)}{L_p T} \tag{5.8}$$

and

$$\mathrm{LsF} = \frac{t_1}{T} + \frac{Ls_1(T - t_1)}{Ls_p T} \tag{5.9}$$

Now, if we consider the peak load of a very short duration or $t_1 \ll T$,

$$\frac{t_1}{T} \to 0 \text{ and } \frac{(T - t_1)}{T} \to 1 \tag{5.10}$$

Hence,

$$\mathrm{LF} = \frac{L_1}{Lp} \text{ and } \mathrm{LsF} = \frac{Ls_1}{Ls_p}. \tag{5.11}$$

With $Ls_1 = k(L_1)^2$ and $Ls_p = k(L_p)^2$, we get

$$\mathrm{LsF} = (\mathrm{LF})^2 \tag{5.12}$$

This approximation becomes better as T becomes larger, such as one year, because the peak load typically exists only for an hour or two. Also, the square relationship between the load factor and the LsF is an extreme situation because in reality all the off-peak loads are not equal. Other approximations are also available in the literature. However, for planning purposes, it is acceptable to use this relationship.

5.5 Design Criteria and Standards

In addition to minimizing the cost of the equipment needed for a distribution system, the job of a planning engineer is to ensure that the selected equipment, layout, and installation meet the design criteria and standards. Voltage and service quality, equipment loading, and safety are the three issues relevant to the planning and design of distribution systems. These criteria are either specified by equipment manufacturers and standards organization or set by the utilities for their own use. Sometimes, standards set by utilities are more stringent than those set by manufacturers or the standards organization. Manufacturers specify equipment loading values to prevent the damage to the equipment. Safety is related to providing proper insulation and adequate clearance between equipment. Voltage and service quality affect the customers directly.

5.5.1 Voltage Standards

According to ANSI C84.1 standard, utilities in the United States are required to provide service voltage to customers in the 114–126 V range (on a 120-V nominal). This is Range A, or the preferred range, and is based on +/−5% fluctuations around the nominal. During emergencies, the utilities are allowed to use Range B, which is from 110 to 127 V. With 4 V drop in the building wiring allowed by ANSI C84.1, the customers receive utilization voltage in 110–126 V range under normal operation. The equipment manufacturers use this range to design their equipment. The life of equipment is usually shortened if operated outside of this range. Usually, utilities try to keep the service voltage for all customers at or above 120 V to have a larger safety margin.

Providing voltage within the American National Standards Institute (ANSI)-specified values to all the customers is a big challenge for the planners. The challenge is compounded by the fact that the flow of power over the conductors from the substation to the customers will result in voltage drop, and this voltage drop fluctuates with the fluctuation in load throughout the day. Also, the customers close to the substation see a lower voltage drop compared to the customers away from the substations. In addition to proper system design, utilities use load tap changers, line regulators, capacitors, or a combination of the three to provide service voltage to the customers within the ANSI-specified range.

5.5.2 Conservation Voltage Reduction

It is widely recognized that reducing voltage reduces both the real and the reactive power demand and energy consumption. Therefore, with proper control and monitoring, the utilities can operate the system to provide service voltage in the lower side of the ANSI range, with some customers having service voltage of 114 V, to capitalize on this phenomenon. Several experiments were conducted by various utilities in the 1970s and 1980s to determine the effects of voltage reduction on the system load. Despite some success of a few utilities in implementing voltage control specifically to reduce energy (or *conservation voltage reduction* (CVR)) and peak power, such control presently is practiced only by a few utilities. Uncertainties in the level of load reduction and fear of voltage-related complaints led to the lack of interest among utilities on this issue. Interest in this topic has grown lately due to national interest in energy conservation and concerns about environmental impacts and to efforts to reduce carbon emissions. Implementation of voltage control for energy and power reduction allows utilities to participate in energy conservation efforts, which makes them good corporate citizens and promotes their "green image."

Various studies conducted on this issue show that real load-to-voltage sensitivities (% change in real power to 1% change in voltage) may range from 0.4 to as high as 2.5, but the average is in the 0.8–1.0 range. The reactive load-to-voltage sensitivity is larger with a range of 3–5 with an average of about 4. The energy sensitivity reported for a period of 24 hours ranges from 0.6 to 1.237. However, for a given system, the energy-to-voltage sensitivity is usually lower than the real load-to-voltage sensitivity. The exact load-to-voltage sensitivity for any substation/feeder will be impossible to predict and will vary each day, throughout the day, and throughout the year. The commonly acceptable number by industry for real load-to-voltage sensitivity is 1.0, and for reactive load-to-voltage sensitivity is 3.0. The general practice of utilities regarding voltage reduction is to either use it on a continuous basis to reduce energy or only during emergency to reduce power, or a combination of both.

5.6 Distribution System Design

5.6.1 Substation Design

Distribution substation is where transmission or subtransmission system ends and distribution system begins. Power comes into distribution substation between 34.5 and 230 kV range, and it is reduced to a voltage level between 2.4 and 34.5 kV. The voltage level at the distribution level is determined by the load density and the distance to which power must be delivered from the substation. Usually, the substations are located to uniformly cover the whole service area without having excessively long distances between some customers and the substation. However, this rule may not apply in rural settings. The layout of an example substation is shown in Figure 5.16. In this figure, there are three 115-kV incoming lines, and they are connected to the 115-kV bus. The 115-kV bus is normally operated as one contiguous bus with four normally closed (NC) switches. In emergencies, some of these switches can be opened to operate this bus in a different configuration. Note that each incoming line has its own circuit breaker. This is the simplest

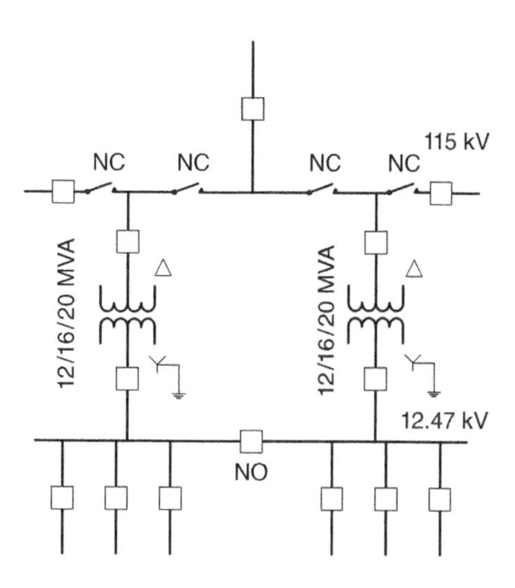

Figure 5.16 Typical layout of a substation for a semiurban area (NC, normally closed and NO, normally open).

bus arrangement. Other arrangements, such as main and transfer bus, ring bus, and breaker-and-a-half bus schemes, can be used to enhance reliability. However, these schemes are more expensive, and some of them also complicate the protection system.

Distribution substations can have one to three transformers. The substation of Figure 5.16 has two transformers, each rated 12/16/20 MVA with 115 kVΔ/12.47 kV Y-grounded connections. It is preferred to have both transformers of the same capacity, which is selected based on the total load in the service territory of the substation. The tie breaker on the 12.47-kV bus is normally open to avoid parallel operation of transformers, but can be closed under emergencies. Parallel operation of transformers is not recommended because that creates circulating currents and increases losses. Typically, the transformers are loaded to 50% of their highest rating during peak load conditions. This is done so that if one of the transformers fails, the other one is able to pick up the whole load. Thus, if the substation has three transformers, they can be loaded to 67% of their highest rating. In this case, the other two transformers will pick up the load of the failed transformer. Several transformers are needed in the substation to further increase the loading on transformers. If the substation has only one transformer, its failure will result in sustained loss of service to the customers. This was the old way to select sizes of distribution substation transformers. In the new approach, the distribution systems can be designed with automation to receive support from other substations in case of failures. Automation allows quick operation of switches and thus load transfer to other transformers. The distribution substation transformers usually have Δ/Y configuration to provide neutral on the low-voltage side for distribution of single-phase power to customers using one of the phases and the neutral. The rule of keeping Y-connection on the high-voltage side of the Δ/Y transformers is overruled in this case to make the neutral available on the low-voltage side.

The low-voltage side of the transformers are connected to the 12.47-kV bus, which is operated in two segments with the normally open (NO) switch in between. Closing the switch would parallel the two transformers, which is not desirable. Each segment of the low-voltage bus has a number of feeders, which can range from 1 to 4 per segment. Deciding on how many feeders to use is a subjective process, and it depends on the load to be served and the capacity of the transformer. A typical feeder is designed to carry 4–8 MVA. In the substation shown in Figure 5.16, there are three feeders per transformer, with a total of six feeders.

5.6.2 Design of Primary Feeders

Primary feeders carry power from substations to customers and thus are the backbone of the distribution systems. Except for a very densely populated service area,

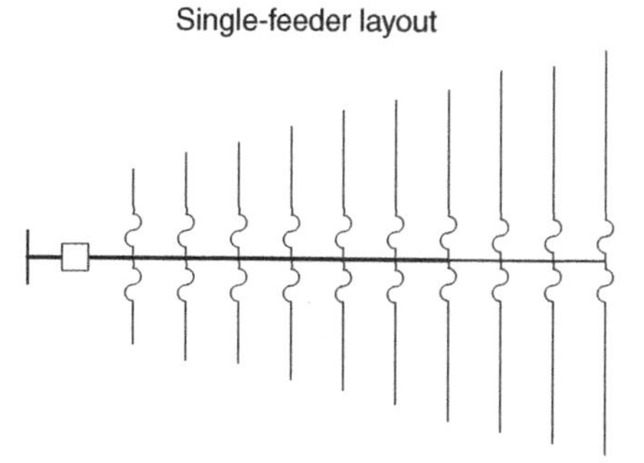

Figure 5.17 Single-feeder layout of distribution feeders.

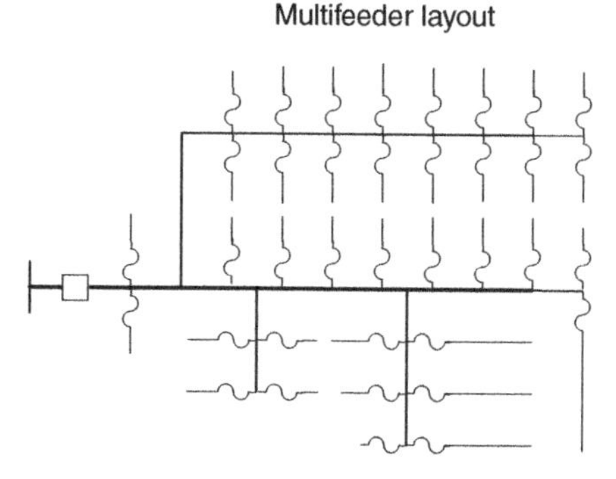

Figure 5.18 Multifeeder layouts of distribution feeders.

such as downtowns of large cities, distribution systems have a radial structure. In a radial structure, the power from the utility to customers flows only in one direction. Although the radial structures provide lower reliability, they are substantially cheaper to build and much easier to operate and protect. Therefore, distribution systems are predominantly radial with the feeders arranged in a tree-like structure. The root of this tree is at the substation. From this root the main three-phase branch of the primary feeder or the trunk comes out. This primary feeder may split into multiple subfeeders or remain as one throughout its service territory. Examples of single and multifeeder layouts are shown in Figures 5.17 and 5.18, respectively. Due to the radial nature, the load on the primary feeders reduces as we move away from the substation. Therefore, in some cases, conductor sizes for the primary feeders are tapered as we move away from the substation.

Each primary feeder leaving a substation is assigned a part of the service area of the substation so that each primary feeder has roughly equal service areas. In an ideal situation, a substation with four feeders will have the division of the service area as shown in Figure 5.19. Similarly, a substation with six feeders will have the service area division as shown in Figure 5.20. In real life such clean demarcation between service areas is not possible, but the planners try to maintain this division as much as possible. If a substation has a higher number of primary feeders, the service territory is divided equally.

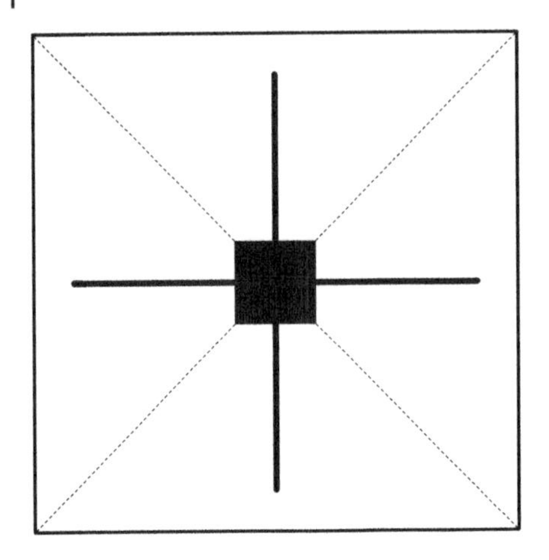

Figure 5.19 Service territory division for four primary feeders.

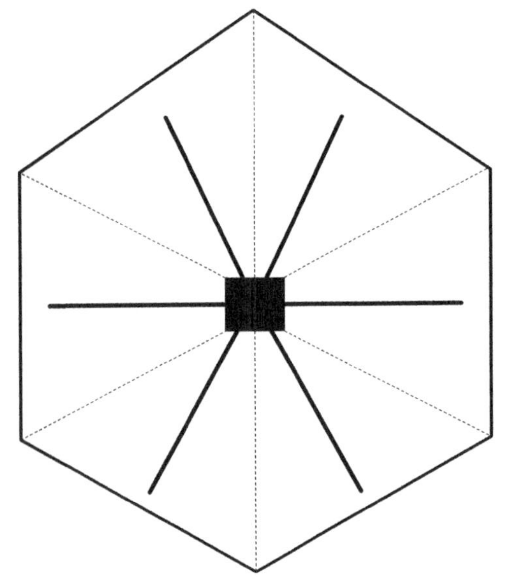

Figure 5.20 Service territory division for six primary feeders.

Although distribution systems operate in a radial configuration, most of the primary main feeders have a NO (Normally Open) switch or circuit breaker, called the tie switch, at their ends. When a permanent fault occurs on a primary feeder so that it loses connection to the substation, the device at the end of the feeder is closed to restore power to some of the customers receiving power from the faulted feeder. The feeder on the other side of this device could be connected to the same transformer, or to a different transformer from the same substation or to a transformer from a different substation. Each of them provides a different level of reliability and has different costs associated with it. Therefore, careful planning of the feeder layout is needed to achieve the desired results. For example, if two feeders are connected to the same transformer, the loss of that transformer will result in no backup possibilities for customers connected to these feeders. Power will have to be restored at the substation by connecting these feeders to a different transformer. On the other hand, if they are connected to different transformers, the loss of a transformer will result in power loss to only one of the feeders. Power to this feeder can be supplied by connecting to the other feeder by closing the tie switch.

Connected to the primary main feeders are the primary lateral feeders, which are usually single phase, with phase and neutral conductors. They carry power from the primary main to customers. Primary main feeders are typically along the city streets, but the lateral feeders are mostly in the utility easement behind houses in residential subdivisions in cities and towns. The phase selection (A, B, C) for specific laterals branching off from the primary main is done to maintain a rough balance among the three phases. Figure 5.21 shows the feeders and streets in a typical city in the United States.

5.6.3 Design of Secondary Systems

The secondary system is that part of the distribution system which is closest to the customers. It includes the distribution transformers, secondary feeders, and service drops. The secondary system is single phase for residential customers, but it could be three phase for large industrial and commercial customers.

Distribution transformers are small transformers, ranging in size from 5 to 100 kVA with 120 V/240 V secondary. The size of the transformer used in a specific location depends on the number and the class of customers served by it. Typically, a transformer serves four to eight medium-sized single-family homes. If homes are very large, a transformer may serve only one or two homes. On the other hand, a single transformer may serve several apartments in a complex. Typical layouts of secondary systems are shown in Figure 5.22.

5.6.4 Underground Distribution Systems

A large percentage of distribution systems all over the world is overhead. Cost is the main factor for this because the cost of an underground system is 5–10 times more than the overhead system. However, in densely populated areas and in business districts, the distribution systems are underground. In other areas, aesthetics is the motivation for underground distribution systems. Many cities are mandating underground systems within new housing developments. Even if cities do not mandate, some developers use underground distribution in high-cost subdivisions. In these situations, laterals and secondary systems are underground, but the primary main is still overhead.

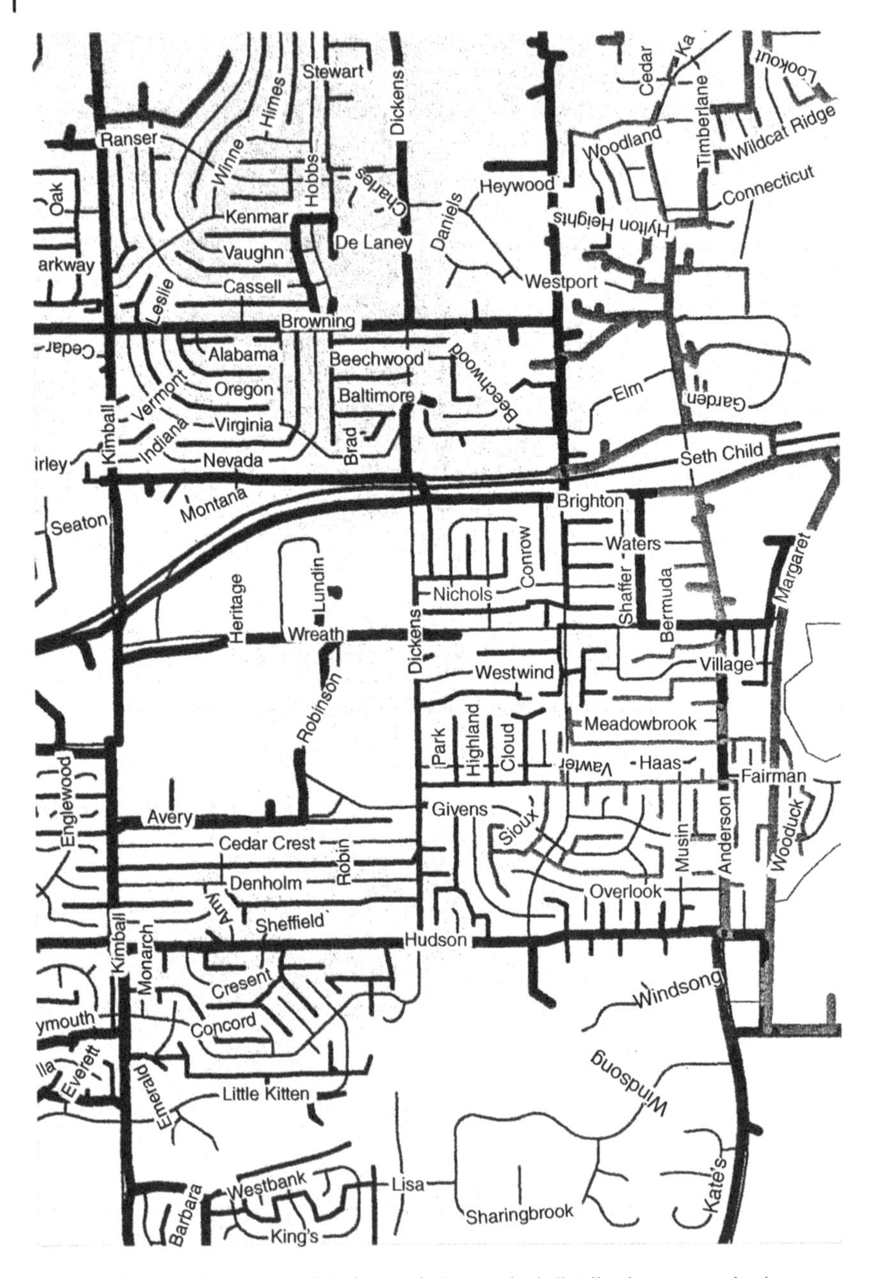

Figure 5.21 Feeders along with the roads in a typical distribution system in the United States.

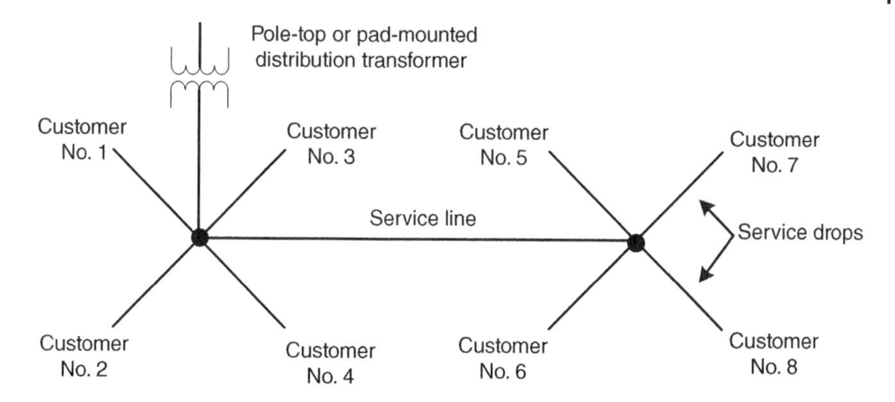

Figure 5.22 Illustration of a secondary system for service to eight customers.

Overhead systems are more exposed to the environment and therefore more prone to failure. Several studies have shown that more than 50% of the outages on overhead distribution systems are caused by environmental factors including weather, trees, and animals. One advantage of overhead systems is that the location of faults and repairs are relatively easy. In contrast, fewer faults occur in underground systems; but if a fault does take place, it is very difficult and time consuming to locate that fault. The repairs are more difficult and take a longer time.

5.6.5 Rural vs. Urban Systems

In general, the distribution systems can be divided into three types based on the load density: urban, semiurban, and rural. Urban systems have very high load density and are designed with lower substation spacing, higher size substation transformers, and bigger conductors for feeders. Since the substation spacing is small, the total distance from the substation to the farthest customer served from the substation is not very long. Each substation also has two or more power transformers. Rural systems have very low load density and the loads are spread out. Therefore, the substation spacing is very large, the transformer sizes are small, and the conductors for the feeders are relatively small in size. Further, the substation may have only one transformer. Generally, sufficient number of NO tie points are available in the urban systems, but very few are available in the rural systems. The tie points improve reliability of the system significantly and are easy to locate in urban systems. It is very difficult and expensive to locate tie points in rural systems, and also, as the number of customers in rural systems is low, the impact on reliability is only marginal. As a consequence of this, the reliability of urban systems is substantially higher than the rural systems. The semiurban systems are in between these two systems in all respects.

In urban systems, since the total load served by feeders is higher, and feeders are shorter in length, by design, and they are mostly thermally limited. That is, the total load that a substation can supply is based on the thermal loading capacity of the transformers and the feeders. Voltage drop on the feeders is usually not a problem. The rural systems, on the other hand, are voltage-drop limited and that is due to the length of the feeders, which makes the voltage drops on them are quite large. Hence, they can only carry a limited amount of load to ensure that the last customer on the feeder gets service voltage within the limits. The semiurban systems are in between. They may have both problems, depending on the size and type of equipment selected while designing the system.

5.7 Cold Load Pickup (CLPU)

5.7.1 CLPU Fundamentals

Loads on a predominantly residential feeder can be divided into two categories those that are thermostatically controlled and those that are manually controlled. Thermostatically controlled devices such as air conditioners, heaters, heat pumps, and refrigerators provide the largest contribution to the total load in a typical house. Manually controlled loads are switched on and off by occupants of the house in a random fashion based on their need. The aggregate load of a group of houses is less than the connected load under normal operation due to diversity among loads. However, after an extended interruption, some or all thermostatically controlled devices would be on as soon as the power is restored. Similarly, the aggregate load of manually controlled devices immediately after restoration would be higher than normal because more people may want to use different devices. The lifestyle of the occupants of the house has a significant influence on the contribution of these loads to the total load of the house.

The load phenomenon that is experienced upon restoration after an extended power interruption is called cold load pickup (CLPU). CLPU can be categorized into four phases according to the current levels and durations. These phases are inrush, motor starting, motor running, and enduring phases. The first three phases last approximately less than 15 s, and the current may reach 5–15 times of the preoutage current. The enduring phase follows the third phase and continues until the normal diversity among the loads is reestablished. The load in this phase may vary from two to five times of the diversified load level, and this phase may last for several hours, depending on the outage time, outside temperature, and the type and ratings of devices. Several utilities also experienced this phenomenon during implementation of direct load control on thermostatically controlled devices.

CLPU appeared first in the literature in the 1940s as a problem related to high inrush currents that last a few seconds and prevent the circuit from being reenergized after extended outages [3, 4]. Application of very inverse characteristic relays and sectionalizing the distribution systems were some of the solutions engineers used to overcome this problem. Since then, penetration of thermostatically controlled devices in distribution systems has increased. These types of loads may cause restoration problems during CLPU before they cause serious overload problems in normal operation. Therefore, sustained or enduring load after restoration becomes an important issue for loading limitations of distribution equipment.

5.7.2 CLPU Models

With increasing penetration of thermostatically controlled loads, the enduring components of CLPU grabbed the attention of researchers in the 1970s and 1980s. Some experimental studies were conducted to predict the magnitude and the duration of the peak demand upon service restoration following a long power outage [5]. Substantial work on physically-based modeling of this phenomenon of loads took place [6]. Some of it was initiated under the sponsorship of the U.S. Department of Energy. The main idea is to develop a physically-based load model for individual loads as a function of weather and human use patterns and then use aggregation of these loads to find the total demand. Aggregated load behavior, when a large number of customers are considered, can be determined based on these models either by use of numerical techniques to solve partial differential equations derived from individual load model or by Monte Carlo simulation based on the stochastic difference equations. Some simplified models have also been used to find the effect of CLPU on substation and distribution transformers [7]. Because the thermal response of a transformer to a load is slow, simplified models are sufficient to analyze its loading capabilities. Simplified models in which postoutage load is constant for some time and then decreases linearly and piece-wise linear CLPU models have been used to find the overloading capability of transformers.

Figure 5.23 shows an example of data recorded by a utility in the east coast of the United States after restoration following an interruption of 1 h and 39 min around midnight during winter. The load prior to interruption was 749 kW. The graph clearly shows that the load in the first 15-min period immediately after restoration is 1451 kW, which is approximately two times the load prior to interruption. Also, the load returns to normal in the third interval. The value of the initial load after restoration and the time taken to return to normal would depend on the load characteristics, duration of interruption, time of interruption, and ambient temperature. Since the data was recorded for 15-min intervals, it is hard to say more

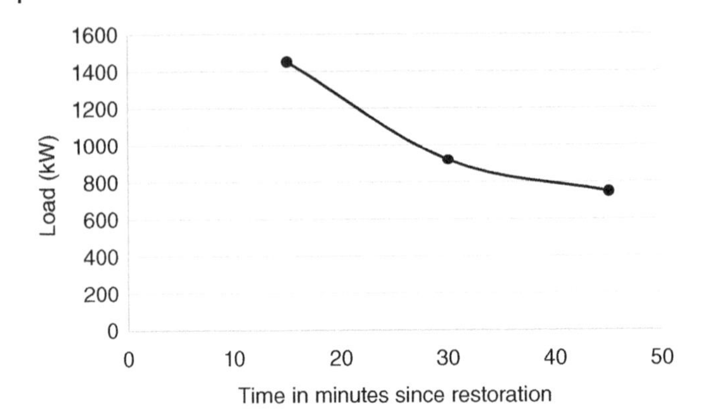

Figure 5.23 Load upon restoration following a long outage during winter recorded by an east-coast utility in the United States. The dots show the average load for 15 min prior to that time.

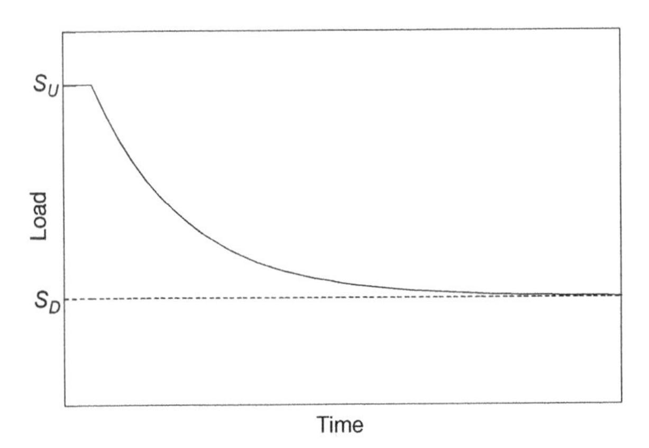

Figure 5.24 Delayed exponential model for cold load pickup. S_U is the undiversified load, and S_D is the diversified load.

about the cold pickup characteristics based on this graph. Data recorded on 5- or 1-min resolution could provide more information about CLPU.

Simulations as well as experimental data suggest that aggregate load on a residential feeder in distribution system during CLPU can be represented by a delayed exponential model [7], as shown in Figure 5.24.

5.7.3 Impacts of CLPU

CLPU can impact both the operation and design of distribution systems. These impacts are dependent on the type of system. Since load during CPLU is much higher than normal load, it causes overloading of the transformer and conductors and excessive voltage drops on the feeders. The first factor is obviously more crucial for the urban systems, whereas the second factor is more crucial for the rural systems. In either case, it may not be possible to restore the entire load in one step. One possible approach is to install remotely operated sectionalizing switches on the main feeder to split the system into several sections. During restoration, the sections can be restored in steps. Load behavior of sections and restoration sequence of these sections play an important role in the restoration procedure. Based on the load dynamics of each section, the restoration procedure should be chosen in such a way that the selected restoration objectives are met while the loading of transformers and feeders is within limits, and there are no voltage-drop violations. For example, one of these objectives is to minimize customer interruption duration because it directly affects the reliability of the system. The shorter the customer interruption is, the more reliable the system will be. We will discuss additional details on CLPU when we revisit it in a later chapter.

5.7.4 Operating Limits

According to ANSI/IEEE PC57.91-1994 standard, the load on a transformer can exceed the rated load under an emergency situation, but the loss of life of the transformer should not exceed 4% [8]. Further, the maximum top-oil temperature should not exceed 110 °C, the maximum hottest spot winding temperature should not exceed 180 °C, and the maximum short time loading should not be more than two times the maximum normal rating. Similarly, short time emergency loading of conductors is 133% of the normal full load.

At all times, the supply voltage to all the customers must be within the limits specified by ANSI C84.1. Of the total voltage drop from the substation to the service entrance, approximately 2.5% under normal operation and 3.54% under emergency operation can be assumed for the laterals, and the rest for the main feeder. Hence, if the voltage at the substation is considered to be at the upper end of the zone, the total permissible voltage drop on the main feeders would be 7.5% under normal operation and 10.62% under emergencies. Thus, during restoration under CLPU conditions, the voltage drop should not exceed 10.62%.

5.8 Asset Management

Asset management, as the name implies, is an approach for utilization of assets in an optimal way. This applies both for planning and operation. During the planning process, decisions have to be made on replacing or updating the existing equipment while keeping in mind the age and current operating status of these equipment. Waiting for too long can have disastrous results, but at the same time replacing equipment too early will cost utilities more money. Therefore, decisions have to be made while accounting for risk associated with selected actions. Similarly, during operation, real-time decisions on loading of equipment have to be made based on their conditions and current operating status. Sustained overloading could result in equipment failure, but on the other hand, without overloading service to some customers would have to be curtailed. Hence, risk of overloading has to be balanced with loss of power to some customers.

The issue of asset management has come into limelight because a large part of the existing distribution and transmission infrastructure is very old, with some equipment with age higher than 40 years. Complete replacement of these equipment is cost prohibitive. So a planned replacement strategy is required.

Problems

5.1 Power demands of a house recorded for a house on a 15-min basis for a duration of two hours are given in Table 5.3.

Table 5.3 Data for Problem 5.1.

Period	Power demand (kW)
1	1.2
2	1.4
3	2.0
4	1.8
5	1.6
6	1.4
7	1.2
8	1.0

Find the average demand for each of the two hours and for the total duration of two hours.

(a) What is the total energy consumed in the first hour?

(b) What is the total energy consumed in the total duration of two hours?

5.2 Energy consumptions in a house recorded on five-minute basis for a one-hour duration are shown in Table 5.4.

(a) Find the power demand for each period.

(b) What is the average power demand for the hour?

(c) What is the total energy consumption for the hour?

Table 5.4 Data for Problem 5.2.

Period	Energy consumed (kWh)
1	0.5
2	0.6
3	0.8
4	1.0
5	0.9
6	0.8
7	1.0
8	0.8
9	0.7
10	0.6
11	0.4
12	0.5

5.3 Energy consumptions recorded on a one-hour basis for four hours for five houses are shown in Table 5.5.

Table 5.5 Energy consumption data of houses for Problem 5.3.

	Energy consumption (kWh)				
	House number				
Hour	1	2	3	4	5
1	2.0	4.8	3.2	1.2	1.6
2	2.4	1.6	1.2	3.0	2.0
3	3.2	2.0	1.6	1.0	0.8
4	4.0	0.8	2.0	2.2	3.0

(a) Find the combined power demand of five houses for each hour.
(b) What is the coincidence factor for these houses?
(c) What is the diversity factor?

5.4 The hourly load power demands for a primary distribution feeder of a power company for a typical winter day are given in Table 5.6.

Table 5.6 Data for Problem 5.4.

	Load (kW)		
Hour	**Street lighting**	**Residential**	**Commercial**
1	200	200	50
2	200	200	50
3	200	200	50
4	200	200	50
5	200	200	50
6	180	300	50
7	150	400	50
8	0	600	50
9	0	500	200
10	0	500	600
11	0	500	600
12	0	500	600
13	0	500	600
14	0	500	600
15	0	500	600
16	0	500	600
17	100	600	600
18	200	800	600
19	200	800	600
20	200	600	600
21	200	600	600
22	200	300	300
23	200	300	300
24	200	300	300

(a) Find the DFs for each load type with the total connected loads of 200, 1200, and 800 kW for street lighting, residential, and commercial loads, respectively.

(b) Find the diversified maximum demand and the coincidence factor of the feeder.

(c) Find the total combined energy consumption in kWh for the day. Use this information to find the load factor of the feeder for this day.

(d) If the losses on this feeder at peak load are 5% of the peak load, find the average hourly loss using LsF.
(e) Find the total cost due to losses on this day with an energy cost of 3 cents/kWh.

5.5 Table 5.7 shows the hourly load on a substation transformer for a 24-h duration.

Table 5.7 Data for Problem 5.5.

Hour	Load (MW)	Hour	Load (MW)	Hour	Load (MW)	Hour	Load (MW)
1	4.1	7	3.5	13	10.4	19	17.8
2	3.5	8	5.1	14	12.1	20	15.1
3	3.1	9	6.2	15	14.2	21	12.8
4	2.3	10	7.4	16	15.9	22	8.8
5	2.6	11	8.9	17	17.8	23	6.5
6	3.2	12	9.8	18	20.2	24	5.2

(a) Draw a plot showing load in MW on y-axis as a function of time.
(b) Draw a load duration curve for the given data with a step size of 1 MW.
(c) Draw a load duration curve for the given data with a step size of 0.5 MW.
(d) What is the load factor for parts (b) and (c)?

5.6 A 100 sq. mile (10×10 miles) area has a load density of 1.5 MW/square mile with 250 customers per square mile. The load power factor is 0.9 lagging. Each primary feeder can carry load in 4–8 MVA range, and the maximum length allowed for the feeders is six miles. Find the substation spacing, number of substations, substation capacity, number of transformers needed per substation and their capacity, and the number of primary feeders per substation. (Note: This is a challenge problem and does not have a unique solution.)

References

1 Willis, H.L. (2002). *Spatial Load Forecasting*. Boca Raton, FL: CRC Press.
2 U.S. Energy Information Administration (2015). U.S. Residential Energy Consumption Survey (RECS). https://www.eia.gov/consumption/residential/index.php (accessed 28 April 2022).
3 Hartay, C.E. and Couy, C.J. (1952). Diversity: a new problem in feeder pickup. *Electric Light and Power* 142–146.

4 Ramsaur, O. (1952). A new approach to cold load restoration. *Electrical World* 101–103.
5 McDonald, J.E., Bruning, A.M., and Mahieu, W.R. (1979). Cold load pickup. *IEEE Transactions on Power Apparatus and Systems* PAS-98: 1384–1386.
6 Ihara, S. and Schweppe, F.C. (1981). Physically based modeling of cold load pickup. *IEEE Transactions on Power Apparatus and Systems* PAS-100: 4142–4150.
7 Lang, W.W., Anderson, M.D., and Fannin, D.R. (1982). An analytical method for quantifying the electrical space heating component of a cold load pick up. *IEEE Transactions on Power Apparatus and Systems* PAS-101: 924–932.
8 Transformers Committee of the IEEE Power and Energy Society (1996). IEEE Guide for Loading Mineral-Oil-Immersed Transformers, *IEEE Std C57.91-1995*, doi: 10.1109/IEEESTD.1996.79665.

6

Economics of Distribution Systems

6.1 Introduction

Most engineering decisions require the commitment of financial resources. Some of these resources will be spent in the present, and more will be needed for the future. In addition, there could be benefits that accrue on an annual basis. For example, consider a utility that decides to install capacitors in a distribution system to improve power factor and reduce losses. This will require spending money to purchase capacitors and related hardware, and for their installation. The utility will have to spend a certain amount of money annually to maintain the capacitors; but reduction in losses will result in an annual benefit. As expenses take place over different years, we cannot simply add all the costs and all the benefits to obtain the net expenses for the project. This is because the value of money is not constant but decreases from one year to the next. We need to examine some of the factors that are commonly used in engineering economic analysis and arrive at simple approaches to determine costs and benefits of projects over a period of time [1, 2]. In this chapter, we present the basic economic concepts and provide a few examples of economic analysis related to distribution systems.

6.2 Basic Concepts

6.2.1 Interest Rate

Interest rate is the rate that is applied to a financial transaction. The rate can be different, depending on the type of transaction. For example, if you deposit money in your checking account, you get a very small interest, but if you deposit the money in a certificate of deposit (CD), you will get higher interest. However, you will not be able to use the money that you deposited in a CD until it matures. Similarly, the interest rate that you have to pay to borrow money to buy a car is very different

Electric Power and Energy Distribution Systems: Models, Methods, and Applications, First Edition.
Subrahmanyam S. Venkata and Anil Pahwa.

Companion website: www.wiley.com/go/Pahwa/ElectricPowerDistributionSystems

from the interest rate that you would pay to buy a house. The rates are different due to risks, flexibility, and other factors related to the transaction.

6.2.2 Inflation

Inflation is the rate at which the price of goods and services increases with time. Do you remember your parents telling you that when they were your age certain item used to be so cheap and that it has become too expensive now? Whether we like it or not, inflation in fact is an integral part of the modern economy. It varies from one year to another and from one country to another. Nobody seems alarmed if inflation stays at a small value, but everyone gets worried if its value becomes very large. Inflation, however, is a complex phenomenon, and its value at a given time depends on the spending behavior of people, policies of the government, international events, and many other factors. It is almost impossible to predict inflation for the future years.

6.2.3 Discount Rate

Discount rate is the rate that is used by businesses to discount money with time to make financial decisions that span several years. Different businesses use different discount rates based on their specific conditions. This topic is rather complex and we will not go into the details here. Most people consider the discount rate to be related to the opportunity cost of investing capital. In other words, if an investor considers receiving $100 today as equivalent to receiving $112 one year from today, then the discount rate is 12% for that investor. Discount rate is usually much higher than the interest rate that you would get from the bank since it includes risk, cost of capital, government policies, and other business factors. Generally, the discount rate used for engineering economic analysis does not include the effect of inflation. However, if inflation rates for future years are available, they can be included in the discount rate. If the discount rate does not include inflation, all the future costs and benefits are represented in real dollars (today's dollar), but if they do include inflation, then all the future monies are represented in inflated or nominal dollars. It turns out that either way the results are the same. Therefore, most people prefer to leave inflation out to keep things simple.

6.2.4 Time Value of Money

As we have discussed in previous sections, money loses value with time; that is, the more the investor waits to receive the money, the higher the amount he will expect. It is a common practice in financial analysis to convert all the future financial outlays to a present value by defining the present worth factor or discount factor, which is a function of discount rate (d) and the number of years from the present (n):

$$DF_n = \frac{1}{(1+d)^n} \tag{6.1}$$

Thus, a discount rate of 12% and $n = 1$ gives a present worth factor of 0.893, which implies that \$100 one year from today will have the present worth of \$89.30. Similarly, $n = 5$ with the same discount rate gives a present worth factor 0.567, and therefore, \$100 five years from today have a present worth of \$56.70.

6.2.5 Annuity

Annuity is a term that defines equal financial outlays that occur at fixed interval of time. For example, if we take a loan for three years from the bank to buy a car, we will be paying the bank a fixed amount of money every month for the following three years. The amount of payment will depend on the amount of loan and the interest rate. Similarly, in engineering applications, the same amount of expense or benefit may take place on a yearly basis, such as annual cost of maintaining equipment and annual benefits accrued due to the implementation of a new technology.

6.2.6 Present Worth of Annuity

An annuity of \$A over N years can be converted to the present worth by first finding the present worth of the financial outlay for each year and then summing the present worth,

$$PW(A)_N = \frac{A}{(1+d)} + \frac{A}{(1+d)^2} + \cdots \cdots + \frac{A}{(1+d)^N} \tag{6.2}$$

We have considered that payments for each year are made at the end of the year. Now add and subtract A from the right-hand side of this equation, and let $\frac{1}{(1+d)} = a$. Then,

$$PW(A)_N = A(1 + a + a^2 + \cdots \cdots + a^N) - A \tag{6.3}$$

The quantity within the parenthesis is a finite geometric series equal to $\frac{1-a^{N+1}}{1+a}$. Upon substitution and simplification, we get

$$PW(A)_N = A\frac{(1+d)^N - 1}{d(1+d)^N} \tag{6.4}$$

6.2.7 Present Worth of Geometric Series

In some situations, certain quantities will increase at a certain rate annually. An example of such a quantity is losses in distribution feeders. Since losses are dependent on load, and if the load grows at a certain rate, the losses will also grow annually. With j as the rate of load growth, we get a geometric series,

$$A_k = A_{k-1}(1 + j) \text{ for } k = 2 \text{ to } N \tag{6.5}$$

where A_k is the cost due to losses in year k. The present worth of such a series over N years with the total cost at the end of the first year equal to A is

$$PW(A)_N = A(a + a^2(1 + j) + a^3(1 + j)^2 + \cdots\cdots + a^N(1 + j)^{N-1}) \tag{6.6}$$

or

$$PW(A)_N = aA(1 + s + s^2 + \cdots\cdots + s^{N-1}) \tag{6.7}$$

where $s = a(1 + j)$.

Now, the quantity within the parenthesis is $\frac{1-s^N}{1-s}$. Substituting this in the above expression and simplifying gives

$$PW(A)_N = \frac{A\left(1 - \left(\frac{1+j}{1+d}\right)^N\right)}{(d - j)} \tag{6.8}$$

6.3 Selection of Devices: Conductors and Transformers

Economics is one of the major factors in the selection of devices and equipment for distribution systems. The decision should include the total cost including the capital cost, the operating cost, and the cost of losses over the selected life of the equipment. In this section, selection of conductors for distribution feeders and selection of distribution transformers are discussed.

6.3.1 Distribution Feeder Conductors

The selection of conductors for design and upgrade of distribution systems is an important part of the planning process. After taking all the factors into consideration, utilities select four or five conductors to meet their requirement, which is done based on life-cycle economic analysis and engineering judgment. Historical factors also play a role in the selection process, i.e., if a company has been using a particular size of conductor, they would want to continue to use that size unless there are compelling reasons not to do so. Some of the important factors to consider are conductor ampacity, conductor resistance, feeder cost, power cost, energy cost, and load growth.

6.3.1.1 Conductor Economics

Every conductor has a unique cost vs. load characteristic. Thus, for the proper selection of conductors, it is necessary to analyze these characteristics.

Initial installation, annual operation and maintenance, and losses are the three components of the total cost. Initial installation is a one-time cost, and it is incurred whenever the line is built. This cost is different for different conductor sizes since heavier hardware is needed for larger conductors, and also the handling cost is higher for larger conductors. Annual operation and maintenance costs are also higher for lines with larger conductors. This is mainly due to the fact that utilities would spend more money on maintenance of those lines that carry higher load since the failure of these lines would impact a larger number of customers. This cost could have an annual escalation, but in most planning studies such escalation is neglected. Thus, the present worth of fixed annual expense of $\$A$/year for operation and maintenance over a period of N years at a discount rate of d can be determined by multiplying it by the present worth factor given by Eq. (6.4).

Unlike transmission systems, the feeders and distribution systems are not transposed. Thus, each phase has a different resistance and reactance value. However, for planning purposes, it is prudent to consider them to be equal to simplify computations. One option is to consider the feeders to be transposed and determine the phase impedance matrix. In the next step, transform the matrix to sequence matrix and use the positive sequence resistance and reactance in the calculations. The other option is to ignore the mutual impedance values in the phase impedance matrix and take the average of the phase a, b, and c values to determine resistance and reactance of conductors.

Losses in the lines, which are a function of the peak load, contribute to the variable part of the cost. If there were any load growth, this cost would increase every year. Thus, if the peak load in the first year of operation of the line is P MW, losses due to this load can be computed by first determining the current, which is

$$I = \frac{P \times 10^3}{\sqrt{3} V\, pf} \text{ A} \tag{6.9}$$

where V is the line-to-line voltage in kV, and pf is the power factor. Hence, for an annual loss factor (ratio of average loss to loss at peak load) L_{sf}, the total energy losses in the first year for one mile of a three-phase line with a resistance of r ohms per mile are

$$\text{Loss}_1 = 3I^2 r = \frac{8760 \times 10^3 L_{sf} P^2 r}{(V\, pf)^2} \text{ kWh/mile} \tag{6.10}$$

$$= DP^2 \text{ \$/mile} \tag{6.11}$$

where $D = \frac{8760 \times 10^3 L_{sf} r}{(V\, pf)^2}$ is a constant for the given values of various quantities.

Multiplying Eq. (6.10) by c (cost of energy in \$/kWh) gives the cost of losses in \$/mile for the first year.

Now, consider that peak load grows at a rate m per year. Therefore, the peak load in the nth year is given by

$$P_n = P(1+m)^n \tag{6.12}$$

Therefore, losses in the nth year (Loss_n) with the peak load of P_n and the assumption that all the quantities in D are fixed are

$$\begin{aligned}\text{Loss}_n &= DP_n^2 = DP^2(1+m)^{2n} \\ &= DP^2(1+m^2+2m)^n\end{aligned} \tag{6.13}$$

Now, if we let $j = m^2 + 2m$

$$\text{Loss}_n = DP^2(1+j)^n \tag{6.14}$$

$$= \text{Loss}_1\ (1+j)^n \tag{6.15}$$

Hence, losses will grow at the rate j. We have assumed that all the other terms in Eq. (6.10) remain fixed. The present worth of a quantity that escalates with a rate j is obtained by multiplying the value in the first year by a present worth factor given by Eq. (6.8). As a variation to this issue, we can consider that the load would grow at a certain rate for the first N years and then remain constant. Such a situation can be very easily included in the calculations by splitting the present worth calculation into two parts. In the first part, the present worth of geometric series for the first N years is determined with load increasing with the rate of m per year. In the second part, the present worth of the remaining years is determined with load fixed at the value determined for the Nth year.

To obtain the total present worth, the cost of a line with a specific size of a conductor, we add the installation cost, the present worth of operation and maintenance, and the present worth of losses and obtain the following expression of the total present worth cost:

$$PW_{\text{Cost}} = a + b\,P^2\ \$/\text{mile} \tag{6.16}$$

where $a = \text{Cost}_{\text{ins}} + (PW_1 \times \text{Cost}_{\text{O\&M}})$;

$b = \frac{8760 \times 10^3 L_{sf}\ r\ c\ PW_2}{(V\ pf)^2}$;

Cost_{ins}, initial installation cost in \$/mile;

$\text{Cost}_{\text{O\&M}}$, Operation and maintenance cost in \$/year/mile, PW_1, present worth factor for fixed annual O&M cost/year, and PW_2, present worth factor for losses changing yearly with changing load.

Thus, the total present worth as a function of peak load in the first year turns out to be a quadratic function. We can similarly find the present worth cost of lines of different conductor sizes, and when the total present worth cost vs. peak load characteristics of different conductors are plotted on the same graph, we get a graph of the type shown in Figure 6.1. An important thing to note is that the plots of different conductors intersect with each other, and thus, for every value of peak load there is a conductor with the least cost. Hence, the peak load range for which

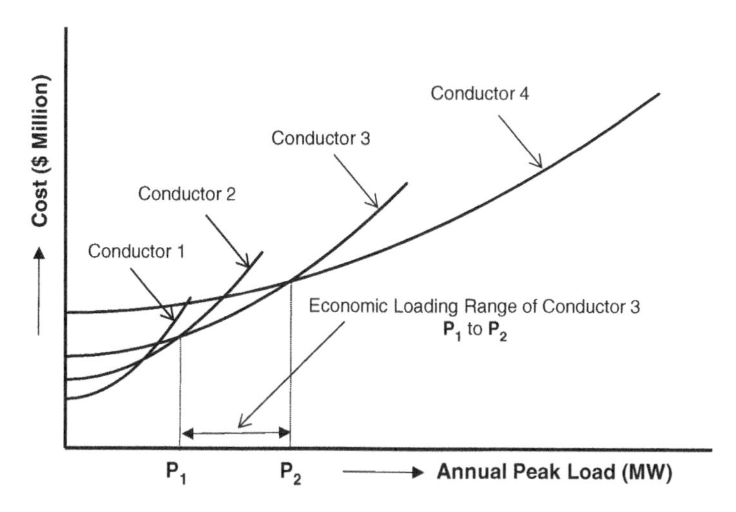

Figure 6.1 Economic characteristics of a set of four conductors.

a conductor is the most economical is defined as the economic loading range of that conductor. Figure 6.1 illustrates the economic loading range of Conductor 3. The largest conductor in the set has a lower economic loading limit, but the upper economic loading limit is equal to the thermal loading limit of that conductor.

6.3.1.2 Reach of Feeders

The reach of a feeder built with conductors of a specific size is defined as the distance up to which a certain load can be delivered over the feeder without violating the voltage limits at the end of the feeder. Thus, we define the following reaches [3]:

- *Thermal reach* of a feeder is the distance up to which the feeder can move power at its thermal loading limit (maximum allowed current or I_{max}) while maintaining the voltage within the specified limits.
- *Economic reach* of a feeder is defined as the distance up to which the feeder is capable of carrying power equal to that determined by the upper limit of the economical loading range of the selected conductor without violating the voltage limits.

The reach of a feeder with a specific conductor size at a given load is determined by first finding the voltage drop on the feeder for the current corresponding to that load as discussed in Chapter 4 (Eq. 4.52). We are assuming that the load is balanced and all the phases have the same impedance; thus, the voltage drop will be identical on all the phases. The percent voltage drop per mile for a distribution feeder is obtained by dividing the voltage drop by the rated line-to-neutral voltage and multiplying the result by 100, or

$$\%V_{drop}/\text{mile} = \frac{\sqrt{3}\,I(r\cos\emptyset + x\sin\emptyset)}{V \times 1000} \times 100 \tag{6.17}$$

where V is the rated line-to-line voltage of the system in kV. If the allowed voltage drop is $\Delta V\%$, the reach, R, of the feeder is given by

$$R = \frac{(\Delta V/100)V \times 1000}{\sqrt{3}I\,(r\cos\emptyset + x\sin\emptyset)} \text{ miles} \tag{6.18}$$

or

$$R = \frac{(\Delta V/100)V^2 \cos\emptyset}{P(r\cos\emptyset + x\sin\emptyset)} \text{ miles} \tag{6.19}$$

When $I = I_{max}$, the computed reach is the thermal reach, and when $I = I_{econ}$ or current corresponding to power at the upper limit of the economic loading range of the conductor, the reach is called economic reach. I_{econ} is the current carried by the feeder for power at the point of intersection of the selected conductor's curve with that of the next higher conductor as shown in Figure 6.1. This point gives the economic loading limit of the conductor; i.e., the conductor is not economical for use beyond this limit. If a reach higher than the economic reach is desired from a feeder, the feeder has to be derated, or the upper loading limit reduced to keep the voltage drop within the limits. In distribution system design, it is a recommended practice to keep the reach of all the feeders in a system the same by derating all the conductors in the selected set corresponding to the desired reach [3].

A plot showing variation of reach (R) with peak load (P) for any given conductor and power factor can be obtained using Eq. (6.19). Note that reach must be computed using the final peak load on the feeder. For example, if a conductor has a certain peak load in the first year and it increases over the years to the final peak load in year N, the peak load in N must be used for reach calculation. Figure 6.2 shows reach as a function of peak load for four conductors. It is very obvious that reach decreases with increase in peak load. Now, if we desire a common reach for these conductors, we draw a horizontal straight line at the desired reach value, which is R_c miles in this figure. Intersections of the horizontal line at R_c with the reach characteristic of different conductors provide the maximum peak load of the respective conductors for the desired reach. Therefore, the maximum peak load for a common reach of R_c miles is P_{max1} for Conductor 1, P_{max2} for Conductor 2, P_{max3} for Conductor 3, and P_{max4} for Conductor 4.

Now, we use the maximum peak load values obtained from Figure 6.2 and derate the conductors for a reach of R_c miles to get the revised cost characteristics as shown in Figure 6.3. Since the cost characteristics are based on the peak load in the first year, the maximum peak load values from Figure 6.2 must be scaled down to the first year. For example, the maximum first-year peak load for Conductor 1 will be

$$P_1 = \frac{P_{max1}}{(1+m)^N} \tag{6.20}$$

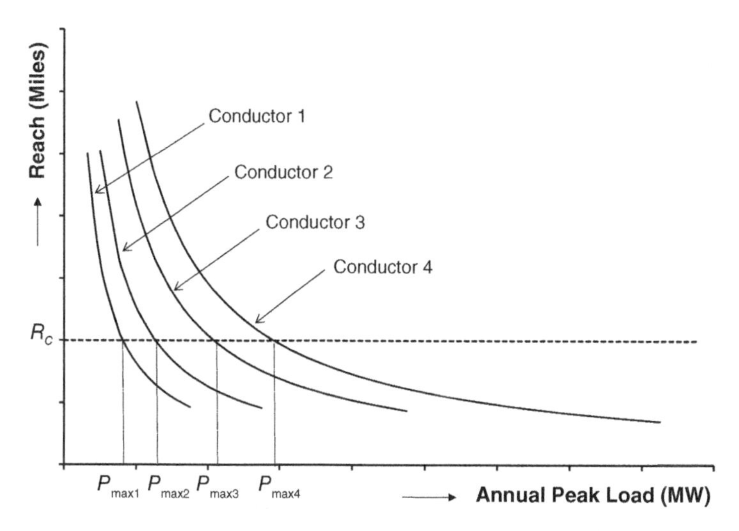

Figure 6.2 Reach at different peak loads for selected conductors.

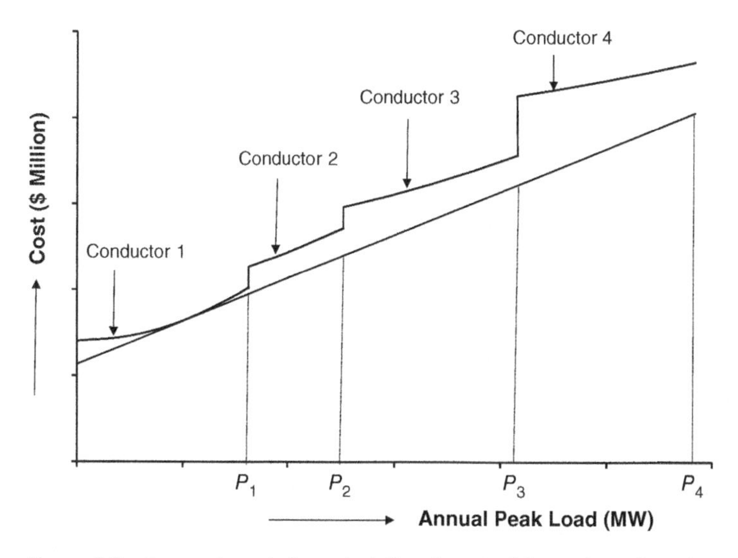

Figure 6.3 Economic cost characteristics of a set of four selected conductors with a common reach of R_c miles.

Similarly, we can obtain values of P_2, P_3, and P_4. Because we are derating the conductors, the cost characteristics do not intersect each other any longer. Therefore, we see a jump from one conductor to the next at the peak loads P_1, P_2, P_3, and P_4. Although the most economical characteristics are those given by the graphs of different conductors in a conductor set, a straight line can be a simple linear approximation for these characteristics. This line could be a line tangent to the smallest and the largest conductor as shown in Figure 6.3. Note that the full graph of the largest conductor is not shown in this figure, but the straight-line approximation does meet its graph at higher values of peak load.

Calculations shown in this section, as well as in the previous section, are for three-phase feeders. The analysis can be modified for single- or two-phase feeders. Such modifications are important because a large percentage of feeders in a distribution system carry load below 1 MW, and often, these feeders are single-phase laterals. Further, the voltage drop calculations were done with an assumption that the power flowing on the feeder remains the same at the beginning and end of the feeder. But in reality, the power flow decreases as we move along the feeder because the feeder will have laterals tapped off of it to supply loads. This in turn decreases the voltage drop or increases the reach of the feeder. A factor to account for the reduction in voltage drop for such phenomenon can be included in the calculations. Specifically, a precise value of this factor can be determined if the feeder serves a uniformly distributed load in a rectangular or triangular service area [4]. However, in reality, these conditions do not exist. So computing a factor to account for reduction in voltage drop is difficult. On the other hand, it is alright if we do not consider this factor because, in emergencies, the feeders may be required to support the additional load of an adjoining feeder. The extra load and reach margin available on the feeders would be valuable in such situations.

6.3.1.3 Optimal Selection of Conductors for Feeders

Selection of a good conductor set is very important for proper planning of distribution systems. An ideal conductor set should have roughly equal economic loading range for each conductor, sufficient thermal capacity in the largest conductor to take care of situations with very high load, and should be the most optimal in terms of cost. Generally, it is very difficult to select a conductor set that will meet all the criteria of an ideal set. Most utilities choose conductors for their system based on experience and historical applications. Usually, the number of conductors in a set is limited to four or five for proper management of inventory. Some authors have addressed this issue to provide a systematic procedure for selection of a good conductor set [5–8].

An approach to determine the optimal conductor is by minimizing the aggregate area under the plots of different conductor sets between the minimum and the maximum loads as discussed in [8]. For example, in Figure 6.1 we would integrate from zero to the maximum value of annual peak load on the graph of the lowest cost characteristics considered for different load ranges. The conductor set with minimum value for this area would have an average minimum cost over the whole range of loading. A simpler approach is to compute the area enclosed by the conductor curves and the tangent line to the smallest and the largest conductors in the set as shown in Figure 6.3. We call this line the ideal line. Also, the conductor characteristics shown in this figure account for derating of conductors to accommodate the selected feeder reach. During various studies we found that none of the conductors had plots that would go below the tangent line. Hence, minimizing the total area under the curves is equivalent to minimizing the area between the curves and the ideal line. The equation of the ideal line can be obtained by finding the two points where the ideal line would touch the smallest and the largest conductors in the set. These two points are given by the following equations:

$$x_1 = \sqrt{\frac{b_4(a_1 - a_4)}{b_1(b_4 - b_1)}} \tag{6.21}$$

and

$$x_2 = \frac{b_1}{b_4} x_1 \tag{6.22}$$

where $y_1 = a_1 + b_1 x^2$ and $y_4 = a_4 + b_4 x^2$ are the equations of the smallest and the largest conductors in the set. Let the ideal straight line be

$$y = mx + c \tag{6.23}$$

where m is the slope of the line, and c is the intercept on the y-axis.

However, instead of using this area for the selection of the optimal conductor set, a better approach is to give different weights to areas in different loading ranges, since in most distribution systems a large percentage of feeders carry very small load, and a few feeders carry large load. For example, in a system discussed in [3], out of 24 800 miles of primary voltage line, close to 56% of the total length of feeders carry power less than 0.5 MW. Around 11% of the total length of feeders carry power between 0.5 and 1 MW. Similar information is given for every increment of 0.5 MW. Of these feeders, the ones that carry less than 1 MW are usually single-phase laterals, and the remaining are three-phase feeders. This shows that two-thirds of the feeders in a distribution system are single-phase feeders. These numbers are for an example system and will be somewhat different for different systems. A utility can very easily obtain such information from their existing system.

Data obtained from [3] was modified slightly to get Table 6.1, which shows the percent of the total feeder length with the given peak load in an example distribution system. The modification included adjusting the percentage for each range to make the sum of the percentages equal 100. Since this table includes single-phase laterals, it must be modified to determine weights to compute the weighted average deviation (WAD). If we consider that a very small percentage of three-phase feeders will carry peak load below 1 MW, we can discard the first two rows and scale up the percentages for the rest of the rows such that their sum is equal to 100.

Table 6.1 Percentage of conductors with different loading ranges in an example distribution system.

Loading range (MW)	% Of conductors
0–0.5	56.18
0.5–1.0	10.99
1.0–1.5	7.94
1.5–2.0	5.50
2.0–2.5	3.66
2.5–3.0	3.05
3.0–3.5	2.44
3.5–4.0	1.83
4.0–4.5	1.53
4.5–5.0	1.22
5.0–5.5	1.22
5.5–6.0	1.22
6.0–6.5	0.92
6.5–7.0	0.61
7.0–7.5	0.61
7.5–8.0	0.46
8.0–20	0.31

Following that, the WAD of the cost of a conductor set from the tangent line can be computed as follows:

$$\text{WAD} = \frac{\text{Total Weighted Area}}{\sum \text{Loading range} \times \text{Adjusted Weight}} \tag{6.24}$$

To explain this further, let $P_1 = 1.6$ MW and $P_2 = 2.4$ MW. Now, we want to compute the weighted area between Conductor 2 and the ideal line. Since the weights are given for 0.5 MW ranges, we divide the range of integration in 0.5 MW increments. Therefore, the weighted area between Conductor 2 and the ideal line is given by

$$\text{WAD}_2 = w_2 \int_{1.6}^{2} (a_2 + b_2x^2 - mx - c)dx + w_3 \int_{2}^{2.4} (a_2 + b_2x^2 - mx - c)dx \tag{6.25}$$

The characteristics of Conductor 2 are given by $y_2 = a_2 + b_2x^2$, where w_2 is the weight for the loading range from 1.5 to 2 MW, and w_3 is the weight for the loading range from 2.0 to 2.5 MW. The process is repeated for all the conductors in their respective loading ranges up to the maximum load of the largest conductor for the selected reach to determine the WAD using Eq. (6.24).

6.3.1.4 Example [8]

In this example, 17 conductors spanning from conductor #1 to 795 kcmil are considered for a 12.47-kV distribution system. The discount rate is 8%, the load growth rate is 0.5% over a period of 30 years, the loss factor is 0.46, the power factor is 0.9, the cost of energy is 3 cents/kWh, and the planning horizon is 30 years. The smallest and the largest conductors are preselected to be #1 and 795 kcmil, respectively, and the remaining two or three conductors are determined using the procedure described in the previous section. In Method A, the reach of the conductors is not considered for determining the optimal conductor set, and in Method B, a reach of 4.7 miles and allowed voltage drop of 7.5% are considered. The results for a four-conductor set and a five-conductor set are given in Table 6.2.

Now, we want to compare the selected sets for a reach of 4.7 miles. To be able to do so, we have to derate the conductors determined with Method A appropriately. The adjusted average-weighted deviations of the two conductor sets from the ideal cost characteristics are $17 049/mile and $15 160/mile, respectively, for the four-conductor sets. Thus, the latter conductor set offers an average saving of $1889/mile for a reach of 4.7 miles. Similar analysis for the five-conductor set gives WAD of $13 471/mile with Method A and $12 691/mile with Method B, thus yielding an average savings of $780/mile.

When the common reach was changed to 3.6 miles, the same conductor set #1, 266, 477, and 795 kcmil were found to be the best whether Method A or Method B was used. This is a very interesting result, which can be explained as follows. We already know that to increase the reach of a conductor higher than its economic reach, the conductors need to be derated to carry maximum load lower than its economic loading limit. Thus, when the common reach is fixed at 4.7 miles after selecting the conductors, the conductors get derated significantly. This is because of the fact that the economic reaches of these conductors vary from 3.1 to

Table 6.2 Best conductor sets.

Number of conductors	Method	Best conductor set
4	A	#1, 266 kcmil, 477 kcmil, 795 kcmil
4	B	#1, 2/0, 300 kcmil, 795 kcmil
5	A	#1, 1/0, 266 kcmil, 477 kcmil, 795 kcmil
5	B	#1, 2/0, 266 kcmil, 477 kcmil, 795 kcmil

4.7 miles. So, when the reaches of all the conductors are adjusted to be 4.7 miles, all except one of them (266 kcmil, which has the economic reach of 4.7 miles) get derated by a large value. But when the reach is fixed at 3.6 miles, the conductors are not derated by a big margin, which implies a smaller deviation from the already selected characteristics. Hence, for smaller reach, the results are the same whether Method A or Method B is used. Also, the average weighted deviation from the straight line for a reach of 3.6 miles for the four-conductor set turned out to be \$10 633/mile, which is lower than that for a reach of 4.7 miles. This can again be explained by the fact that for a lower reach, derating of the conductors is lower than that for higher reach. Also, the optimal conductor set was found to be the same using both methods for the five-conductor set, giving a WAD of \$9317/mile.

The results show that Method B is better than Method A if reach is known ahead of time. But both methods give the same result for shorter reaches. However, another question that needs to be answered is whether it is better to select five or four conductors. Five-conductor set offers savings of \$2468 over a four-conductor set for a common reach of 4.7 miles. It offers savings of \$1316/mile for a common reach of 3.6 miles. These savings can be very significant, since distribution systems are very extensive and cover large areas. However, extra cost associated with maintaining an inventory of five conductors in comparison to four conductors has to be considered while making decisions to stock four or five standard conductors in the inventory.

6.3.2 Economic Evaluation of Transformers

Similar to conductors, transformers have to be selected while considering both the capital cost and the operating cost. We use the economic principles presented in the previous section to determine the best size of the transformer. Let C_c be the capital cost of the transformer, A_r be the annual charge rate (note that it is inverse of the present worth factor), and OPEX are the annual operating expenses associated with the transformer. Now, we can write an equation for the total annual cost AC_T of the transformer as

$$AC_T = C_c A_r + \text{OPEX} \tag{6.26}$$

The annual operating cost includes the cost due to the exciting current (C_{ex}), core losses (C_{cl}), and copper losses (C_{cu}). Therefore,

$$\text{OPEX} = C_{\text{ex}} + C_{\text{cl}} + C_{\text{cu}} \tag{6.27}$$

These costs can be estimated as follows:

$$C_{ex} = I_{ex}\,\text{kVA}\,C_{sc}\,A_r \tag{6.28}$$

where I_{ex} is the per-unit exciting current of the transformer, kVA is the name plate rating of the transformer, and C_{sc} is the installed cost per kvar of the primary shunt capacitors.

In other words, we are computing the cost of the capacitors needed to compensate the inductance caused by the exciting current in the transformer.

$$C_{cl} = (C_{kW}\,A_r + c_{gen}h\,)L_{cl} \tag{6.29}$$

where C_{kW} is the cost of the investment per kW load, c_{gen} is the cost of energy generation per kWh, L_{cl} are core losses of the transformer in kW, and h are the number of hours/year or 8760 hours.

$$C_{cu} = (C_{kW}\,A_r + c_{pgen}\,h\,\text{LsF})P^2\,L_{cu} \tag{6.30}$$

where c_{pgen} is the incremental cost of the peak generation per kWh, LsF is the loss factor in pu, P is the peak load of the transformer in pu, and L_{cu} are the copper losses of the transformer at the rated load.

Note that the cost of the core losses is treated differently than the cost of the copper losses. The core losses are constant at all loads, but the copper losses increase with increasing load. Now, we define an index for the transformer or

$$\text{Economic Performance Index} = \text{AC}_T/\text{kVA} \tag{6.31}$$

Example

For a 100-kVA pad mount transformer, the data are given in Table 6.3.

Compute the total annual cost and the economic performance index of the transformer.

Solution

Substitute the values in Eqs. (6.26)–(6.31).

$$C_{ex} = 0.015 \times 100 \times 5 \times 0.25 = \$1.875$$

$$C_{cl} = (600 \times 0.25 + 0.025 \times 8760)0.37 = \$136.53$$

$$C_{cu} = (600 \times 0.25 + 0.05 \times 8760 \times 0.36)\,(1.0)2 \times 1.17 = \$359.98$$

$$\text{OPEX} = 1.875 + 136.53 + 359.98 = \$498.39$$

Total Annual cost: $\text{AC}_T = 3000 \times 0.25 + 498.39 = \$1248.39$
Economic Performance Index = 1248.39/100 = \$12.48/kVA

Table 6.3 Data for a 100-kVA pad mount transformer.

C_c	\$3000
A_r	0.25
I_{ex}	0.015 pu
kVA	100 kVA
C_{sc}	\$5/kvar
C_{kW}	\$ 600/kW
c_{gen}	\$0.025/kWh
c_{pgen}	\$0.05/kWh
h	8760 h
L_{cl}	0.37 kW
LsF	0.36
P	1.0 pu
L_{cu}	1.17 kW

6.4 Tariffs and Pricing

Unlike the transmission systems, which are regulated by the Federal Energy Regulatory Commission, distribution systems are regulated at the state levels. Electric distribution utilities have an obligation to serve all the customers in their service territory while maintaining a certain level of reliability. The utilities build, upgrade, and maintain the system to fulfill the obligation. The state corporation commissions allow utilities to include some of these expenses into the rate base. In other words, the utilities recoup these expenses over time through revenues collected through bills. Every state has a different model, but there are many common elements to the electricity bills, and we will explore them in this section.

6.4.1 Electricity Rates

6.4.1.1 Energy

Energy consumed by customers is measured in kilowatt-hours (kWh) by a meter installed at the service entrance. The simplest form of rate charges for energy at a fixed rate irrespective of the time when it was consumed. For customers with large requirements, some utilities have block rates in which the first block of energy consumed has a rate and the next block of energy has a lower or higher rate. Subsequent blocks have lower or higher rates.

6.4.1.2 Demand

Demand is defined as the largest power demand (kW) measured for a specified duration during the billing period. For example, customers on a typical demand rate will pay for the energy consumed over the month and will also pay for the maximum one-hour demand during the month. The demand can be applicable for all the months or for certain months of the year.

6.4.1.3 Time of Use (TOU)

As the name implies, under this plan varying electricity rates for different blocks of time are usually defined for a 24-hour day. TOU rates generally reflect the average cost of generating and delivering power during those time periods. TOU rates often vary with the time of the day and by the season, which are typically predetermined by the utilities. This type of pricing requires meters that register cumulative usage during different time blocks. As an example, the utility can define 2:00 p.m. to 7:00 p.m. as the peak period with a higher rate for electricity. The rate for the off-peak hours is lower than the flat rate to provide incentive to customers to sign up for this rate plan. In addition to the peak period, some utilities use three blocks of rates. These blocks are peak, shoulder, and off-peak. Thus, the off-peak period could be from 10:00 p.m. to 10:00 a.m., shoulder period from 10:00 a.m. to 2:00 p.m. and from 7:00 p.m. to 10:00 p.m., and peak period from 2:00 p.m. to 7:00 p.m. In certain cases, the TOU rates are applied only during the summer months. As we can see, there are multiple combinations for time of use rates. Some utilities have implemented TOU rates and some have experimented with them, but they are not used very widely in the utility business.

6.4.1.4 Critical Peak Pricing (CPP)

This type of pricing includes a prespecified high rate for usage during periods designated by the utility to be the critical periods. The CPP events are generally triggered by system contingencies or high prices faced by the utility in procuring power in the wholesale market. CPP rates may be superimposed on either a TOU or time-invariant rate and are called on a relatively short notice for a limited number of days or hours per year. CPP customers typically receive a price discount during non-CPP periods. CPP rates have been tested in pilots for large and small customers in several states [9]. While they have not been implemented at a wide scale, they are available to certain class of customers in some areas.

6.4.1.5 Critical Peak Rebates (CPRs)

Under this plan, the utilities give a predetermined amount of refund to the customers for reducing their consumption when they observe or anticipate high wholesale market prices or power system emergency conditions. The utilities may call critical events during prespecified time periods (e.g. 3 p.m. to 6 p.m. summer weekday afternoons). Although the price for electricity during these time periods remains the same, the customer is refunded for any reduction in consumption relative to what the customer was expected to consume.

6.4.1.6 Interruptible Rates

These rates are available to large industrial, commercial, or government customers. Those under this plan agree to curtail their load by a certain percent under declared emergencies.

6.4.1.7 Power Factor-Based Rates

Since lower power factor results in higher current for the same amount of real power, some utilities charge large customers a penalty for low power factor. For example, the utility may require the customer to maintain the power factor between 0.95 lagging and unity. Deviations outside of that are assessed a penalty.

6.4.1.8 Real-Time Price

Real-time price is commonly used in bulk power systems. For example, power pools such as Southwest Power Pool have the price of electricity in its systems declared every 5 minutes in real time. Such implementation has not been done in distribution systems. However, in the future, the FERC Order 2222 will allow third-party aggregators to pool resources from customers in distribution systems and bid in the market. The resources would be in the form of customer-owned solar or other generation and reductions in customers load in response to price changes.

6.4.1.9 Net Metering

Net metering is an idea that came into existence to address issues arising from individual customers owning their own generation, such as rooftop solar photovoltaic generation, and injecting the excess power generation into the grid. In the early days, the utilities compensated the customers at the same rate at which they were selling it to the customers. However, this arrangement was flawed because the rate that utilities charge includes not only the cost of generation but also the cost of delivery. In addition, the utility has invested in the infrastructure to deliver electricity to customers while maintaining a certain level of expected reliability. In this arrangement, they are not able to recover money for their investments. Hence, the new net metering rates in place compensate customers at the wholesale rate for electricity in the market. To make this possible, the utilities install two separate meters, one for flow of energy from the grid to the customers and the other from customers to the grid. The compensation for a given month is credited in the same month or carried forward to the next month. Accounting for the load offset behind the meter by local generation is done in different ways by utilities. Some utilities may consider compensation for it at the full rate, while others may have a reduced rate. In addition, some states are allowing utilities to charge a monthly fee in dollars per kW based on the capacity of the customer-owned system. This is a complex and controversial subject with many variations, and it is still evolving. With increased number of homes owning their own generation and participating in the market through aggregators, net metering will evolve further. The readers are encouraged to check websites of utilities to obtain details of different net metering policies in place for their customers.

6.4.2 Understanding Electricity Bills

The rates that a utility charges to its customers are approved by the state regulators of the state in which the utility provides services. Therefore, each utility has its own schedule of rates, and thus, no two utilities are identical in this regard even within the same state. While it is not possible to cover all possible billing options, we show an example of a typical utility bill based on the rates for residential customers given in Table 6.4 for a large utility company. The rates have been slightly modified for this example.

6.4.2.1 Monthly Rate

In addition, the utilities include fuel charge on the bill, which covers the cost for the fuel needed to generate electricity and any purchased power acquired to meet customer needs. Further, some utilities offer the option to purchase renewable energy certificates (RECs) from wind, solar, or biomass facilities. In that case, the purchase amount associated with them is included in the bill.

Table 6.4 Example of rate schedule for residential customers.

1. Basic customer charge: $7.28 per month		
2. Distribution charge	June to September	October to May
First 800 kWh	2.1086¢ per kWh	2.1086¢ per kWh
Over 800 kWh	1.1943¢ per kWh	1.1943¢ per kWh
3. Electricity supply service charges		
a. Generation charge		
First 800 kWh	3.5826¢ per kWh	3.5826¢ per kWh
Over 800 kWh	5.4500¢ per kWh	2.7632¢ per kWh
b. Transmission kWh charge		
All kWh	0.970¢ per kWh	0.970¢ per kWh

Example

Consider that a customer consumes 1125 kWh in the month of July. This customer has purchased RECs to buy electricity from solar facilities. Compute the monthly bill for this customer based on the rate schedule given in Table 6.4. The customer has also volunteered to purchase 500 kWh of electricity per month from solar resources and pay a surcharge of 2.055¢ per kWh. In addition to kWh usage, the utility has determined a fuel charge of $21.65 and taxes of $6.13.

Solution

Since the month is July, we use the schedule to compute the cost for distribution and energy supply.

Distribution:

$$800 \times 0.021086 = \$16.86$$

$$325 \times 0.011943 = \$3.88$$

Generation:

$$800 \times 0.035826 = \$28.66$$

$$325 \times 0.054500 = \$17.71$$

Transmission:

$$1125 \times 0.00970 = \$10.91$$

Green Power:

$$500 \times 0.020550 = \$10.27$$

$$\text{Monthly Bill} = 7.28 + (16.86 + 3.88) + (28.66 + 17.71) + 10.91 + 21.65 + 6.13 = \$113.08$$

Therefore, the average cost of electricity for this customer is 113.08/1125 = 10.051¢ per kWh.

6.4.3 Rural Electric Cooperatives (RECs)

United States is a large country with many rural communities. Since investor-owned utilities were not catering to the needs of the rural communities, the government passed legislation in 1936, which resulted in the formation of RECs. The RECs are member-owned nonprofit organizations that provide electricity to rural communities. Large RECs own their own generation and transmission, but smaller ones only engage in the distribution business.

6.4.4 Municipal Utilities

Several small cities in the United States operate their own distribution system. Some of them own small generation within the city limits, but many of them buy electricity from the market to serve their customers. Since individual cities do not have large market power, typically several cities cooperate to form power purchase pools for better rates.

Problems

6.1 Find the present worth of $100 eight years from today with a discount rate of 9%.

6.2 Find the present worth of an annuity of $500 per year for a period of 15 years.

6.3 A utility spends $250 000 on a project which is financed for a period of 20 years at a discount rate of 10%. What will be the yearly payments for repaying the loan?

6.4 A distribution utility supplying power to customers with a peak load of 500 MW lost approximately $4 million this year due to losses in the distribution system. This amount is expected to grow at 2% annually due to load growth. Find the present worth of losses over a period of 10 years at a discount rate of 8%.

6.5 A three-phase 12.47-kV feeder with resistance of 0.6 ohms and reactance of 0.8 ohms per mile carries a peak load of 2 MW at a power factor of 0.9.

(a) Find the annual losses in kWh/mile for a loss factor of 40%.

(b) Consider that the load for the feeder will grow beyond 2 MW at a rate of 5% per year over the next 10 years. Find the present worth of losses over a period of 10 years in $/mile if the cost of energy is $0.03 per kWh and discount rate is 8%.

(c) Find the voltage drop per mile at the peak load.

(d) Find the reach of the feeder at the initial peak load and also after 10 years if 7.5% voltage drop is allowed from the substation to the end of the feeder.

6.6 For the feeder specified in the previous problem, consider that the load for the feeder will grow beyond 2 MW at a rate of 5% per year over the next 10 years and after that it will stay fixed for the next 10 years. Find the present worth of losses over a period of 20 years in $/mile if the cost of energy is $0.03 per kWh.

6.7 The objective of this problem is to select a set of four conductors for distribution systems design. 556 kcmil and #1 are the two conductors that have

Table 6.5 Conductor data for Problem 6.7.

Size	r/mile (Ohms)	x/mile (Ohms)	Current carrying capacity	Installation cost ($/mile)	O&M Cost ($/year per mile)
#1	1.50	1.50	200	50,000	2500
1/0	1.30	1.48	230	55,000	2600
2/0	1.10	1.45	270	65,000	2700
3/0	0.90	1.40	300	75,000	2800
4/0	0.75	1.35	340	85,000	2900
336 kcmil	0.45	1.30	530	105,000	3200
556 kcmil	0.35	1.20	730	130,000	3600

already been selected, and we want to pick two additional conductors in between from Table 6.5. Data for the problem are given below.

Voltage = 12.47 kV

Power factor = 0.9

Load factor = 60%

Cost of energy = 3 cents/kWh

Discount rate = 8%

Initial peak load on the feeders = 60% of final peak load

Load growth = 5.0%/year

Allowed voltage drop = 7.5%

Duration of study = 30 years

(a) Choose any two conductors from the given list for a load range of 0–6 MW.
(b) Find the number of years for the peak load to reach the final value.
(c) Determine the cost equations for the given scenario for each of the selected conductors and plot these characteristics on the same graph.
(d) Determine the thermal reach of each conductor, economic loading range of each conductor, and economic reach of each conductor using the voltage drop equations.
(e) Compute the reaches of each conductor for loads ranging from zero to the highest based on the thermal capacity of the conductors and draw plots with load in MW on the x-axis and reach on the y-axis for all the conductors. Highlight the economic loading range for each conductor on this graph.
(f) Adjust the loading of the conductors such that the reach of all the conductors is three miles. Redraw the conductors' costs with adjusted reaches.

(g) Determine the straight-line approximation for the cost vs. load characteristics of the feeders based on the discussion in Section 6.3.1.3. Draw this line of your graph.

(h) Compute the weighted area between your conductors and the straight line with the weights shown in Table 6.6. Find the average weighted deviation between your conductors and the straight line.

(i) Give a critical summary of the results.

Table 6.6 Loading ranges and weights for conductors for Problem 6.7.

Loading range (MW) (MW)(MW)	% Of conductors
0–1.0	60.00
1.0–2.0	20.00
2.0–3.0	10.00
3.0–4.0	5.00
4.0–5.0	3.00
5.0–6.0	2.00

6.8 A transformer is rated 15/20/28 MVA, 115 kV/12.47 kV, $Z = 7.5\%$. Let the core losses be 20 kW and copper losses be 25 kW at a rated load of 28 MVA. The rest of the information is given in Table 6.7.
Compute the total annual cost and the economic performance index of the transformer for the given data.

Table 6.7 Transformer data for Problem 6.8.

C_c	\$200,000
A_r	0.15
I_{ex}	0.012 pu
C_{sc}	\$5/kvar
C_{kW}	\$800/kW
c_{gen}	\$0.03/kWh
c_{pgen}	\$0.06/kWh
h	8760 h
LsF	0.3
P	0.8 pu

References

1 Khatib, H. (1997). *Financial and Economic Evaluation of Projects in the Electricity Supply Industry*. London, UK: The Institution of Electrical Engineers.
2 Sigley, R.M. Jr., (1991). Engineering economic analysis overview. In: *Tutorial on Engineering Economic Analysis: Overview and Current Applications*. Piscataway, NJ: IEEE Publication No. 91EH0345-9-PWR.
3 Willis, H.L. (1997). *Power Distribution Planning Reference Book*. New York: Marcel Dekker, Inc.
4 Kersting, W.H. (2012). *Distribution System Modeling and Analysis*, 3e. Boca Raton, FL: CRC Press.
5 Funkhouser, A.W. and Huber, R.P. (1955). A method for determining economical ACSR conductor sizes for distribution systems. *AIEE Transactions on Power Delivery* 74 (3): 479–484.
6 Lakervi, E. and Holmes, E.J. (1997). *Electricity Distribution Network Design*, 2e. London, UK: The Institution of Electrical Engineers.
7 Leppert, S.M. and AllenA.D. (1995). Conductor life cycle cost analysis. In Rural Electric Power Conference. C2-1–C2-8.
8 Mandal, S. and Pahwa, A. (2002). Optimal selection of conductors for distribution feeders. *IEEE Transactions on Power Systems* 17(1): 192–197.
9 Faruqui, A. and Sergici, S. (2010). Household response to dynamic pricing of electricity: a survey of 15 experiments. *Journal of Regulatory Economics* 38: 193–225.

7

Distribution System Operation and Automation

7.1 Introduction

Distribution system operation typically deals with some form of switching. The switching operations cover activities such as restoration following outages, substation maintenance and clearance, project construction, load transfers to alleviate overloads, capacitor switching, and system reconfiguration. Other activities include disabling automatic reclosing and turning on instantaneous tripping for live line work and temporary switching for other work and planned outages. Operations center also request system changes, such as phase balancing projects. Usually, all planned switching operations are done without any interruption of service. There are exceptions of course, but they are a small percentage.

For a typical midsize utility in the United States, the number of switching orders executed in a distribution system varies between 10 and 20 per working day. The length of these orders varies quite a bit from just a few steps to hundreds of steps. Most of these are planned operations, which have written instructions to respond to a switching request, but there are always some that are outage related. This number is highly dependent on the weather; on severe weather days, the number of orders could spike into the hundreds that are processed and executed in real time. A switching order could be outstanding (not returned to normal) for days or in some cases months, but most are returned to normal on the same day.

As reported by a utility, in 2008, the total number of switching orders was in the neighborhood of 4500 for the year. Most of them involved substation equipment. This number has escalated in recent years in part due to increased capital and maintenance spending. Also, the utility has experienced more outages due to severe weather and an aging system.

Personnel in the distribution control center also monitor substation alarms that are reported to supervisory control and data acquisition (SCADA) and alert substation personnel of problems. Outage management and troubleshooting-dispatch

Electric Power and Energy Distribution Systems: Models, Methods, and Applications, First Edition.
Subrahmanyam S. Venkata and Anil Pahwa.

Companion website: www.wiley.com/go/Pahwa/ElectricPowerDistributionSystems

personnel stay within the control center in most circumstances. It is the control center's obligation and responsibility as the operating authority to properly analyze the system, equipment ratings, circuit loads, and other switching requests and determine whether requests for switching can be granted. It may be that some temporary work needs to be done, or a mobile substation is needed in order to allow work on a particular piece of equipment. Or they may need to wait for other switching requests to be finished, or they may have to avoid doing the work when the load is high, such as in the summer.

Historically, the utilities have executed most of the distribution system operations manually. In fact, many utilities do not have any real-time information available beyond the distribution substation. With systems becoming bigger and with advancements in communications and computer technology, gradually many of them are becoming automated, which has given rise to the concept of *distribution automation* (DA). DA is a precursor to *Smart Grid* in relation to distribution systems. In this chapter, we look at the history of DA and several common operation functions and examine the impact of automation on these functions. A case study on cost/benefit of DA is included.

7.2 Distribution Automation

Deregulation and restructuring of the electric utility business have directed the attention of utilities toward providing higher supply reliability and quality to customers at the distribution level. Although higher reliability and quality are the goals of the utilities, they would like to accomplish this while optimizing the resources. Another goal for utilities is improving system efficiency by reducing system losses. DA provides options for real-time computation, communication, and control of distribution systems and thus opportunities for meeting the goals mentioned above. The concept of DA first came into existence in the 1970s [1], and, since then, its evolution has been dictated by the level of sophistication of the existing monitoring, control, and communication technologies as well as performance and economic factors associated with the available equipment. Evolution of SCADA systems, which have been in use for monitoring the generation and transmission systems, have also helped progress in the field of DA.

Although progress in computer and communication technology has made DA possible [2–7], advances in distribution control technology have lagged considerably behind advances in generation and transmission control. The progress of DA has been relatively slow due to reluctance of utilities to spend money on automation as many utilities have found it difficult to justify automation based purely on cost–benefit numbers. However, DA provides many intangible benefits, which should be given consideration while deciding to implement it. Now, with the emergence of the smart grid concept and integration of distributed energy resources, DA has come to the forefront with many utilities looking at automation of different aspects of distribution systems. Automation allows utilities to implement flexible control of distribution systems, which can be used to enhance efficiency, reliability, and quality of electric service. Flexible control also results in more effective utilization and life extension of the existing distribution system infrastructure and in enhancing resiliency of the distribution system.

In general, functions that can be automated in distribution systems can be classified into two categories, namely, monitoring functions and control functions. Monitoring functions are those needed to record (i) meter readings at different locations in the system, (ii) status of switching devices at different locations in the system, and (iii) events of abnormal conditions. The data monitored at the system level are not only useful for day-to-day operations but also for system planning. SCADA systems perform some of these monitoring functions. The control functions are related to switching operations, such as switching a capacitor, or reconfiguring feeders. In addition, system protection can also be a part of the overall DA schemes. Some functions specific to service at customer locations, such as remote load control, demand response, automated meter reading, and remote connect/disconnect, may also be considered as DA functions. However, automated meter reading has evolved significantly as part of the advanced metering infrastructure (AMI) as a separate area. Now, with the emergence of smart grid concept, both distribution automation and AMI can be considered part of the smart grid.

The functions mentioned above are performed in a relatively slow time frame (minutes to hours). These devices are not designed to endure frequent switching. Recently, several new devices have been developed, which allow rapid control. Application of distribution-level power electronic devices such as static condenser (STATCON) for distribution system control has already been demonstrated [8]. In addition, smart inverters are being deployed with inverter-based resources (IBRs) to manage large-scale integration of solar photovoltaic (PV) generation in distribution systems. These devices are continuously controlled and respond in real time to system changes. Coordination of a STATCON with load tap changer (LTC) and mechanically switched capacitors reduces fluctuations in system voltage, which improve the quality of service. Similarly, smart inverters manage reactive power to manage voltage magnitude and fluctuations [9].

Electric power quality has become an increasingly problematic area in power system distribution systems. Power quality may be defined as "the measurement, analysis, and improvement of bus voltage, usually a load bus voltage, to maintain that voltage to be a sinusoid at rated voltage and frequency [10]." A direct correlation exists between the lack of electric power quality delivered to the customer and the number of complaints received from the customer. As a result, Electric

Power Research Institute (EPRI) has directed substantial research efforts into the development of advanced technologies to improve the performance of utility distribution systems. The technology, called custom power, seeks to integrate modern power electronics-based controllers such as the solid-state breaker (SSB), the STATCON, and the dynamic voltage restorer (DVR) with DA and integrated utility communications to deliver a high grade of electric power quality to the end user [11]. Although extremely useful, custom power devices have been used in distribution systems only on a limited basis. Detailed study of these devices and their applications is a separate subject by itself and is beyond the scope of this book.

Demonstration of the feasibility of DA through various pilot projects increased the interest of the technical community in this field. Some of the early pilot projects include the Athens Automation and Control Experiment [12] sponsored by the U.S. Department of Energy and EPRI-sponsored projects at Texas Utilities and Carolina Power & Light [13]. A list based on an IEEE survey of other projects is available in a report that was prepared by the author [2] in the 1990s. Since then, the number of manufacturers offering DA equipment has increased substantially.

Until the early part of the 1990s, equipment reliability was a major concern. The equipment available now is more reliable and robust compared to that of the older generation. But there are still several issues which are obstacles to widespread implementation of DA. These issues include cost of the equipment, hardware and software standards, and interoperability. Standardization allows the users of DA systems to mix and match components manufactured by different manufacturers and also to port software from one platform to another.

Implementation of DA requires careful thinking and planning. As discussed in a presentation [14], the utilities can either adopt the "top-down" approach or the "bottom-up" approach. The top-down approach is the revolutionary approach in which a large-scale fully integrated automation system is installed to automate all or most of the functions performed by various individual devices in the distribution system. The bottom-up approach is evolutionary in the sense that automation devices to perform only a selected function are installed, or only a small part of the system is automated. Other functions and other parts of the system are automated gradually.

The top-down approach is expensive and requires major modifications in the utility operation, and so it is suitable for only a few utilities. The bottom-up strategy is more suitable for a majority of utilities. This approach allows utilities to adjust to changes at a more measured pace and to install automated systems for the most immediate needs. The most difficult task for a utility contemplating DA is to identify the functions to be automated [3, 15]. The needs of each utility are dependent on geographic location, operating philosophy, and financial situation. A careful screening of all the possible control functions is imperative before implementing any of them.

7.3 Communication Infrastructure

DA system requires a communication infrastructure to communicate with individual customer locations and control points in the distribution system on the one side, and the energy management system on the other side. Generally, such a communication infrastructure is a hybrid system utilizing different communication mediums for different parts. Some of the earlier DA systems used telephone for communication between the control center and the substation, and communications from substation to the customers and control points were based on power-line carrier (PLC) or radio. PLC was an obvious choice because a wired link to all the points of communications was available. It was only a question of installing the right equipment to accomplish communication. However, PLC-based communications suffered from heavy attenuation in certain parts of the distribution system, specifically transformers. Thus, gradually the popularity of PLC has decreased, but some variations of it are still being used successfully for AMI systems.

Earlier, radio systems also had a problem because of limited range and their inability to send signal across obstacles, such as tall buildings. Advances in the radio technology and availability of 900 MHz frequency spectrum to electric utilities have made radio a very popular communication medium. Currently available radio systems can communicate very reliably with points in a large area.

Developments in the fiber-optic technology have made it a viable communication medium for certain applications, and its use in distribution systems has been increasing steadily. One popular approach is to use fiber optic cables along the feeders up to specific points defined as concentrators and radios at that those points to reach customers and other metering locations on the system.

Cellular telephone technology can also be used for some applications in distribution systems. Satellite, as a communications medium, has been experimented with by some utilities, but it is not used very widely.

7.4 Distribution Automation Functions

DA functions can in general be separated into two groups: customer- and system-level functions. Customer-level functions are those functions that require installation of a device with communication capability at customers' premises. These include meter reading, demand response, time-of-use rates, outage alert, and remote connect/disconnect. System-level functions are related to system operations. Control and communication devices for these functions are installed at different locations in the system, such as substations and feeders. These functions include fault detection and service restoration, feeder reconfiguration, voltage/var control, and equipment monitoring. In addition to system operation functions, digital protection of substations and feeders can be considered a part of DA in some situations.

This chapter deals only with system operation-related functions. Customer-related functions can be implemented as a part of AMI, which has been implemented by many utilities as a part of smart metering initiative. AMI has been mainly focused on recording energy consumption in homes for billing and to get alerts on loss of power. AMI evolved independently of DA, and as a result of that, there is no coordination between them. For instance, the utilities have smart meters in homes, but the metered information is not available to the distribution system operators in real time. So they are not able to leverage this information for system operations. Most of the smart meters record energy consumption every 15 minutes, and this information is communicated to the utility either every 15 minutes or stored locally and communicated to the utility once in a day. Finally, it is posted to a website once in a week. Consumers can see this information through their personal account. Several utilities have also implemented additional features as part of smart metering initiative where the consumers see a display of their energy usage, which is updated periodically, inside their home. The hypothesis is that consumers are likely to make adjustments in their usage to conserve energy if they can get feedback about their consumption. AMI is also useful in locating outages. We explore this feature in a later chapter. Except for outage location, AMI has not been integrated well with DA for the control of distribution systems. AMI has a built-in latency, which is not suitable for faster controls in distribution systems. In contrast, micro-phasor measurement units (μPMUs) are being installed in distribution systems. μPMUs are very expensive, and they have not yet seen widescale adoption. Integration of fast data from μPMU (microseconds) with slow data from SCADA (seconds) and AMI (minutes) is a challenge, which has not been addressed adequately.

The operation-related functions can be divided into two classes, namely substation- and feeder-related functions [1]. In fact, some people consider the domain of DA to include only feeder-level functions, and substation-level functions are covered by a separate field called substation automation [16]. Although we mainly focus on feeder-level functions in this chapter, such division of functions are not considered. Each function selected may be applicable for both substation and feeders. In some situations, the functions at substation and feeder levels may be performed in a coordinated fashion; for example, switching of capacitors on the feeders may be coordinated with switching of capacitors at the substation. The following sections provide information of four functions, which are the most important for distribution system operations. These are (i) outage management, (ii) feeder reconfiguration, (iii) voltage and reactive power management, and (iv) equipment monitoring and control. The discussion includes only an overview. Specific formulations and details on implementation are discussed in a separate chapter.

7.4.1 Outage Management

Distribution systems, particularly with a large percentage of overhead feeders, are susceptible to various types of faults due to exposure to the environment. While underground feeders are not exposed to several environmental factors, they also experience faults. Also, a large majority of distribution systems around the world are overhead due to cost considerations. Since underground feeders typically cost 5–10 times more than overhead feeders, so they are used only in special situations where aesthetics or higher reliability is of concern.

The system can have either a temporary or a permanent fault. For a temporary fault, the affected customers experience interruption of power for a very short duration, which is typically in seconds. In the event of a permanent fault, the protective devices are expected to operate and isolate the faulted section, which results in longer service interruption for the customers. However, if the fault is of a high impedance type, the protective devices might not operate to isolate the faulted section. In such situations, locating faults becomes more difficult. In both cases, some customers experience a power interruption. If no information on the status of various devices in the distribution system is available to the distribution system dispatchers, there is no direct way to find out about the outages. The dispatchers depend on telephone calls from customers or a sudden change in power flow at a metered location upstream in the system to come to know of the outages. Calls from customers only provide an approximate location of the outage. Moreover, in the case of a major storm, multiple outages can be widespread and difficult to locate. Once the approximate location of outages is known, line crews are dispatched to drive along the lines to look for damage. After the damaged area is located, it has to be isolated from the rest of the system if the protective devices have not done that already. The next step is to restore power to those parts of the system that are undamaged but have power interruption due to problems elsewhere in the system. The power to these parts may be provided from alternates routes. The dispatchers determine such possible routes and ask the line crew to operate switches. Many of the switches cannot be operated under load; therefore, the substation breaker has to be opened before operating them. If the whole process is done manually, it takes a long time.

Automation of this function requires installation of remotely controlled sectionalizers and sensors on the feeders and leveraging AMI at customer locations to detect interruption of service. Upon loss of power, the meter automatically sends a message, or the outage is detected when the metered information is not received at the end of the metering duration. In the latter case, there can be a delay up to the metering duration, for example 15 minutes, if the outage happens at the beginning of the metering duration. Since the activity of locating outages has historically depended on customer calls, it is called *Trouble Call Analysis* in the utility jargon.

Once sufficient calls are received, locations of the outages are determined based on the escalation of data from the customer level to the substation level. To aid the operators in an outage location, the calls are automatically mapped on the system map. However, simple escalation can lead to wrong diagnosis, specifically during multiple outages. Therefore, advanced inferencing techniques [17] are used to determine possible locations of faults. Installation of sensors on the feeders to obtain additional information also increases the possibility of correct diagnosis. One approach requires recording the time of service interruption [18] at different locations in the system. Techniques based on voltage sags are also used for location of outages. In general, it is easier to locate the main feeder outages but more difficult to locate lateral outages.

The next step is the restoration of service to customers. If the fault happens on the main feeder, power to the unfaulted section of the feeder is restored by automatically closing a tie switch, which routes power from a neighboring feeder. Following that, the crew is directed toward the faulted section to fix the problems, and to restore service. If the fault is on a lateral, restoration requires replacing the fuse manually after the crew has made repairs. Laterals usually do not have a tie switch from a neighboring lateral. Also, replacing fuse by an automated sectionalizer is a very expensive proposition because a distribution system of a small city may have hundreds of laterals. With the cost of an automated single-phase sectionalizer about $10,000 and the cost of a fuse for the same lateral about $50, it is difficult to justify replacing the fuse by an automated sectionalizer.

As automation of this function provides more precise location of the outages, the crew goes to the identified location instead of looking for problems in a general area. Thus, with automation, the whole process of outage location and service restoration can be accomplished more efficiently by less people in much less time with minimum interruption to the customers.

7.4.2 Feeder Reconfiguration

Load in a distribution system varies by hour, by day, and by season. For every load level, the system has an optimal configuration of feeders. In the past, optimality had been defined in terms of minimum loses, but now service reliability has become an equally important criterion for system operation. Several different objectives can be included in a multiobjective distribution system reconfiguration problem. These objectives may include loss minimization, balancing load on transformers for reliability, balancing load on feeders, maximum load on feeders, and deviation of voltages from nominal. Reconfiguration requires opening of a closed switch and closing of an open switch such that the radial structure of the system is maintained, and no parts of the system are without power supply. In a manual system, the reconfiguration of a system is typically done on a seasonal basis, perhaps a few times in a year or during system expansion. Since such reconfiguration may require several manual switching operations and as it causes short-term interruption of service to some customers, it is not feasible to do it more frequently.

Reconfiguration of the system can be accomplished in an automated system using the same sectionalizers that are used for fault isolation and service restoration. As the operation of the sectionalizers is controlled remotely, system reconfiguration can be done as frequently as the dispatcher desires. From a practical point of view, reconfiguration poses several challenges. First, reconfiguration requires closing the normally open switches, which will temporarily create loops in the radial system. This has to be followed by opening of a closed switch to bring the system back to radial configuration. If creation of loops is not desirable, a closed switch has to be opened first before closing the open switch, which will cause short-term interruption of service to some customers. Additionally, opening and closing of switches causes wear and tear of switches, which will reduce the life of these switches. The benefits of frequent reconfiguration must be evaluated against the cost of replacing switches.

7.4.3 Voltage and var Management

This function is very important for utilities to provide customers with proper service voltage under different operating conditions and for minimizing losses in the system. Also, most of the customer loads are inductive, which requires utilities to place capacitors in the distribution system to provide reactive power compensation. Proper management of voltage and reactive power requires coordination between LTC at the substation transformer, switched capacitors in the distribution system, and line regulators. In this section, the role of each of these devices is discussed.

7.4.3.1 Transformer LTC Operation

The substation transformers have LTC control, which changes the tap position in response to load. Since higher load results in higher voltage drop on the feeders, the tap moves to a higher position to maintain the voltage at a proper level at all points on the feeders. Similarly, under low load conditions, the tap moves to a lower setting to compensate for lower voltage drop on the line. The LTC operates in response to the voltage measured on the feeders at the substation. Another scheme based on line-drop compensation is used if voltage at a specific point on the feeder has to be regulated. In this scheme, a voltage-regulating relay is used to control the taps. The voltage across the regulating relay is equal to the voltage measured at the substation minus the estimated voltage drop on the feeder. Since the actual measurements are not available, line impedance from the regulator to the regulated point and measured current at the regulator are used to estimate the voltage at the regulated point. None of these schemes use voltage measurements on the feeders. Therefore, the control is only approximate without any idea of its effects on the customers at the end of the feeder. Now, utilities are installing voltage measurements at selected locations, typically at the end of the feeder, for LTC control, which provides precise and optimal control of LTC. This would also allow implementation of conservation voltage reduction (CVR) effectively to reduce demand on the system.

7.4.3.2 Capacitor Operation

Capacitors are used in the distribution systems for voltage and reactive power support. The capacitors can be of fixed or switched types and are placed at strategic locations to improve the overall system operation. The switched capacitors switch on or off upon the receipt of a signal from a controlling device attached to the capacitor. This control device may be a timer, a temperature-sensitive relay, a voltage-sensitive relay, a current-sensitive relay, a reactive power-sensitive relay, or a combination of the above. The timers are set based on an assumed load curve. However, on a given day, the load may not be the same as the assumed load curve. Moreover, the timer does not discriminate between a working day and a holiday. The temperature-sensitive device is set based on an assumption that the load is high when the temperature is high because the air-conditioning demand goes up during hot weather. This type of control does not work very well because there is a lag of few hours between the outside temperature and the corresponding air-conditioning load due to thermal inertia of the houses. Other types of controls also have problems, which are discussed in the available literature [19].

To alleviate some of the problems associated with the abovementioned schemes of controlling capacitors, microprocessor-based controllers can be used. Often, these controllers have the facility included for communication with the central station. These controllers can be programmed to use a combination of several factors to switch the capacitors. These controllers perform significantly better than the conventional controllers. However, they do not provide the most optimal capacitor configuration for the system. A major drawback of these controllers is that they respond to the local conditions existing at the location of the capacitor. They do not consider the impacts of switching a capacitor on other parts of the system.

An optimal capacitor configuration can be obtained by implementing a higher level of automation where switching of all the capacitors can be coordinated under any load condition. In such a scheme, meters are needed at various locations to measure the real and reactive power, voltage, and current. The metered data and the status of capacitors are sent to the central computer via communication lines. The computer then determines the optimal switching configuration of the capacitors for the measured system conditions. Under the optimal configuration, the system losses are kept at a minimum value with all the voltages within the specified limits. Since the system has real-time measurement capabilities, the switching configuration can be changed as frequently as desired. From a practical point of view, it is not desirable to switch capacitors too frequently to prevent failure of switches or capacitors and due to power quality concerns. However, in the future, power electronics-based schemes will be available for the control of reactive power, which will eliminate the abovementioned concern associated with mechanical switching of capacitors.

7.4.3.3 Regulator Operation

Voltage regulators are autotransformers that are used in distribution systems for additional control of voltage, particularly on long distribution lines where voltage drops are high. These regulators are set to maintain voltage within a specified band, and when the voltage becomes lower than the low setting, the tap on the regulator moves to increase the number of turns of the output side. Similarly, when the voltage goes above the high setting, the tap moves to reduce the number of turns on the output side. The regulators are set to regulate the voltage at a specified point on the downstream side using line drop compensation.

It is quite obvious that if DA is implemented, actual voltage at the regulated point can be metered and can be used in control of regulators. Also, operation of regulators can be coordinated with capacitor switching to reduce losses and to obtain better voltage profile on the feeders under different load conditions. Better voltage profile on the feeders will result in lesser low voltage complaints from the customers.

7.4.3.4 Smart Inverters

With increased proliferation of IBRs into the distribution systems, such as solar PV, voltage and var management has become more challenging. Injecting real power from these resources into the system causes voltage rise. In addition, intermittent clouds change the power output of solar PV rapidly, which in turn causes voltage fluctuations. The previous generation of inverters interfacing IBR did not have options for reactive power control. However, the current generation of inverters is smart with options to control reactive power. Such controls can be implemented locally or in a coordinated manner. Local controls are easier to implement and may be acceptable in many situations. Coordinated control is more challenging because it requires communications with all the devices for data gathering, and the data have to be processed quickly to make control decisions. For a system with hundreds or thousands of such distributed devices, centralized control is infeasible. A hybrid approach with distributed data management and computing is a viable approach to manage such systems. Another complicating issue is the ownership of data associated with distributed resources. Most of the customer-owned rooftop solar PV are behind the meter, and, as a result of that, the utilities do not have access to that data and do not have control jurisdiction over those devices. Any such control will require a special contract between the utility and the customers. The recent Federal Energy Regulatory Commission (FERC) order 2222 allows third-party aggregators to coordinate behind the meter resources for participation in the market. In such a scenario, the aggregators will have access to data and will be able to control these devices. However, the control actions should not have adverse impacts on the distribution systems to which they are connected. Thus, some coordination between the aggregators and the utility is imperative. The situation becomes more complex with large number of aggregators simultaneously engaged with customers in a distribution system. A possible architecture to leverage smart inverters for transactive energy is shown in Figure 7.1. This subject is still being investigated with no unique solution to the problem.

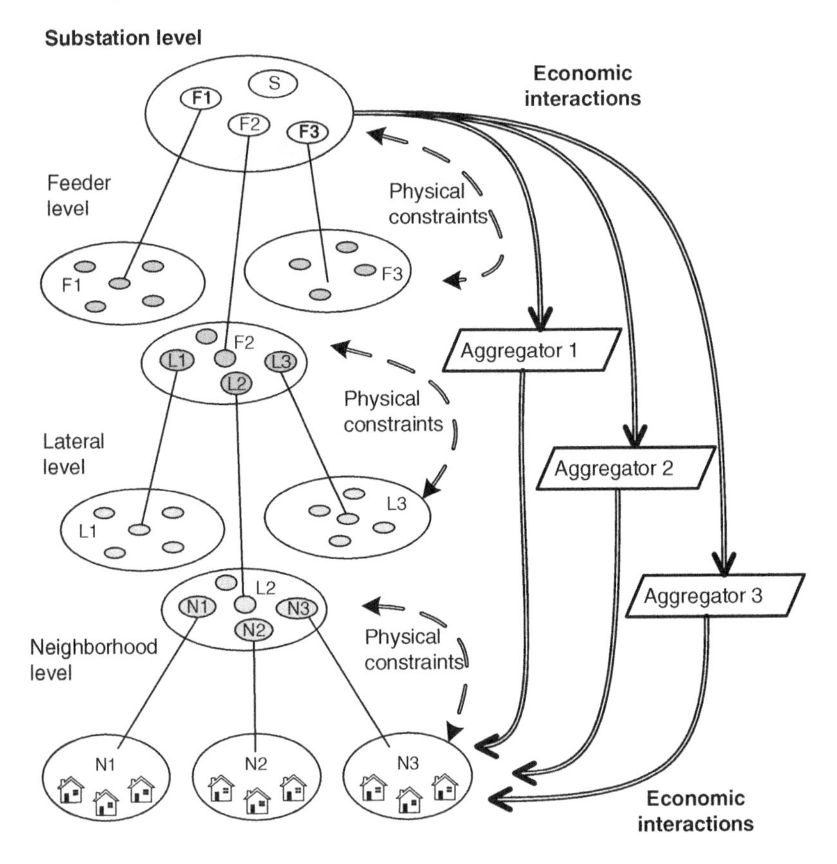

Figure 7.1 An architecture for transactive energy in distribution systems.

7.4.4 Monitoring and Control

The purpose of distribution system monitoring is very similar to SCADA in the traditional sense. Monitoring is necessary to acquire data for many of the distribution functions. Some of these functions require real-time data from the system to make control decisions. Real-time data is also useful in providing information to operators on abnormal system conditions in the form of alarms. In addition to the real-time data, system data can be gathered and archived for later use. Such data can then be used for forecasting and planning purposes. Monitoring is also useful for the management of major assets, such as substation transformers and circuit breakers, and in distribution systems as described in the following sections.

7.4.4.1 Transformer Life Extension

The substation transformers normally operate at loads lower than their capacity. However, during emergencies, such as the failure of another transformer, they can be operated at loads higher than the rated capacity. But overloading can be done only for a limited time without jeopardizing the life of the transformer. The higher the overloading, the lower is the time allowed for overloading. In a manual process, the dispatcher has to rely on trial and error to get proper level of loading. The dispatcher would close the switch to supply additional load with an expectation that the total load would be less than a certain value. But if the load after switching happens to be higher than expected, he would have to open the switch, drop a few feeders, and then close the switch again. The process would have to be repeated until the load is at a desired level. The switching on and off of load can stress the transformer significantly and thus reduce its total life. Proper monitoring allows this task to be performed with minimum switching operations and without stressing the transformer.

Automation of this function requires equipment for monitoring the transformer including oil and winding temperature. Equipment for monitoring the health of the transformer based on dissolved gas analysis are also available. Data and measurements from the feeders connected to the transformer are needed too. The oil and winding temperatures determine the level of overloading possible under the given loading conditions. Then, the feeders can be selected such that there is a balance between the desired loading and the loads of the feeders. Overloading of the transformers can be controlled precisely without too many unwanted switching operations. In this way, stress on the transformers can be avoided, and life extension of transformers can be achieved.

Such monitoring and control of distribution transformers is very challenging due to a large number of such transformers in the system. Since these transformers are inexpensive, additional expense to monitor them is not justified at present.

However, in the future with increased proliferation of electric vehicles and rooftop solar PV, monitoring of these transformers could become essential to avoid transformer failures.

7.4.4.2 Recloser/Circuit Breaker Monitoring and Control

Typically, no remote monitoring and control is available on circuit breakers and reclosers. The settings of the relay and the recloser timings can be changed only by physically going to the location of the equipment. In case of pole-mounted reclosers, it is extremely time consuming to change settings. Further, with no monitoring, the recloser and the circuit breaker contacts are refurbished at fixed intervals whether it is necessary or not. This maintenance frequency is usually based on the duty level the recloser or the circuit breaker is expected to perform. Generally, the maintenance interval is estimated conservatively (i.e. refurbishment is done, on the average, sooner than is necessary). Hence, in many cases, the contacts are serviced before it is necessary.

The advantages of automating this function are many. In an automated scenario, first, the relay settings and recloser timings can be set remotely. This will allow for better control and protection of the system whenever the system configuration changes. Moreover, the labor needed to reset the relay and recloser timings can be saved because these settings can be done remotely instead of going to the location. Further, monitoring of the energy interrupted by the recloser and the breaker provides a precise estimate of health of the contacts. Using this information, the contacts can be scheduled whenever necessary. Hence, too early or too late servicing of the reclosers and the breakers contacts can be avoided.

7.5 Cost–Benefit of Distribution Automation

Determining costs and benefits associated with different DA functions is very important for making decisions regarding implementation of these functions. Cost for automation includes cost for communication infrastructure, hardware, monitoring and control equipment, installation, software, training, and operation and maintenance. Benefits of automation are obtained in four broad categories: higher efficiency, higher reliability, higher quality, and higher resiliency. Higher efficiency includes lower system losses, lower peak power requirement, and reduced manpower needs. Higher reliability includes lesser and shorter sustained power interruptions to the customers, failure avoidance of equipment, and balanced loading of equipment. Higher power quality includes lesser momentary interruptions, lesser voltage sags and swells, and lesser incidences of voltage out of range at the customer ends. Now with increased frequency of extreme weather events, resiliency or the ability to recover from these events has become an important topic, and automation can assist in the recovery. Benefits are usually accrued annually after implementation of DA, whereas cost has two components: capital cost and annual operating and maintenance expenses. Some benefits and cost may increase at a certain rate annually.

A suitable approach for determining benefits is to map DA applications to operational benefits. Following the examples presented by Northe-Coate Green [20], the functions discussed in the previous section are shown in Table 7.1 along with the associated operational benefits. Definitions of System Average Interruption Frequency Index (SAIFI) and System Average Interruption Duration Index (SAIDI) are given in Chapter 9. The next step is to map operational benefits to monetary benefits. The outcomes of automation must be compared with the status quo or a traditional improvement scheme to obtain benefits or costs. If actual values of some of the variables to determine monetary benefits or costs are not available, the planners have to use an estimated value for these parameters. It is not necessary to go into too many details to obtain a very precise value. In case of uncertainty, different values of some variables can be tried to get a range of answers. Also, sensitivity analysis of the outcomes with respect to the input variables gives an indication of the importance of certain variables on the cost–benefit analysis. Thus, planners should focus on those variables to which the cost–benefit analysis is more sensitive.

To illustrate the computation of monetary benefits of different operation benefits, let us consider a few examples of operational benefits of outage management.

Table 7.1 Mapping the benefits of distribution automation functions.

DA functions	Higher energy sales	Lower SAIFI	Lower SAIDI	Life extension of equipment	Lower losses	Reduced peak demand	Reduced manpower	Lesser low-voltage complaints
Outage management	X	X	X	X			X	
Feeder reconfiguration			X	X	X	X		X
Voltage/var management					X	X	X	X
Monitoring and control		X	X	X			X	

7.5.1 Higher Energy Sales

With faster restoration, the customers receive service sooner than with the traditional approach. This in turn results in additional revenues for the utility for selling electricity to the customers.

$$\begin{aligned}\text{Annual Benefit} = {} & (\text{Cost of energy in \$/kWh})(\text{Average Customer Demand in kW}) \\ & (\text{\% of System Lost}/100)(\text{Number of Customers}) \\ & (\text{Line Failures per Year per Mile})(\text{Circuit Miles}) \\ & (\text{Average Decrease in Outage Hours per Event}) \end{aligned} \tag{7.1}$$

7.5.2 Reduced Labor for Fault Location

With automation, faults can be located sooner, which saves time spent on the task for the field crew as well as the operators. This in turn results in labor cost savings.

$$\begin{aligned}\text{Annual Benefit} = {} & (\text{Line Failures per Year per Mile})(\text{Circuit Miles}) \\ & \begin{bmatrix} (\text{Line Crew Hours per Fault})(\text{Line Crew \$/Hour}) \\ +(\text{Operator Hours per Fault})(\text{Operator \$/Hour}) \end{bmatrix} \end{aligned} \tag{7.2}$$

7.5.3 O&M of Switches and Controllers

Automated switches will have different failure rates and repair cost than the manual switches. Difference in these costs must be accounted. This quantity will be a benefit if it is positive and will be additional cost if it turns out to be negative.

$$\begin{aligned}& \text{Annual Benefit or Cost} \\ & = \begin{bmatrix} (\text{Manual Switches per Feeder})(\text{Failure Rate of Manual Switches}) \\ (\text{Repair Cost per Manual Switch}) - (\text{Automated Switches per Feeder}) \\ (\text{Failure Rate of Automated Switches})(\text{Repair Cost per Automated Switch}) \end{bmatrix} \\ & \qquad (\text{Number of Feeders}) \end{aligned} \tag{7.3}$$

7.5.4 Lesser Low-Voltage Complaints

Automation is expected to provide better voltage at customer ends, which will result in lower low-voltage complaints. This will result in savings because the utility has to spend time and resources to investigate the cause of such complaints.

$$\text{Annual Benefit} = (X - X')(\text{Cost per Complaint})(\text{Number of Customers})/1000 \tag{7.4}$$

where X is the complaints without automation per thousand customers per year and X′ is the complaints with automation per thousand customers per year.

Similar relationships can be developed for all the other benefits or costs. The intent here is not to go through all them, but to give an overview of the process associated with determining benefits and costs. The readers are referred to a report on this subject [21].

7.6 Cost-Benefit Case Studies

Two example systems representing an urban distribution system and a rural distribution system are considered for implementing DA. All of the functions are considered for automation in the rural system, but in the urban system all except control of voltage regulator are considered. Since voltage drops are not very large in urban systems, many utilities do not use voltage regulators. Data for these systems, taken from [22], are given in Table 7.2. The duration of the economic analysis is 15 years in both cases. A software PC-ADAM [23], developed by one of the authors and his colleague, was used for analysis. Results shown in this section are for illustration only and do not apply to all the systems. Every utility must evaluate their options based on their own data to determine the cost-benefit of selected functions.

Results of the benefit/cost analysis of the examples considered are summarized in Tables 7.3 and 7.4. The overall benefit/cost ratio is 2.93 for the urban system and 3.29 for the rural system, which implies that DA is beneficial in both cases but slightly more for the rural system. However, detailed analysis of the cost/benefit ratio for individual functions provides a clearer picture. Outage management has a benefit/cost ratio of 1.16 for the urban system and 4.11 for the rural system. The difference in this ratio is due to the fact that rural systems are spread out, and thus, it takes much longer to locate faults and to restore service. The field crew has to drive over longer distances to perform switching. Hence, most of the benefits

Table 7.2 Data for the two example systems.

Parameter	Urban system	Rural system
Number of substations	5	4
Number of transformers per substation	3	3
Number of feeders per transformer	3	1
Total length of primary feeders	288	292
Customer density (per sq. mile)	525	3.5
Power consumption (kW/customer)	6.19	3.77
Total area served (sq. mile)	90	1000
Total number of customers	47,250	3500

Table 7.3 Summary of all costs and benefits for the urban system.

No.	Description	Benefits ($)	Costs ($)	B/C ratio
	Initial planning and equipment cost	—	339,829	—
	Intangible benefits	197,208	—	—
1	Outage management	1,195,716	1,032,612	1.16
2	Option 1 + feeder reconfiguration	3,236,062	1,038,393	3.12
3	Voltage and var management			
	Transformer LTC operation	18,192	19,654	0.93
	Capacitor operation	996,435	200,299	4.97
4	Monitoring and control			
	Distribution system monitoring	1,680,210	500,512	3.36
	Transformer life extension	114,086	70,046	1.63
	Recloser/circuit breaker monitoring and control	157,199	86,709	1.81
All selected options		8,079,603	2,755,953	2.93

Table 7.4 Summary of all costs and benefits for the rural system.

No.	Description	Benefits ($)	Costs ($)	B/C ratio
	Initial planning and equipment cost	—	174914	—
	Intangible benefits	98,604	—	—
1	Outage management	1,800,333	437,790	4.11
2	Option 1 + feeder reconfiguration	1,918,271	443,571	4.32
3	Voltage and var management			
	Transformer LTC operation	16,920	16,186	1.05
	Capacitor operation	366,483	152,608	2.40
	Capacitor and regulator operation	385,967	165,326	2.33
4	Monitoring and control			
	Distribution system monitoring	821,173	181,436	4.53
	Transformer life extension	12,697	57,193	0.22
	Recloser/circuit breaker monitoring and control	54,784	33,528	1.63
All selected options		4,129,591	1,253,588	3.29

are due to savings in hours needed for this task. When feeder reconfiguration is included with the previous option, the benefit/cost ratio changes to 3.12 for the urban system and 4.32 for the rural system. This implies that addition of this option is beneficial in both cases, but it is significantly more for the urban system. The load in urban system is much larger than the rural system; therefore, even small savings per feeder due to reconfiguration result in large overall savings. In particular, load reduction at peak results in substantial savings because of capacity release.

The benefit/cost ratio for transformer TLC operation for the urban system is 0.93, and for the rural system it is 1.05. In both cases it is very close to 1, which implies that the benefits derived from automating this function are equal to the cost of additional equipment and software. The existing LTC control works very well; therefore, the benefits of further automation of this function are not many. Voltage and var management with switched capacitors have a benefit/cost ratio of 4.97 for the urban system and 2.4 for the rural. In both cases, the benefits are substantial but significantly more in the urban system. This is again mainly due to loss reduction, which results in significant capacity release under peak conditions, specifically for the urban system. The benefit/cost ratio of automation of voltage regulator operation in the rural system turned out to be 1.53. Hence, when it was combined with capacitor control, the overall benefit/cost ratio reduced to 2.33.

The benefit/cost ratio for distribution system monitoring is 3.36 for the urban system and 4.53 for the rural system. Slight extra benefit in rural systems is associated with reduction in cost for visiting substations to collect data. Transformer life extension option has a benefit/cost ratio of 1.63 for the urban system and 0.22 for the rural system. Thus, it is useful for the urban system but not for the rural system. The reason for this result is that the transformers in urban systems are larger in size and thus very expensive. On the other hand, transformers in rural systems are small in size and not very expensive. Since the monitoring equipment cost the same irrespective of the size of the transformer, it is not beneficial to automate monitoring of small transformers. The benefit/cost ratio for monitoring of reclosers and circuit breakers is 1.81 for the urban system and 1.63 for the rural system. Thus, it is more or less equally beneficial in both cases.

References

1. Clinard, K. and Redmon, J. (eds.) (1998). Distribution Management Tutorial. IEEE PES Winter Meeting, Tampa, FL (February 1998).
2. Pahwa, A. and Shultis, J.K. (1992). Assessment of the Present Status of Distribution Automation. *Report No. 238*, Engineering Experiment Station, Kansas State University, Manhattan, KS (March 1992).

3 Bassett, D., Clinard, K., Grainger, J., Purucker, S., and Ward, D. (1988) Tutorial Course: Distribution Automation. IEEE Publication 88EH0280-8-PWR.
4 Moore, T. (September 1984). Automating the distribution network. *EPRI Journal* 9: 22–28.
5 Moore, T. and Bunch, J.B. (1984). Guidelines for Evaluating Distribution Automation. *EPRI Report EL-3728* (November 1984).
6 Kendrew, T. (January/February 1990). Automated distribution. *EPRI Journal* 46–48.
7 Bunch, J.B. (1984). Guidelines for Evaluating Distribution Automation. *EPRI Report EL-3728* (November 1984).
8 Paserba, J.S., Miller, N.W., Naumann, S.T., Lauby, M.G., and Sener, F.P. (1993). Coordination of a distribution level continuously controlled compensation device with existing substation equipment for long term Var management. Paper No. 93 SM 437-4 PWRD, IEEE PES Summer Meeting in Vancouver, Canada (July 1993).
9 Malekpour, A. and Pahwa, A. (2017). A dynamic operational scheme for residential PV smart inverters. *IEEE Transactions on Smart Grid* 8 (5): 2258–2267.
10 Heydt, G.T. (1991). *Electric Power Quality*. West Lafayette, IN: Stars in a Circle Publications.
11 Douglas, J. (1994). Power quality solutions. *IEEE Power Engineering Review* 14 (3): 3–7.
12 Gnadt, P.A. and Lawer, J.S. (ed.) (1990). *Automating Electric Utility Distribution System: The Athens Automation and Control Experiment*, Prentice-Hall Advanced Reference Series. Upper Saddle River, NJ: Prentice-Hall.
13 L.L. Mankoff, S.L. Nilsson, T.J. Kendrew (1990). Proceedings: Transmission and Distribution Automation Systems. *EPRI Report EL-6762* (March 1990).
14 Undren, E.A. and Benckenstein, J.R. (1990). Protective relaying in integrated distribution substation control systems. Presentation for Panel Session on Integration of Demand-Side Management and Distribution Automation, IEEE Power Engineering Society Winter Meeting, Atlanta, Georgia (February 1990).
15 Davis, E.H., Grusky, S.T., and Sioshansi, F.P. (19 Jan 1989). Automating the distribution system: an intermediary view for electric utilities. *Public Utilities Fortnightly* 22–27.
16 Block, D. (1996). Utility automation technology. In: *Electric Power Industry Outlook and Atlas 1997 to 2001* (ed. PennWell Power Group Magazine). Tulsa, OK: PennWell Books.
17 Subedi, L., Pahwa, A., and Das, S. (2015). Trouble call analysis in radial distribution feeders using an idiotypic immune system. *Electric Power Components and Systems* 43 (17): 1990–1998.

18 Rodrigo, P.D., Pahwa, A., and Boyer, J.E. (1996). Location of outages in distribution systems based on hypotheses testing. *IEEE Transactions in Power Delivery* 11: 546–551.

19 Coughlan, B.W., Lubkeman, D.L., and Sutton, J. (1990). Improved control of capacitor bank switching to minimize distribution system losses. The Proceedings of the Twenty-Second Annual North American Power Symposium (October 1990), pp. 336–345.

20 Northcote-Green, J. and Wilson, R. (2007). *Automation and Control of Electrical Power Distribution Systems.* Boca Raton: Taylor & Francis.

21 Shultis, J.K. and Pahwa, A. (1992). Economic Models for Cost/Benefit Analysis of Eight Distribution Automation Functions. *Report No. 234.* Engineering Experiment Station, Kansas State University, Manhattan, KS (June 1992).

22 Willis, H.L. (1997). *Power Distribution Planning Reference Book.* New York, NY: Marcel Dekker, Inc.

23 Shultis, J.K. and Pahwa, A. (1992). A User's Guide for PC-ADAM. *Research Report 237*, Kansas State Eng. Expt. Stn. (June 1992).

8

Analysis of Distribution System Operation Functions

8.1 Introduction

Implementation of distribution automation functions requires thoughtful planning and analysis. Specific details, such as the data requirements and algorithms to arrive at a decision are important. In this chapter, we examine in detail some functions that were presented in the previous chapter.

8.2 Outage Management

Outage management is a key activity for a utility company since reputation of the utility depends on its performance related to outage management. This is one area that comes into focus for close scrutiny by the public, the press, and governing bodies if a utility's performance falls below a certain standard. Hence, utilities want to ensure that power is restored without undue delay after an interruption. Loss of power means loss of revenues to industry and businesses, and overall inconvenience to customers. Although loss of power to residential customers does not directly result in significant financial loss (except for those customers who use their home as the primary office). Several studies based on customer surveys also suggest a cost associated with loss of power even for residential customers [1]. It could be due to the loss of food in refrigerator and freezer, or perceived cost of inconvenience. Note that the interruptions that we are discussing in this chapter are sustained interruptions that are caused by permanent faults and could last from several minutes to hours.

Restructuring of the power industry that has taken place over the past 30 years has also contributed to the importance of outage management. Various state corporation commissions require utilities in their states to submit an annual report providing a statistical summary of outages in their distribution systems. Examples of such reports can be found on websites of corporation commissions of

Electric Power and Energy Distribution Systems: Models, Methods, and Applications, First Edition.
Subrahmanyam S. Venkata and Anil Pahwa.

Companion website: www.wiley.com/go/Pahwa/ElectricPowerDistributionSystems

California, Illinois, Michigan, and New York. In the future, it is anticipated that all the states will require utilities in their state to submit such reports. J.D. Power and Associates also conducts an annual customer satisfaction survey of large utilities in the United States. Further, several states have already implemented, and others are contemplating, performance-based rates for sale of electricity. Thus, if a utility's performance related to outages is below a specified standard, they will have to pay a penalty. On the other hand, if their performance is better than a specified value, they would be paid a premium. Moreover, better performance leads to better publicity and more customer loyalty.

Outage management system includes three specific tasks: fault location, fault isolation, and service restoration. Fault location can be based on real-time measurements from protective devices or other devices installed at a strategic location on the system. In addition to the communication link with the control center, these devices have peer-to-peer communication in some cases. Newer techniques based on voltage sag and current rise are being developed. It can also be based on information gathered from customers' premises from smart meters or customer calls. Activity related to outage location based on customer calls is also called trouble call analysis, as discussed previously.

Outage management takes on a different meaning for utilities in the event of major storms, such as tornadoes, hurricanes, and ice storms. These storms can result in outage to thousands of customers. Thus, whenever such a storm hits the service territory of a utility, normal operation of the utility ceases to exist. All the resources, personnel as well as equipment, are diverted toward repairing the system and restoring power to the customers.

Outage location, fault isolation, and restoration can be illustrated with the help of an example system shown in Figure 8.1. As described in the figure all the sectionalizers and circuit breakers have communication and automated control.

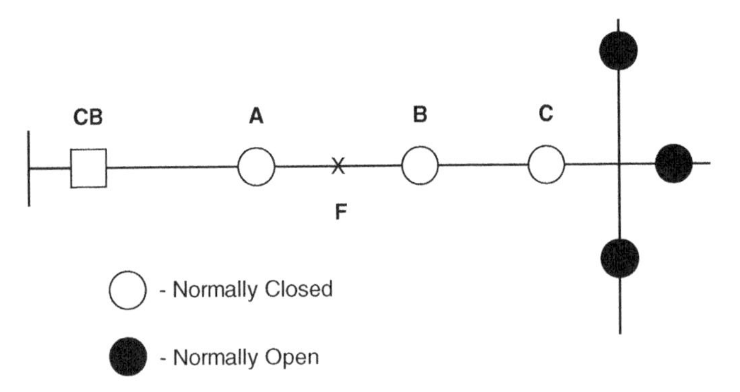

Figure 8.1 An automated distribution feeder with outage location, fault isolation, and restoration capabilities. Light circles are closed sectionalizers, and dark circles are open sectionalizers.

Let us consider that a fault takes place at F. Since the feeder is radial, the circuit breaker and sectionalizer A see this fault, but sectionalizers B and C do not see it. The circuit breaker and sectionalizer A open, and sectionalizer A sends a signal to the next downstream device, that is sectionalizer B, to open. Following this, the circuit breaker closes to restore power to the line section between it and sectionalizer A. However, line sections between sectionalizer B and the normally open sectionalizers D, E, and F are healthy but do not have power. This initiates a control action to close one of the open sectionalizers to restore power to the healthy parts of the feeder. Several rules should be followed to decide which open sectionalizer to close:

- Look for a path from the same transformer or another transformer, whichever has higher spare capacity
- Feeders in the path should not be overloaded
- Voltages should stay within the American National Standards Institute (ANSI) limits
- Minimize switching operations

Additional load transfers may be required to mitigate feeder and transformer overload. Several papers are available in the literature on determining the optimal path for restoration of service to the unfaulted parts of the feeders while considering the abovementioned rules.

While this approach works well for faults on three-phase main feeders on distribution systems, it is impractical to apply it for faults on laterals for numerous reasons. Laterals are typically single-phase feeders, which are protected by fuses. Replacing fuses by an automated remote-controlled sectionalizer for every lateral in the system is cost prohibitive. Also, laterals do not have a tie point at the end to connect to other feeders. Hence, techniques based on information from customers, such as trouble call analysis, are the best to locate faults on laterals.

8.2.1 Trouble Call Analysis

An issue with trouble call analysis that depends on customer calls is that the number of calls received from the customers depends on the time of day and weather conditions. For example, if the outage is in the night, lesser numbers of people are likely to call. According to Northcote-Green and Wilson [2], after an outage under stormy conditions, few people call initially, but if outage continues, the number of calls keep increasing gradually for about 30 minutes. If the outage continues beyond that, the calls start decreasing with almost zero calls after approximately 90 minutes.

Advances in the telecommunications technology have made it possible for utilities to dedicate several lines for receiving calls. This is a significant improvement compared to the 1980s when it was very difficult to make a call to the local utility after an outage. The authors recall making such calls to the local utility on several occasions without much success. The main problem was that the utility had only a few lines to receive such calls, and thus, most people got a busy signal whenever they made the call. After a few attempts, most people gave up. So the information received from customer calls would be very minimal. The utilities now use several telephone lines for customers to report power outages. Customers can enter their address into a database through phones or a web interface. The data gathered is processed and displayed on system maps to pinpoint possible locations of outage. In addition to receiving calls from the customers, trouble call analysis system monitors the progress of restoration and provides customers feedback on expected length of the outage.

Automated retrieval of outage information from advanced metering infrastructure (AMI) can take this a step further. Not many customers actually make attempts to call unless the interruption becomes very long. However, the AMI deployed by utilities is primarily focused on metering with no real-time feedback for operations. In a typical AMI, the meter at home communicates with a nearby data concentrator periodically through a communication medium to report accumulated energy consumption data over a period of time. The data concentrator conveys this information to a billing services company's server. The billing services company processes the data and sends them to the utility. The utility in turn makes it available to the customers through its website. In some cases, the feedback time from the meter to the customers can be much shorter, especially if the utility is expecting customers to modify their usage pattern based on real-time display of energy consumption. To use AMI for outage location, it will require a feature where the outage can be reported either directly to the utility or through the third party. In any case, reporting outages through the AMI could cause communication bottlenecks with some loss of data if large number of meters report service interruption simultaneously. We will look at some simple methods for locating outages based on outage report escalation.

8.2.1.1 Outage Location Using Escalation Methods

Outage report escalation method uses all the reported interruptions to identify a common point of electrical connectivity [3]. The analysis begins from the extreme downstream points in the system and moves upstream. Let us look at Figure 8.2, which shows a part of a typical distribution system, to illustrate this concept. The number downstream of the transformers shows the number of calls received from the customers served by those transformers. This is part of an actual system in which devices have numbers that are used by the utility. Thus, following the utility practice, we have used the same numbers. The number in parenthesis in front of the devices in this figure is the total number of downstream devices that are one

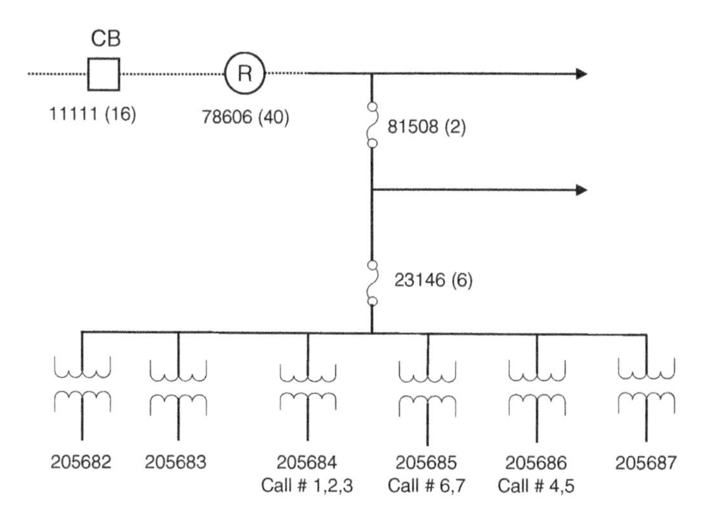

Figure 8.2 One-line diagram of part of a typical distribution system showing the primary feeders, laterals, and distribution transformers.

level below these devices. For example, device 78606, which is a recloser and a Level 1 device, has 40 Level 2 devices (lateral fuses) below it.

At first glance, all the calls associated with transformer 205684 suggest that there was a problem with this transformer. Similarly, calls associated with transformers 205685 and 205686 point to problems with those respective transformers. If these calls are considered independently, the conclusion is that all the three transformers have separate problems, which is unlikely. Therefore, the other possibility is that there is a problem in the feeder upstream, which is causing all the customers connected downstream from it to lose power even though no calls were received from customers connected to 205682, 205683, and 205687. Since ad-hoc approaches can lead to a wrong diagnosis, a systematic approach is necessary to process data for identifying the correct fault location. In the following section, a simple rule-based approach for processing the data is presented. The readers are referred to a paper inspired by human immune system for additional information [4].

8.2.1.2 Rule-Based Escalation

To implement rule-based escalation of outage calls, the protective devices within the system are designated different levels based on their location within the distribution system. All primary feeder circuit breakers and reclosers are designated as Level 1 devices. Fuses connected to the primary side of the distribution transformer are designated as Level 3 devices. Consequently, all the protective devices that are located between these two levels are designated as Level 2 devices. This generally includes the lateral fuses serving the laterals. Hence, with reference to Figure 8.2, 11111 and 78606 are Level 1 devices, 81508 and 23146 are Level 2 devices, and 205683, 205684, 205685, and 205686 are Level 3 devices.

In the next step, we define the state of the protective devices in the distribution network. Each protective device in the system can be either in a "0" state for a normal device or in a "1" state if a device is suspected to have operated. The states of all the protective devices in the distribution feeder are assumed to be "0" initially. The state of a Level 1 device, which could be a recloser or a circuit breaker, is changed to "1" if the percentage of devices connected directly beneath it that have their state changed to "1" is equal to or higher than a specified threshold (θ_1). Level 2 devices are the lateral fuses, and their state changes to "1" only if the percentage of devices connected directly beneath it that have their state changed to "1" is equal to or higher than a specified threshold (θ_2). Level 3 devices are the distribution level transformer fuses, and a single call from a customer served by it changes its state to "1."

The process of call escalation starts by changing the state of Level 3 devices that are associated with the received calls to "1." Any subsequent calls that are received from customers served by the same device will not affect its state. After processing all the registered calls, the state of the Level 2 devices is determined, which is followed by the state of Level 1 devices. Once the states of all devices are determined, calls are grouped based on the state of the devices that have their state changed to "1" and they do not have any other device upstream with a state of "1."

An important aspect of the method is selection of the right value for the threshold percentage for grouping the calls. While selecting these values it has to be realized that not all customers can be relied upon to make a call in the event of an outage. Further, the probability of two separate outages in a time frame is lower than one outage and the probability decreases as larger number of outages are considered. Hence, for Level 2 devices (lateral fuses), which typically do not have many devices downstream, using a high value for threshold could result in calls not getting grouped together and instead would be assigned to several lower level devices.

Level 1 devices include the main feeder circuit breakers and reclosers. In a radial distribution feeder, they will have larger number of immediate downstream devices. Therefore, a small value of threshold could result in calls being grouped into the same group, indicating a single outage. On the other hand, if it is high, calls would not get grouped together, and an outage on the main feeder could be considered as multiple outages on laterals. This number seems paradoxical, as it should be low as well as high. Important factors on which this threshold could depend are weather conditions. Under normal weather conditions, the chance of more than one outage occurring within the same time period is relatively low. During such conditions, if calls that are associated with multiple devices are received, it is very likely that a main feeder circuit breaker upstream of the call origin would have operated. Hence, a lower threshold value for Level 1 devices would make it possible to classify the incoming calls in a single group. On the other hand, during stormy conditions, the likelihood of multiple outages increases. Therefore, under such conditions, a relatively higher threshold value for Level 1 devices would ascertain that the calls are not classified into a single group. Thus, the threshold value of Level 1 devices has to be chosen as a trade-off between allowing multiple outages to be classified into different groups during rough weather conditions while also allowing calls to be grouped together during normal conditions. The best way to achieve this would be to set a variable threshold that is changed between certain ranges depending on the weather conditions.

Several simulations presented in Ref. [4] suggest a threshold value of 20% for Level 2 devices and 50% for Level 1 devices, which will be used for the test cases presented discussed in this chapter.

8.2.1.3 Test Cases

The distribution circuit configuration and call record data for the test cases were obtained for a feeder from a utility in Kansas. In addition, a list of operated protective devices that were discovered by the field restoration crew associated with each call sequence was obtained. All calls that were received within a period of one and a half hours after receiving the first call are considered for processing. Calls received after this period are considered to belong to separate instances of outages. The period is determined based on engineering judgment and examination of various outage records, but it can be different for different utilities.

The first case considered is for the system shown in Figure 8.2. Table 8.1 shows the log of calls, and Table 8.2 shows escalation of calls and their impacts on the status of the devices. 205683, 205684, 205685, 205686 are fuses for the distribution transformers and thus are Level 3 devices, which change status even with one call. Note that 205682 and 205687 did not have any calls and are not listed in Table 8.1. 23146 is a Level 2 device, and it will change status after more than 20% of devices

Table 8.1 Calls log of 18 June 2010.

Device ID	Call number	Date/time of the call
205684	1	18 June 2010 8:19
205684	2	18 June 2010 8:20
205684	3	18 June 2010 8:25
205686	4	18 June 2010 8:26
205686	5	18 June 2010 8:28
205685	6	18 June 2010 8:33
205685	7	18 June 2010 8:35

Table 8.2 Escalation of calls and status of device.

	Device									
	205683		205684		205685		205686		23146	
Call number	**Call count**	**Status**	**Call count**	**Status**	**Call count**	**Status**	**Call count**	**Status**	**Device count**	**Status**
1	0	0	1	1	0	0	0	0	1	0
2	0	0	2	1	0	0	0	0	1	0
3	0	0	3	1	0	0	0	0	1	0
4	0	0	3	1	1	1	0	0	2	1
5	0	0	3	1	2	1	0	0	2	1
6	0	0	3	1	2	1	1	1	3	1
7	0	0	3	1	2	1	2	1	3	1

below it change their status to 1. After the fourth call, two devices below it become 1, which exceeds the threshold of 20%, and its status changes to 1. Additional calls do not change the status of any other Level 2 or Level 1 devices. This implies that all the calls belong to one group, which are associated with device 23146. It was also reported as the operated protective device by the utility.

Figure 8.3 shows another example of trouble calls due to the operation of multiple protective devices within the same period with call log given in Table 8.3. Each feeder downstream of the fuse serves multiple transformers, but in this figure only one downstream transformer is shown, to keep the figure simple. A table

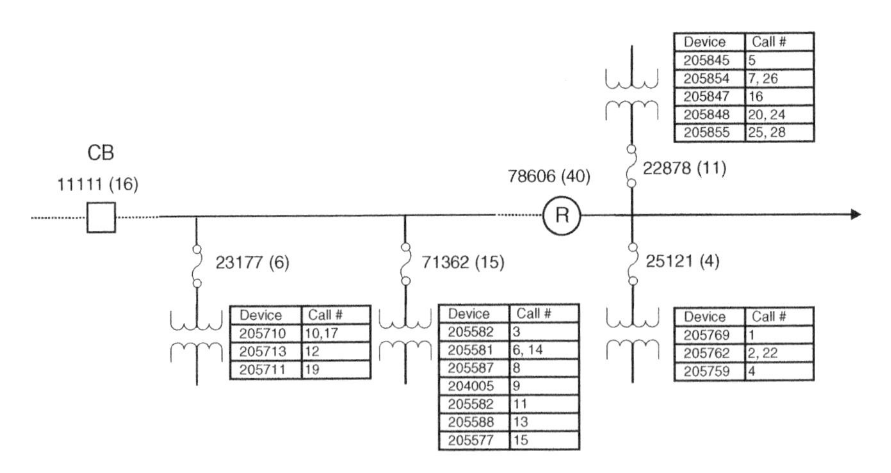

Figure 8.3 Circuit diagram and call scenario for 13 August 2010.

Table 8.3 Calls log of 13 August 2010.

Device ID	Call number	Date/time of the call
205769	1	13 August 2010 16:30
205762	2	13 August 2010 16:33
205582	3	13 August 2010 16:34
205759	4	13 August 2010 16:37
205845	5	13 August 2010 16:37
205581	6	13 August 2010 16:37
205854	7	13 August 2010 16:39
205587	8	13 August 2010 16:39
204005	9	13 August 2010 16:40
205710	10	13 August 2010 16:41
205582	11	13 August 2010 16:47
205713	12	13 August 2010 16:50
205588	13	13 August 2010 16:55
205581	14	13 August 2010 16:58
205577	15	13 August 2010 17:00
205847	16	13 August 2010 17:05
205710	17	13 August 2010 17:14
205742	18	13 August 2010 17:18
205711	19	13 August 2010 17:19
205848	20	13 August 2010 17:22
142557	21	13 August 2010 17:31
205762	22	13 August 2010 17:33
219600	23	13 August 2010 17:35
205848	24	13 August 2010 17:35
205855	25	13 August 2010 17:41
205854	26	13 August 2010 17:48
205838	27	13 August 2010 17:49
205855	28	13 August 2010 17:49

next to each transformer lists the transformers associated with the customer calls. For example, fuse 22878 serves 11 transformers below it, but only five are listed in the table because trouble calls were received only from customers served by these transformers. While applying the escalation rules, we note that customers associated with three of the six devices below 23177 are reporting an outage, and thus, the status of 23177 is changed to "1" because the calls exceed the threshold. Customers from seven of the 15 devices below 71362 are reporting an outage, and thus, the status of 71362 is changed to "1." Similarly, the remaining calls are processed, and the status of 25121 and 22878 are changed to "1." Now, we escalate these outages to the higher level to see if the status of any Level 1 device needs to be changed. Flagged devices 23177 and 71362 are two of the 16 devices below 11111, which give 12.5%. Since this is below the threshold, the status of 11111 remains "0." Similarly, flagged devices 25121 and 22878 do not exceed the threshold of 78606, and their status also remains "0." Hence, the calls are grouped into four groups with devices 22878, 23177, 71362, and 25121 as the operated protective devices. These devices are the same as those reported by the utility as the operated devices. Note that call # 18, 21, 23, and 27 are single calls from other devices, and they do not get grouped with other calls.

8.3 Voltage and var Control

Maintaining voltages within the limits and providing proper utilization voltage to all the customers under varying load conditions is an important task for the utilities. This is a difficult task because it requires consideration of customers located at the beginning of the feeder and also the customers at the end of the feeder. Since the load changes throughout the day, utilities must continuously monitor the voltage and take corrective actions as needed. Typically, this is done at three levels by utilizing the load tap changer (LTC) at the substation transformer, line regulators, and capacitors located at specific locations in the system. Both LTC and regulators, currently used widely by utilities, work well, but they also have some limitations. Firstly, they respond to local conditions and do not have coordination between them. Also, they do not have any direct coordination with capacitors.

8.3.1 Load Tap Changer

LTC is a mechanical device that is built into the transformers to change the number of windings by moving the tap up or down. The LTC moves up and down based on the load on the feeder to increase the voltage under heavy load conditions and decrease the voltage under light load conditions. These are custom designed to meet the needs of individual utilities. As an example, a transformer can have LTC with $\pm 10\%$ control from the nominal in 5/8% steps. This would give 16 steps on either side of the nominal. LTC also has an adjustable bandwidth around the nominal voltage and a time delay for initiating operation of taps. Mechanical tap changer controls had limited options, but modern digital tap changer controls provide bandwidths of 1 or 10 V in 0.1 V increments with time delay of 1–120 seconds in 1-second increments. LTC moves the taps if the voltage goes outside the set band and stays there for the specified time delay. Bandwidth and time delay are included to prevent frequent operation of taps. Both smaller bandwidth and smaller time delay would increase the frequency of operation. Many utilities specify 500,000 operations before contact replacements [4]. So, for a life of 40 years, the LTC will be able to have 34 operations per day without a need to replace contacts. However, in actual systems, much lower number of LTC operations take place.

LTC's operation can be controlled based on either the voltage measured on the feeder gateway at the substation or on an estimated voltage at a specified point on the feeder. The latter is done using a technique called line-drop compensation. It uses a model to represent the impedance of the feeder up to the point of control in conjunction with a voltage regulating relay that controls the taps.

8.3.2 Line Regulators

Regulators are autotransformers and work very similar to the LTC in terms of operation and provide voltage control in ±10% range in 32 steps of 5/8%. The difference is that regulators are physically separate from transformers. They can be either three or single phase and can be located in the substation or on the feeders. The ones that are designed for feeders are small in size and are mounted on the poles. If a system is well designed with proper selection of conductors for the feeders to match the loading, it should not need any line regulators. These are typically used in rural systems with very long feeders or in systems that have had an unexpected load growth in a specific part of the system. In the latter case, line regulators are a cost-effective way to address voltage-related issues without having to upgrade the whole feeder. Control of regulators is done based on the voltage at the regulator or at a specific point in conjunction with line-drop compensation. Since line regulators are designed for finer control of voltage on the feeders, they could see more frequent operation than the LTC. Some of the literature has reported up to 70 operations per day for regulators [5].

A major limitation both with LTC and regulators is that they only respond to voltage either next to them or at the regulated point. There is very little or no coordination between their operation and those of the capacitors. As a result, the system operates in a suboptimal state most of the times. Further, by design, they can only provide coarse control of voltage.

8.3.3 Capacitors

Capacitors are used in distribution systems to inject reactive power, which helps in voltage control, power factor correction, and loss reduction. Capacitors can be either fixed or switched. Usually, the need for capacitors increases with increase in load. Therefore, fixed capacitors are installed based on the reactive power needs under low load conditions. Additional capacitors are switched on in steps when load increases and switched off in steps when load decreases.

Voltage, reactive power, and power factor provide a direct way to control capacitors because an increase in load will create changes in these variables. In the past, voltage, reactive power, and power factor were measured locally near the capacitor, but with distribution automation it is possible to measure these quantities at different locations to provide better control of capacitors. With larger number of measurements from the system, more coordinated and precise control of capacitors can be obtained. Capacitors, however, should not be switched on very often because every switching operation generates a spike of current. Therefore, frequent operation can lead to failure of the switch controlling it or the failure of the capacitor itself. Capacitor switching has also shown to create power quality issues including harmonics. Devices such as STATCOM (static compensator) provide much better and precise control, but they are significantly more expensive. Therefore, they are used only in very special situations.

The number of capacitors as well as their sizes and locations to provide adequate voltage and var control is an important issue to consider. Obviously, not every bus in the system needs to have a capacitor. This topic is explored further in the following section.

8.3.4 Capacitor Placement

Optimal placement of capacitors in a distribution system is a combinatorial optimization problem, which falls into the class of discrete optimization problems. The amount of reactive power compensation required on a distribution system at any given time depends on the amount of load at that time. Though the instantaneous load on the system exhibits large fluctuations, load aggregated over a period of time, such as one-hour, shows smooth characteristics. Typically, capacitors are not switched to follow the rapid fluctuations in the load, but in response to hourly load fluctuations over a period of time. Since load fluctuates every hour, and with the seasons, the best solution for a given hour obviously will not be the best for another hour. So we look for a solution which is the best over a specified period of time covering different ranges of loads. This solution in turn provides the best locations for placing capacitors. A common practice is to use the load duration curve for a year. Specifically, for capacitor placement problem the yearly load duration curve is approximated as a piece-wise step function. It is also assumed that all the loads vary in a conformal way (all the loads go up and down together in the same ratio, and real and reactive power change simultaneously in the same proportion).

The objective function for capacitor placement includes losses, voltage profile improvement, power factor improvement, load balance, and any combination of these factors depending on the priorities of the utility. In the discussion that follows, the problem has been formulated for minimizing the cost of losses. The objective function is modeled to include three cost components:

(i) cost equivalent of peak power losses in the system,
(ii) energy costs for losses in the system at all load levels, and
(iii) cost for installing and maintaining the capacitors.

The total cost is the summation of the above components and is given by the following equation:

$$\text{Cost} = K_p P_0 + K_e \sum_{i=1}^{L} T_i P_i + K_c \sum_{j=1}^{n} C_j \tag{8.1}$$

where K_p, K_e, and K_c are constants, referred to as peak power, energy, and capacitor cost constants. The first term is the cost of the peak power loss. The second term is the cost of the total energy losses, which is the product of the power loss for each load level, P_i, and the duration of that load level, T_i, summed over the L load levels. Lastly, the third term is the cost of capacitor installations. This is evaluated in terms of $/kvar. In this problem formulation, only the cost of capacitor purchase and installation is considered. The recurring cost for capacitor bank switching and maintenance is ignored for the sake of simplicity.

Capacitor placement is a very widely researched subject with many papers published on the topic over the past 30 years. This optimization problem has traditionally been solved using various mathematical programming techniques including nonlinear, integer, and dynamic programming. Many authors have incorporated heuristics in analytic methods to simplify the problem solution. Intelligent techniques for capacitor placement reported in the literature include neural networks, simulated annealing, fuzzy logic, genetic algorithms, particle swarm optimization, ant colony optimization (ACO), tabu search strategy, and other evolutionary techniques [6, 7].

8.3.4.1 Illustrative Example

In this section, the results of applying a multistep ACO [8] on a 30-bus three-phase radial distribution system [9], shown in Figure 8.4, are presented. The intent of this presentation is not to discuss the details of the ACO algorithm but to highlight some of the important issues related to the capacitor placement. The cost values for different items considered for implementation are $120/kW for the peak power (K_p), $0.03/kWh for energy ($K_e$), and $5/kvar for capacitors. K_c of 0.15 is used to convert the capacitor cost to annuity. Four discrete load levels of different time durations as shown in Table 8.4 are considered. The continuous load duration curve is discretized into four load levels, and a multiplying factor of 0.787 is considered for loss calculation in each time duration such that the total area below the

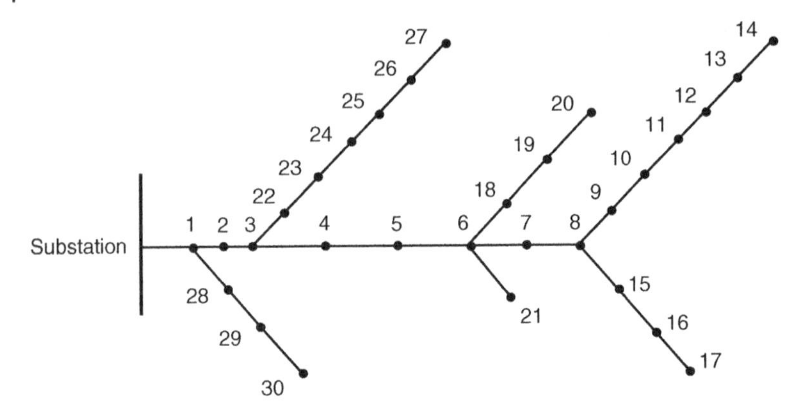

Figure 8.4 Thirty-bus three-phase radial test distribution system.

Table 8.4 Load duration data.

Load level (pu)	1.0	0.8	0.6	0.4
Time interval (h)	952	920	1332	5556

load duration curve remained the same for both the continuous and the discrete load duration curves. Data for the system is shown in Table 8.5 [9] with loads on the bus at the end of the feeder sections. The loads are three phase, and the system is considered to be balanced, and capacitors are selected in discrete steps of 300 kvar.

Variable Number of Locations In this case, the program was allowed to select the number of locations as well as the sizes of capacitors at each location. Since ACO is a stochastic optimization approach, each run gives a different result. Ten runs of the algorithm gave results with the number of locations for capacitors ranging from 11 to 16 with a total capacitance value from 6300 to 8100 kvar as shown in Table 8.6. The lowest cost generated by this method is $212,154 with 7500 kvar capacitors placed at 12 locations. This value is $40,984 better than the cost of the bare system without any compensation. The difference in cost between the best and the worst runs of the algorithm is $1314, which is quite small compared to the total cost.

Predetermined Number of Locations This approach, which gives control over the number of capacitor locations, was also tested on the 30-bus system. Ten runs of the algorithm for nine locations gave results with cost within $477 of each other and the total capacitance value ranging from 6000 to 7200 kvar. The best value obtained is $212,377 with 6900 kvar of capacitance, which is $233 higher than the previous case with 7500 kvar of capacitance and 12 locations. This new solution is

Table 8.5 System data for the test system.

Feeder section	*r* (Ω/mile)	*x* (Ω/mile)	*P* (kW)	*Q* (kvar)
0–1	0.196	0.655	0	0
1–2	0.279	0.015	522	174
2–3	0.444	0.439	0	0
3–4	0.864	0.751	636	312
4–5	0.864	0.751	0	0
5–6	1.374	0.774	0	0
6–7	1.374	0.774	0	0
7–8	1.374	0.774	0	0
8–9	1.374	0.774	189	63
9–10	1.374	0.774	0	0
10–11	1.374	0.774	336	112
11–12	1.374	0.774	657	219
12–13	1.374	0.774	783	261
13–14	1.374	0.774	729	243
8–15	0.864	0.751	477	159
15–16	1.374	0.774	549	183
16–17	1.374	0.774	477	159
6–18	0.864	0.751	432	144
18–19	0.864	0.751	672	224
19–20	1.374	0.774	495	165
6–21	0.864	0.751	207	69
3–22	0.444	0.439	522	174
22–23	0.444	0.439	1917	639
23–24	0.864	0.751	0	0
24–25	0.864	0.751	1116	372
25–26	0.864	0.751	549	183
26–27	1.374	0.774	792	264
1–28	0.279	0.015	792	294
29–30	1.374	0.774	792	294

better from a practical standpoint, as it places capacitors at three lesser locations. Reduction in the number of locations for capacitor placement translates into additional monetary benefits due to the reduction in the capacitor bank switching and maintenance costs.

Table 8.6 Results of 10 runs of the algorithm with variable number of locations for capacitors.

Run number	Number of capacitors	Total capacitance (kvar)	Cost ($/yr)
1	12	7500	212,178
2	14	7200	212,333
3	13	7500	213,496
4	13	7200	212,171
5	13	7500	212,376
6	16	8100	212,588
7	12	7500	212,154
8	11	6300	212,922
9	14	7500	212,555
10	13	8100	212,300

The approach was further tested with different values of m (7, 9, 11, 13, 15, 17, 19, 21, 22), where m is the number of locations for capacitor placement on the system. The algorithm was tested for 10 runs with each value of m. Figure 8.5 shows a plot of the best cost vs. the number of locations where capacitor banks are placed. It clearly shows that as the number of locations for capacitors increases, the value of the cost decreases and reaches a minimum value for 15 capacitor locations. Placement of capacitor beyond 15 locations increases the overall cost.

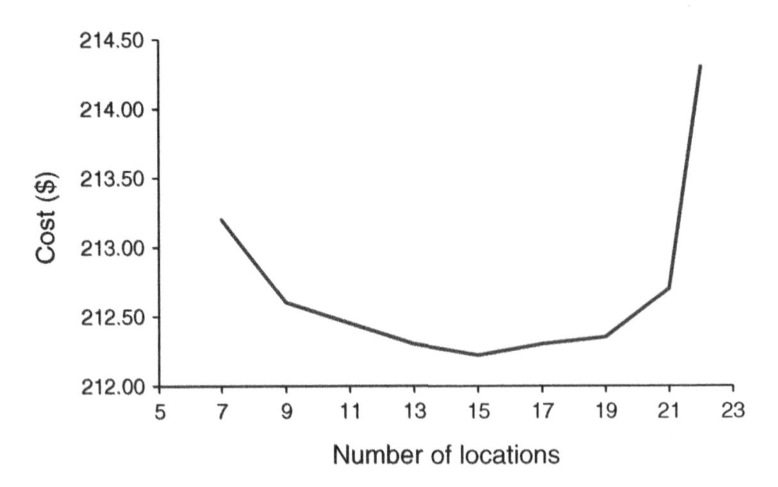

Figure 8.5 Plot showing variation in the best cost with number of locations for capacitor placements on the system.

8.3.5 Capacitor Switching and Control

In the previous section, we examined the problem of placement of capacitors at optimal locations. Since the method was based on load duration curve and conforming loads, the determined locations will not be optimal under all conditions. However, we cannot expect to change locations for different conditions. Therefore, the capacitors installed at predetermined locations must be switched on or off based on the loading conditions to optimize a specified objective, which could be loss reduction, voltage boost, or power factor improvement in the system.

8.4 Distribution System Reconfiguration

Distribution systems are structurally meshed but are operated in a radial configuration. The system has a set of switches, called sectionalizing switches that normally remain closed, and another set of switches, called tie switches that are normally open. The configuration of a distribution network can be modified by changing the status of the tie and the sectionalizing switches. The process of changing the topology of distribution systems by altering the open/closed status of these switches is called distribution system reconfiguration. Reconfiguration is done in normal operation to operate the system optimally. In emergencies, after loss of power to a certain part of the system, reconfiguration is done to restore power quickly to customers whose power supply has been interrupted. The goals are different, but the process is the same.

Mathematically, distribution system reconfiguration problem is a complex, combinatorial optimization problem involving constraints. The complexity of the problem arises from the fact that distribution network topology has to be radial, and power flow constraints are nonlinear in nature. Since a typical distribution system may have hundreds of switches, an exhaustive search of all possible configurations is not a practical solution. Therefore, many of the algorithms in the literature are based on heuristic search techniques or artificial intelligence techniques.

8.4.1 Multiobjective Reconfiguration Problem

Several different objectives can be included in multiobjective distribution system reconfiguration problem. These objectives may include loss minimization, balancing load on transformers, balancing load on feeders, and deviation of voltages from nominal. Under emergencies, loss minimization is not important, but the number of switching operations to complete restoration could be included as an objective. In the following discussion, we have considered three separate objectives, which are system loss, transformer load balance, and voltage deviation from nominal.

In multiobjective optimization, it is possible to compare two solutions using the concept of dominance. Without loss of generality, if we assume that the optimization problem involves minimization of the objective functions, then a solution $x^* \in \Omega$, where Ω is the set of all x that satisfy all constraints, is said to be Pareto Optimal if and only if there does not exist another solution $x \in \Omega$ such that $f_i(x) \leq f_i(x^*)$ for all $i = 1, \ldots, k$ and $f_i(x) < f_i(x^*)$ for at least one i. When comparing two solutions, a solution u is said to dominate over another solution v, if, and only if, u is at least as good as v for all the objectives, and if there is at least one objective where u is better than v. In a solution space, the set of all nondominated solutions is referred to as the Pareto set. The goal of the multiobjective optimization algorithm is to extract diverse samples from this set.

With the three objectives described above, the multiobjective distribution system reconfiguration problem can be defined as the minimization of the vector:

$$F(G) = [f_1(G) f_2(G) f_3(G)]^T \tag{8.2}$$

where $f_1(G), f_2(G)$, and $f_3(G)$, are described below.

8.4.1.1 Minimization of Real Loss

For a given configuration G, the total real loss is defined as

$$f_1(G) = \sum_i I_i^2 \cdot r_i \tag{8.3}$$

where, $i \in \{1, 2, \ldots, N_{cb}\}$. N_{cb} is the number of connected branches in the system.

8.4.1.2 Transformer Load Balancing

Loading on the substation transformers is balanced only when the load shared by each transformer in a distribution system is proportional to the capacity of that transformer. This loading is called ideal loading of the transformer and is calculated by multiplying the fractional capacity of the transformer with the sum of total loss and load (in MVA) on the network. Fractional capacity of a transformer is equal to the ratio between the transformer capacity and the sum of capacities of all transformers in the system. For a given configuration G of the network, unbalance in transformer loading is measured by calculating the linear sum of absolute value of per-unit deviation from the ideal loading for each transformer. Unbalance in transformer loading is defined as

$$f_2(G) = \sum_j \text{dev}_j \tag{8.4}$$

where $j \in \{1, 2, \ldots, N_T\}$; N_T is the number of substation transformers in the system. The quantity dev_j, for the jth transformer, is defined as the percentage deviation of transformer loading (LT_j) from its ideal loading, IL_j, as shown below.

$$\text{dev}_j = \frac{|LT_j - IL_j|}{IL_j} \tag{8.5}$$

where the ideal loading IL_j is defined as

$$IL_j = \frac{TC_j}{\sum_k TC_k} \cdot T_{LL} \tag{8.6}$$

where $j \in \{1, 2, \ldots, N_T\}$ and

$$T_{LL} = \sum_p \text{Load}_p + \sum_q \text{Loss}_q \tag{8.7}$$

such that $p \in \{1, 2, \ldots, N_b\}$ and $q \in \{1, 2, \ldots, N_{cb}\}$. TC_j is the capacity of the jth transformer, T_{LL} is the total load plus losses on the system, Load_p is the load on bus p, Loss_q is loss on the qth connected branch, N_b is the number of buses, and N_{cb} is the number of connected branches.

8.4.1.3 Minimization of Voltage Deviation

Voltage deviation from 1 per unit is defined as

$$f_3(G) = \max\{|1 - \min(V_i)|, |1 - \max(V_i)|\} \tag{8.8}$$

where $i \in \{1, 2, \ldots, N_b\}$, and V_i is the voltage on the ith bus.

8.4.2 Illustrative Example

In this section, the results of applying a hybrid algorithm based on artificial immune system and ACO (AIS–ACO hybrid algorithm) [10] on a sample system are presented. The focus is not to discuss the methodology but to discuss the characteristics of the distribution systems through these results. The system [11] has a total of 86 buses with three substations, eight feeders, and 96 switches. It is assumed that two transformers, each of capacity 20 MVA, are located at each of the substations. The data for peak loading condition was used for simulation.

Table 8.7 presents the numerical values of the objectives for solutions having the minimum value along each objective from the Pareto set for this problem. L is the total real loss in the system (in per unit), ΔV is the maximum deviation of the voltage magnitude from 1 per unit at the buses, ΔT_b is the sum of the per-unit deviation of loads on transformers from ideal loading, and T1, T2, and T3 are the loadings (in MVA) on the transformer at substations #1, #2, and #3, respectively.

It is clear from the results that not all the objectives can be minimized simultaneously. Therefore, the operators have a choice to pick one of the solutions based on their preference. Another interesting observation is that there are five solutions that give the same minimum voltage deviation. Among these, the one that has the lowest losses has the highest transformer loading unbalance. Further, with increase in losses, the transformer loading unbalance decreases. This example is a good illustration of the trade-off between different quantities in the system, which are important for system operation.

Table 8.7 Minimum solutions for each objective for the system.

L (pu)	ΔV (pu)	ΔT_b (pu)	$T1$ (MVA)	$T2$ (MVA)	$T3$ (MVA)
Minimum loss					
0.3918	0.0303	0.544	10.995	16.808	17.493
Minimum voltage deviation					
0.3933	0.0286	0.616	10.447	16.808	18.045
0.3937	0.0286	0.543	11.000	16.808	17.493
0.3962	0.0286	0.502	11.310	15.953	18.045
0.3966	0.0286	0.429	11.862	15.953	17.493
0.4019	0.0286	0.358	12.406	15.420	17.493
Minimum unbalance in transformer loading					
0.7966	0.1036	0.002	15.401	15.417	15.384

All the quantities that we have considered for reconfiguration are important for operation under normal conditions. During restoration after an outage, some of the quantities are removed from consideration, and some others are relaxed. Typically, loss and unbalance in transformer loading would not be a concern during restoration, and higher voltage deviation is allowed. The speed of restoration and the number of switching operations are of major concern during restoration. Availability of spare capacity from the adjoining feeders and substations is taken into consideration to obtain the best restoration solutions. There are several methods available in the literature to search for the best restoration solution.

8.5 Distribution System Restoration

In Chapter 5, we had introduced the concept of cold load pickup (CLPU) and its impacts on the operation and planning of distribution systems. In this chapter, we revisit the topic and provide more details with a few illustrative examples. Note that if the service interruption is of short duration, the enduring component of CLPU is not important during restoration. However, if the interruption is long, the enduring component of CLPU can have significant impact, particularly in systems with high penetration of thermostatically controlled devices. In such situations, step-by-step restoration, as discussed in the following section, may be required instead of restoration of all of the loads in one step.

8.5.1 Step-by-Step Restoration

A single-line diagram of a typical distribution system is shown in Figure 8.6. It is assumed that the main feeders have remotely controlled sectionalizing switches that can be opened and closed to supply power to the respective sections. We assume that the total substation transformer capacity is not sufficient to switch on all the loads simultaneously due to system undiversified load resulting from long outage duration. In general, loss of diversity depends on both the outage duration and outside temperature. For example, on a very hot summer day, even shorter outage durations may result in a complete loss of diversity because temperature inside the houses will increase faster to reach the ambient temperature, resulting in quicker diversity loss. Typically, we can consider that the diversity will be completely lost if the outage lasts for more than half an hour. Switching power to loads during CLPU is dependent on the transformer loading capacity. If the loading capacity is exceeded, sections need to be restored sequentially. Sections with priority loads such as hospital, fire station, and police station must be supplied power as soon as possible. Priority loads are included in the restoration procedure as the priority constraints. Also, if a section that needs to be supplied power is at the end of the feeder, then all the in-between sections on that feeder must be supplied power first. Thus, the selection of a section to restore power will require energizing the upstream sections on the feeder. These types of constraints are called precedence constraints, which are unavoidable when a feeder is divided into sections with sectionalizing switches. In a study [12], we had determined that diving a feeder into more than three sections did not provide sufficient benefits for restoration.

The flexibility of restoring power can be increased by installing single-phase sectionalizing switches on the laterals. In that case, loads on the laterals can

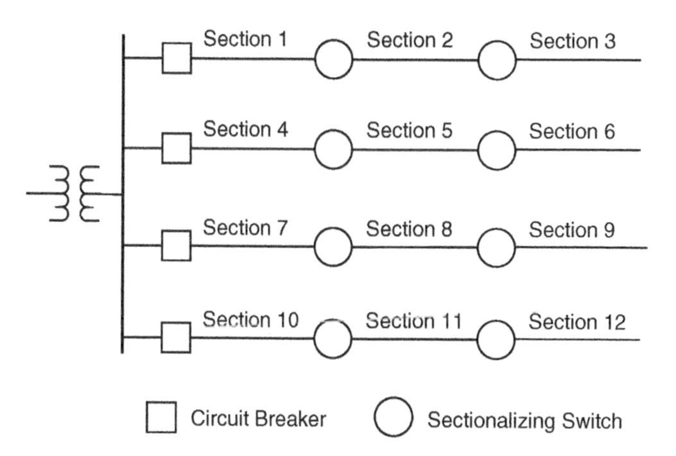

Figure 8.6 One-line diagram of a distribution system with 12 sections.

be restored without any precedence restrictions. The only restriction would be that the main feeders to which laterals are connected must have power before restoration can start. Installing single-phase sectionalizing switches increases flexibility but, on the other hand, complicates the restoration procedure, and the cost becomes prohibitive. Since each lateral is connected to one phase, several switches on the laterals would have to be operated simultaneously to restore a section of the distribution system instead of one three-phase switch on the main feeder. Such a scheme for restoration based on single-phase sections could be complicated and nonpractical. In other words, the outcome of restoration based on this approach may not improve sufficiently to justify additional expenses and increased complexity.

Aggregated load of each section follows the delayed exponential characteristics. The load for section i as a function of time is shown in Figure 8.7 and given below:

$$S_i(t) = \left[S_{D_i} + (S_{U_i} - S_{D_i})e^{-\alpha_i(t-t_i)}\right] u(t - t_i) + S_{U_i}[1 - u(t - t_i)]u(t - T_i) \tag{8.9}$$

where α_i is the rate of decay of load on the ith section, and $u(t)$ is a unit step function, which is $u(t) = 1$ for $t \leq 0$ and $u(t) = 0$ for $t < 0$. For $t < t_i$, there is no diversity among the loads because all the thermostatically controlled devices are in the ON state. The devices start entering the cyclic state after t_i, and the load decreases until full diversity is restored.

All the n sections can be restored simultaneously if the total load does exceed the transformer loading constraint S_{MT}, or

$$S(t) = \sum_{i=1}^{n} S_i(t) \leq S_{MT}\ t \geq 0 \tag{8.10}$$

If the sum of undiversified load of n sections is higher than the transformer maximum allowed loading, restoration of some of the sections will be delayed to

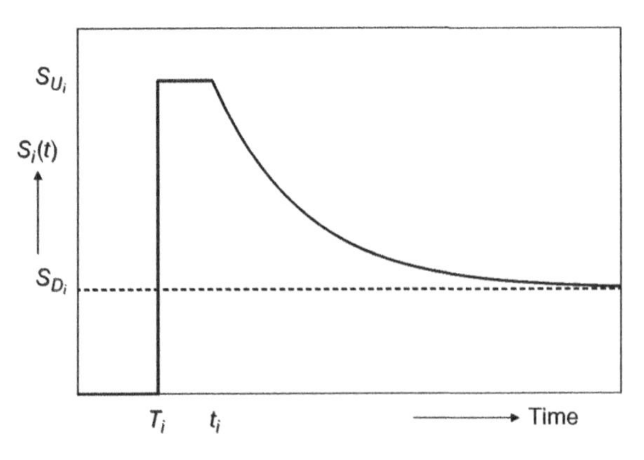

Figure 8.7 Load of section i upon restoration following an extended outage.

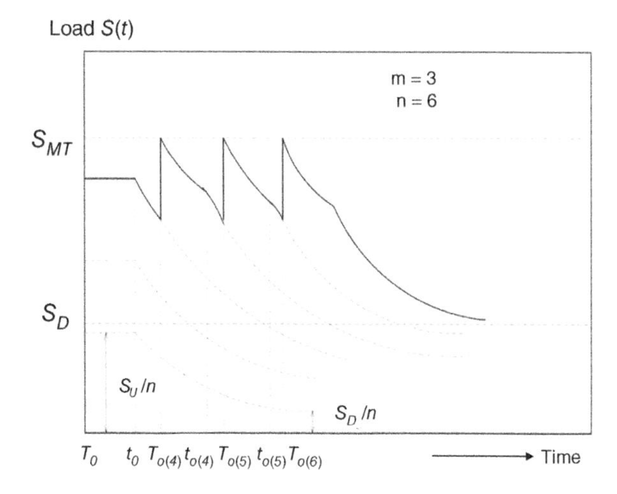

Figure 8.8 An example showing load on the substation transformer as a function of time during restoration. S_{MT} is the maximum load allowed on the transformer to prevent its overheating.

a point where the difference between S_{MT} and total load is larger than or equal to the undiversified load of the next section to be restored. Figure 8.8 shows an example of restoration of a system with six identical sections. Three sections are restored at $t = 0$, and the rest are restored step by step, while maintaining the loading constraints on the transformer.

8.5.2 Restoration Times

The time when a section is restored is called restoration time of that section, which depends on the restoration order. Therefore, to avoid confusion, it is important to distinguish between the section numbers and an index associated with each section, which represents the order of restoration. Each section is represented with a number I, and its restoration order is represented by j, giving a relationship $o(j) = i$. For example, if section number 8 is the sixth in the restoration sequence, $o(6) = 8$. There can be m sections that could be restored in the first step of restoration without violating Eq. (8.10). In that case, the restoration times of sections $o(1)$, $o(2)$, …, and $o(m)$ will be T_0, which is the time when restoration is started. In other words, the first m sections are restored simultaneously, which gives $T_{o(1)} = T_{o(2)} = \ldots = T_{o(m)} = T_0$. The parameter m will change depending on the undiversified load of the sections, maximum allowed loading S_{MT}, and the priority and precedence constraints. The remaining $(n - m)$ sections will be restored in steps at times $T_{o(m+1)}, T_{o(m+2)}, \ldots, T_{o(n)}$. The restoration times of the sections that are restored step by step will obey the inequality $T_{o(j)} < T_{o(j+1)}$ for

$j = (m + 1)$ to $(n - 1)$. In other words, two sections are not restored simultaneously when a step-by-step procedure is considered. Determination of restoration times is important because operators need to know when the sections should be restored.

8.5.3 Derivation of Restoration Times

It can be assumed that the load decay characteristics of each section will be approximately the same if the sections have similar loads and the system outage duration is long enough. For example, after a long outage, house temperatures in the system will reach the ambient temperature, and as a result, their dynamics will be similar upon restoration. Therefore, for a large number of houses connected in each section, the dynamics of aggregate loads are expected to be the same. On the other hand, such approximation cannot be done if house characteristics vary significantly from section to section. The transformer load as a function of time with respect to the restoration order can be expressed using the general representation for the cold load pickup model:

$$S(t) = \sum_{j=1}^{m} S_{o(j)}(t)$$
$$= \sum_{j=1}^{m} \left\{ \left[S_{D_{o(j)}} + \left(S_{U_{o(j)}} - S_{D_{o(j)}} \right) e^{-\alpha_{o(j)}(t - t_{o(j)})} \right] u\left(t - t_{o(j)}\right) + S_{U_{o(j)}} \left[1 - u\left(t - t_{o(j)}\right)\right] u(t - T_{o(j)}) \right\} \tag{8.11}$$

This is a general equation, which is a function of restoration times $T_{o(j)}$. An analytical expression for $T_{o(j)}$ can be derived from the above equation if $\alpha_{o(j)}$ are the same for all sections. With this assumption, the restoration time of each section can be derived while maintaining the maximum loading capacity constraint. It is considered that $(k - 1)$ sections are restored prior to section $o(k)$. According to the restoration procedure, section $o(k)$ should be restored as soon as possible, that is when $S_{MT} - S(T_{o(k)}) = S_{U_{o(k)}}$ is satisfied. Therefore, the restoration time of section $o(k)$ can be found, such that

$$S_{U_{o(k)}} = S_{MT} - \sum_{j=1}^{k-1} \left\{ \left[S_{D_{o(j)}} + \left(S_{U_{o(j)}} - S_{D_{o(j)}} \right) e^{-\alpha(T_{o(k)} - t_{o(j)})} \right] u\left(T_{o(k)} - t_{o(j)}\right) + S_{U_{o(j)}} \left[1 - u\left(T_{o(k)} - t_{o(j)}\right)\right] u\left(T_{o(k)} - T_{o(j)}\right) \right\} \tag{8.12}$$

Since $T_{o(k)} > T_{o(j)}$ for $j = 1, 2, 3, \ldots, k - 1$, the corresponding unit step functions $u(T_{o(k)} - T_{o(j)})$ are equal to 1 for all j. This is true because restoration is done step by step, and simultaneous restoration of two sections after the simultaneous restoration of first m sections is not allowed. The other unit step function $u(T_{o(k)} - t_{o(j)})$ in this equation can be unity or 0 depending on the undiversified load duration of the restored sections. If section $o(k)$ is restored after diversity starts in the previously restored sections, $T_{o(k)} > t_{o(j)}$, and this step function is unity for all j; otherwise, this is not true. However, except for cases where some sections have extremely long undiversified load duration, this condition will be true. This makes the second term under the summation in Eq. (8.12) to go to 0 because $(T_{o(k)} - t_{o(j)}) \geq 0$. Therefore, the equation is reduced to

$$S_{U_{o(k)}} = S_{MT} - \sum_{j=1}^{k-1} \left\{ \left[S_{D_{o(j)}} + \left(S_{U_{o(j)}} - S_{D_{o(j)}} \right) e^{-\alpha(T_{o(k)} - t_{o(j)})} \right] u\left(T_{o(k)} - t_{o(j)}\right) \right\} \tag{8.13}$$

This equation can be solved to find the restoration time of section $o(k)$

$$T_{o(k)} = -\frac{1}{\alpha} ln \left(\frac{S_{MT} - S_{U_{o(k)}} - \sum_{j=1}^{k-1} S_{D_{o(j)}}}{\sum_{j=1}^{k-1} \left(S_{U_{o(j)}} - S_{D_{o(j)}} \right) e^{\alpha(T_{o(j)} + \Delta t_{o(j)})}} \right) \tag{8.14}$$

where $\Delta t_{o(j)} = t_{o(j)} - T_{o(j)}$, which is the delay part or undiversified load duration of section $o(j)$. This equation is valid for $m < k \leq n$. For $k \leq m$, the restoration time of sections $o(1)$, $o(2)$, … , $o(m)$ is T_0, which is the beginning of the restoration and can be in general considered equal to 0.

For cases where the sections have different α, the solution becomes very complex. Interested readers are referred to [13] for additional information.

8.5.4 Optimal Operation and Design for Restoration During CLPU

Both the operation and the design of distribution systems are important to effectively address CLPU. In a system with feeders of substantially different loads, proper sequence of sections to restore becomes important to minimize the restoration time. Thus, an optimal sequence can be determined that requires the least time to restore all of them while meeting all the system constraints. Approaches for finding the optimal sequence are discussed in [14, 15]. In addition, distribution systems can be designed to handle CLPU conditions while considering transformer loading and voltage drop on feeders. In this section, we give two examples to further explain the concepts.

8.5.4.1 Thermally Limited System

The first example deals with a thermally limited distribution system shown in Figure 8.6. It is assumed that feeders are not very long, which would be true for an urban system, and thus voltage drop is not a problem. Also, the thermal limits of the feeders are not considered. The diversified and undiversified loads of the feeder sections are given in Table 8.8. The transformer is rated 30 MVA under forced oil circulation (FOA) conditions, and the maximum loading allowed during restoration is 45 MVA. The delay part in the cold load pickup model for each section is considered to be 20 minutes. The results obtained from a search showed that the minimum time needed to complete restoration without violating the transformer loading limit is 83 minutes, and the maximum time is 93 minutes. The order of restoration of sections for minimum time is 10, 11, 12, 7, 8, 9, 1, 4, 5, 6, 2, and 3.

Table 8.8 Diversified and undiversified loads of sections of the example system.

Section no.	Diversified load (MVA)	Undiversified load (MVA)
1	2.5	7.5
2	2.5	4.5
3	2.0	5.0
4	1.5	3.0
5	2.5	6.0
6	2.0	5.6
7	2.5	6.5
8	2.0	4.0
9	3.0	9.0
10	3.0	6.0
11	1.5	4.8
12	2.0	6.0
Total	27.0	67.9

In another case, the maximum loading on the transformer was reduced to 36 MVA, which changed the maximum and the minimum times to 195 and 168 minutes, respectively. The order of restoration of sections for minimum time in this case is 10, 7, 8, 9, 4, 11, 1, 5, 12, 6, 2, and 3.

8.5.4.2 Voltage Drop Limited System

The second example is based on a semiurban system that has long lines and thus experiences significant voltage drops during restoration due to cold load pickup. Transformer loading is not a concern for such systems because limitations imposed by voltage drop restrict the load such that it does not exceed the transformer's loading limits. Thus, instead of searching for a sequence of sections for switching to minimize the restoration time as in the previous case, the problem is to find the best places to locate the switches based on voltage drop. The data of the system under consideration is given in Table 8.9. The service area was considered to be of triangular shape with uniform load density. The main feeder was considered to have three different conductor sizes, starting with the largest starting at the substation and tapering off away from the substation along with reduction in load. Since such restoration are considered emergency operation, ANSI B voltage ratings were considered for voltage drop calculations. All the feeders are considered to be identical with undiversified load equal to four times the diversified load. Since this is a planning problem, a study duration of

Table 8.9 Input data for the voltage drop limited distribution system.

Voltage level	12.47 kV
Service area	2 sq-miles
Load density	3.61 MVA/sq-miles
Substation spacing	4.56 miles
Main feeder length	3.4 miles
Feeders per transformer	3
Conductors used for feeders	636 kcmil 4/0, and #2
Frequency of CLPU events	0.2/yr

Table 8.10 Results for the voltage drop limited distribution system.

Transformer capacity	30 MVA
Distance of the first switch from the substation	1.86 miles
Distance of the second switch from the substation	2.12 miles
Restoration time of the first switch	30 min
Restoration time of the second switch	59.29 min

30 years was considered for cost calculations. Details of analysis are available in Refs. [16, 17]. Here, we give a summary of the results in Table 8.10.

The examples presented in this section are for illustration and are specific to the systems considered. Results for other systems will be different, but they will follow similar trends. The models of cold load pickup used in the analysis have been postulated based on some field observations and simulations. Obtaining additional real data from the field during restoration following long interruptions is an important issue for further advancing the subject of cold load pickup.

References

1 LaCommare, K.H., Eto, J.H., Dunn, L.N., and Sohn, M.D. (2018). Improving the estimated cost of sustained power interruptions to electricity customers. *Energy* 153: 1038–1047.

2 Northcote-Green, J. and Wilson, R. (2007). *Automation and Control of Electrical Power Distribution Systems*. Boca Raton, FL: CRC Taylor & Francis.

3 Liu, Y. and Schulz, N.N. (2002). Knowledge-based system for distribution system outage locating using comprehensive information. *IEEE Transactions on Power Systems* 17 (2): 451–456.

4 Subedi, L., Pahwa, A., and Das, S. (2015). Trouble call analysis in radial distribution feeders using an idiotypic immune system. *Electric Power Components and Systems* 43 (17): 1990–1998.
5 Short, T.A. (2004). *Electric Power Distribution Handbook*. Boca Raton, FL: CRC Taylor & Francis.
6 Sundhararajan, S. and Pahwa, A. (1994). Optimal selection of capacitors for radial distribution systems using genetic algorithm. *IEEE Transactions on Power Systems* 9: 1499–1507.
7 Annaluru, R., Das, S., and Pahwa, A. (2004). A multilevel ant colony algorithm for optimal placement of capacitors in distribution systems. In IEEE International Congress on Evolutionary Computation, Portland, OR (June 2004), 1932–1937.
8 Annaluru, R. (2004). Multi-level ant colony algorithm for optimal placement of capacitor banks on radial distribution systems. MS thesis. Kansas State University.
9 Civanlar, S. and Grainger, J.J. (1985). Volt/var control on distribution systems with lateral branches using shunt capacitors and voltage regulators part III: the numerical results. *IEEE Transactions on Power Apparatus and Systems* PAS-104 (11): 3291–3297.
10 Ahuja, A., Das, S., and Pahwa, A. (2007). An AIS-ACO hybrid approach for multi-objective distribution system reconfiguration. *IEEE Transactions on Power Systems* 22 (3): 1101–1111.
11 Schmidt, H.P. and Kagan, N. (2005). Fast reconfiguration of distribution systems considering loss minimization. *IEEE Transactions on Power Systems* 20 (3): 1311–1319.
12 Ucak, C. and Pahwa, A. (1992). The effects of number of sectionalizing switches on the total restoration time during cold load pick up. In Proceedings of the 24th North American Power Symposium, Reno, Nevada (October 1992), 42–49.
13 Ucak, C. (1994). Restoration of distribution systems following extended outages. PhD dissertation. Kansas State University.
14 Ucak, C. and Pahwa, A. (1994). An analytical approach for step-by-step restoration of distribution systems following extended outages. *IEEE Transactions on Power Delivery* 9: 1717–1723.
15 Chavali, S., Pahwa, A., and Das, S. (2004). Evolutionary approaches for the optimal restoration of sections in distribution systems. *Electric Power Components and Systems* 32: 869–881.
16 Gupta, V. (2002).Design of power distribution systems for cold load pickup including voltage drop. MS thesis. Kansas State University.
17 Gupta, V. and Pahwa, A. (2004). A voltage drop based approach to include cold load pickup in optimal design of distribution systems. *IEEE Transactions on Power Delivery* 19: 957–963.

9

Distribution System Reliability

9.1 Motivation

The main purpose of distribution systems is to provide electricity to customers to allow them to conduct the activities that require electricity. Assessment of the distribution system to determine the extent to which electricity is made available to the customers without interruptions provides a measure of system reliability. Reliability is defined as the ability of the system to provide electricity without interruptions, and resiliency is defined as the ability of the system to recover from extreme or unplanned events. The subject of system resiliency as well as the various measures associated with it are still evolving. Although resiliency has an impact on reliability, the main focus of this chapter is on reliability. Various relevant definitions and approaches to assess system reliability are presented in this chapter.

The main motivation for reliability assessment is to analyze and improve system performance. Improved performance enhances customer satisfaction and satisfies regulatory requirements. In addition, it provides information for maintenance scheduling, the basis for new or expanded system planning, and it determines performance-based rate making. Various operation, maintenance, and design strategies can be used to enhance reliability. The reliability assessment therefore begins at the design and planning stages to build a new substation, to upgrade existing facilities, to add new feeders, and to identify poorer performing (weak) sections of the system. Operationally, the assessment will be needed to reconfigure the system for reduction in the customers affected, to add tie points to other feeders, or to develop postfault switching plan. To preserve the highest level of reliability, preventive maintenance is essential. These could include trimming trees, inspecting or repairing components, and meeting the codes. In the de- or reregulated utility environment, performance-based rate making aids in forecasting the utility's revenue. These studies also assist in quantifying the "quality" of power delivered to customers, in evaluating the regulatory

Electric Power and Energy Distribution Systems: Models, Methods, and Applications, First Edition.
Subrahmanyam S. Venkata and Anil Pahwa.

Companion website: www.wiley.com/go/Pahwa/ElectricPowerDistributionSystems

requirements specified by Public Utility Commissions (PUCs). It is also useful in benchmarking the system performance in comparison to others.

While the overall performance of a distribution system is measured in terms of system losses and voltage profile, its reliability is measured with respect to probability of experiencing outages and to customer satisfaction. Outages are caused by failure of equipment due to bad quality, aging, human error, or extreme weather events. As all these causes have uncertainties associated with them, we rely extensively on probabilistic analysis to quantify system reliability. Customer satisfaction is measured in terms of the number of momentary and sustained interruptions, the duration of outages, the number of customers affected, and the number of customer complaints. In general, reliability can be improved by hardening or upgrading the entire system. However, such an approach is not cost effective. A targeted approach to selectively harden the system will result in optimal results. Similarly, maintenance techniques can be enhanced to obtain optimal results.

9.2 Basic Definitions

Various standardized indices are used for measuring reliability and associated computations. Definitions are given in this section to aid the readers in understanding the factors that affect the calculation of indices. Many of these definitions were taken directly from *The Authoritative Dictionary of IEEE Standards Terms*, 7th Edition [1] and/or IEEE Standard 1366-2012 [2]. The readers are encouraged to review them very carefully.

- **Connected Load:** Connected transformer kVA, peak load, or metered demand on the circuit or portion of circuit that is interrupted. When reporting, the report should state whether it is based on an annual peak or on a reporting period peak.
- **Customer:** A metered electrical service point for which an active bill account is established at a specific location (e.g. premises).
- **Customer Count:** The number of customers either served or interrupted depending on the usage.
- **Forced Outage:** The state of a component when it is not available to perform its intended function due to an unplanned event directly associated with that component.
- **Interrupting Device:** A device whose purpose is to interrupt the flow of power, usually in response to a fault. Restoration of service or disconnection of loads can be accomplished by manual, automatic, or motor-operated methods. Examples

include transmission circuit breakers, feeder circuit breakers, line reclosers, line fuses, sectionalizers, and motor-operated switches.

- **Interruption:** The loss of service to one or more customers connected to the distribution portion of the system. It is the result of one or more component outages, depending on the system configuration. Note that the outage of a component does not necessarily result in interruption.
- **Interruption Duration:** The time from the initiation of an interruption to a customer until service has been restored to that customer. The process of restoration may require restoring service to small sections of the system until service has been restored to all customers. Each of these individual steps should be tracked, collecting the start time, end time, and the number of customers interrupted for each step.
- **Interruptions Caused by Events Outside of the Distribution System:** Outages that occur on generation, transmission, substations, or customer facilities that result in the interruption of service to one or more customers. While this is generally a small portion of the number of interruption events, these interruptions can affect many customers and may last for an exceedingly long duration.
- **Lockout:** The final operation of a recloser or circuit breaker to isolate a persistent fault or the state where all automatic reclosing has stopped. The current-carrying contacts of the overcurrent protecting device are locked open under these conditions.
- **Loss of Service:** A complete loss of voltage on at least one normally energized conductor to one or more customers. This does not include any of the power quality issues such as sags, swells, impulses, and harmonics.
- **Major Event:** An event that exceeds reasonable design and or operational limits of the electric power system. It includes at least one major event day (MED).
- **Major Event Day:** A day in which the daily system average interruption duration index (SAIDI) exceeds a threshold value, T_{MED}. For the purposes of calculating the daily system SAIDI, any interruption that spans multiple calendar days is accrued to the day on which the interruption began. Statistically, days having a daily system SAIDI greater than T_{MED} are days on which the energy delivery system experienced stresses beyond that normally expected (such as severe weather). Activities that occur on MEDs should be separately analyzed and reported.
- **Momentary Interruption:** A single operation of an interrupting device that results in a voltage zero. For example, two circuit breaker or recloser operations (each operation being an open followed by a close) that momentarily interrupt service to one or more customers is defined as two momentary interruptions.

- **Momentary Interruption Event:** An interruption of duration limited to the period required to restore service by an interrupting device. Such switching operations must be completed within a specified time of five minutes or less. This definition includes all reclosing operations that occur within five minutes of the first interruption. For example, if a recloser or circuit breaker operates two, three, or four times and then holds (within five minutes of the first operation), those momentary interruptions shall be considered one momentary interruption event.
- **Outage (Electric Power Systems):** The state of a component when it is not available to perform its intended function due to some event directly associated with that component. (Note: An outage may or may not cause an interruption of service to customers, depending on the system configuration. This definition derives from transmission and distribution applications and does not apply to generation outages.)
- **Planned Interruption:** A loss of electric power that results when a component is deliberately taken out of service at a selected time, usually for the purposes of construction, preventative maintenance, or repair. (Note: This derives from transmission and distribution applications and does not apply to generation interruptions. The key test to determine if an interruption should be classified as a planned or unplanned interruption is as follows: if it is possible to defer the interruption, the interruption is a planned interruption; otherwise, the interruption is an unplanned interruption.)
- **Planned Outage:** The state of a component when it is not available to perform its intended function due to a planned event directly associated with that component.
- **Reporting Period:** The time period from which interruption data is to be included in reliability index calculations. The beginning and end dates and times should be clearly indicated. All events that begin within the indicated time period should be included. A consistent reporting period should be used when comparing the performance of different distribution systems (typically one calendar year) or when comparing the performance of a single distribution system over an extended period of time. The reporting period is assumed to be one year unless otherwise stated.
- **Step Restoration:** A process of restoring interrupted customers downstream from the interrupting device/component in stages over time.
- **Sustained Interruption:** Any interruption not classified as a part of a momentary event. That is, any interruption that lasts more than five minutes.
- **Total Number of Customers Served:** The average number of customers served during the reporting period. If a different customer total is used, it must be clearly defined within the report.
- **Unplanned Interruption:** An interruption caused by an unplanned outage.

9.3 Reliability Indices

While various indices to define reliability have been used by utilities over the years, IEEE published IEEE Standard 1366 in 1998, with 12 most significant indices, as a guide for utilities to assess the reliability of distribution systems. The IEEE Std 1366 has been revised over the years, and the latest version was published in 2012 [2]. The indices are categorized into system- and customer-level indices as described below.

(a) System-level indices
- Frequency of outages (SAIFI, CAIFI, ASIFI)
- Duration of outages (SAIDI, CAIDI, ASIDI)
- Momentary outages (MAIFI, MAIFI_E)

(b) Customer-level indices
- Frequency (CAIFI, CEMI_n, CEMSMI_n)
- Duration (CTAIDI)

The indices can be classified based on sustained or momentary outages. They can also be classified based on duration and frequency indices. Ultimately, most of these are measures of availability of the system under study or investigation. Unfortunately, there is no one measure that can describe the reliability of the distribution system completely because of its complex nature.

9.3.1 Basic Parameters

The following parameters specify the data needed to calculate the indices for the reporting period:

i An interruption event
r_i Restoration time for each interruption event
I_S Total number of sustained interruption events
K Number of interruptions experienced by an individual customer
CI Customers interrupted
CMI Customer minutes interrupted
CN Total number of distinct customers who have experienced a sustained interruption
$\text{CN}_{(k \geq n)}$ Total number of customers who experienced n or more sustained Interruptions
$\text{CN}_{(t \geq S)}$ Total number of customers who have experienced a sustained interruption of more than S hours
$\text{CN}_{(t \geq T)}$ Total number of customers who have experienced more than T hours of sustained interruptions

$\mathrm{CNT}_{(k \geq n)}$	Total number of customers who have experienced n or more combined sustained and momentary interruption events
IM_i	Number of momentary interruptions
IM_E	Number of momentary interruption events
N_i	Number of interrupted customers for each sustained interruption event
N_{mi}	Number of interrupted customers for each momentary interruption event
N_T	Total number of customers served for the area
L_i	Connected kVA load interrupted for each interruption event
L_T	Total connected kVA load served
T_{MED}	Threshold value for the MED identification

9.3.2 Sustained Interruption Indices

Service interruptions that last more than five minutes are classified as sustained interruptions. Those that are less than five minutes are part of a momentary event. The demarcation of five minutes is due to the fact that many faults that are temporary can be cleared with reclosing operations within one minute. A permanent fault, which causes sustained interruption, needs physical inspections and rarely can be cleared in less than five minutes. Hence, the industry has settled on this demarcation. Indices related to sustained interruptions are defined next.

9.3.2.1 System Average Interruption Frequency Index (SAIFI)

This index gives the average number of times a customer experienced a sustained interruption over a predefined period.

$$\mathrm{SAIFI} = \frac{\text{Total Number of Customer Interruptions}}{\text{Total Number of Customers Served}} = \frac{\Sigma_i N_i}{N_T} = \frac{\mathrm{CI}}{N_T} \tag{9.1}$$

9.3.2.2 System Average Interruption Duration Index (SAIDI)

This index gives the total average duration of interruption experienced by a customer during a predefined period. It is typically measured in customer minutes of interruption.

$$\mathrm{SAIDI} = \frac{\text{Total Customer Minutes of Interruption}}{\text{Total Number of Customers Served}} = \frac{\Sigma_i r_i N_i}{N_T} = \frac{\mathrm{CMI}}{N_T} \tag{9.2}$$

9.3.2.3 Customer Average Interruption Duration Index (CAIDI)

This index gives the average time needed to restore service.

$$\text{CAIDI} = \frac{\text{Total Customer Minutes of Interruption}}{\text{Total Number of Customers Interrupted}}$$

$$= \frac{\Sigma_i r_i N_i}{\text{CI}} = \frac{\text{CMI}}{\text{CI}} \tag{9.3}$$

Also,

$$\text{CAIDI} = \frac{\text{SAIDI}}{\text{SAIFI}} \tag{9.4}$$

9.3.2.4 Customer Total Average Interruption Duration Index (CTAIDI)

This index gives the total average time that customers who experienced at least an interruption were without power. This index is similar to CAIDI except that customers with multiple interruptions are counted only once.

$$\text{CTAIDI} = \frac{\text{Total Customer Minutes of Interruption}}{\text{Total Number of Distinct Customers Interrupted}}$$

$$= \frac{\Sigma_i r_i N_i}{\text{CN}} = \frac{\text{CMI}}{\text{CN}} \tag{9.5}$$

9.3.2.5 Customer Average Interruption Frequency Index (CAIFI)

This index gives the average frequency of sustained interruptions for those customers experiencing sustained interruptions. The customer is counted once regardless of the number of times interrupted for this calculation.

$$\text{CAIFI} = \frac{\text{Total Number of Customer Interruptions}}{\text{Total Number of Distinct Customers Interrupted}}$$

$$= \frac{\Sigma_i N_i}{\text{CN}} = \frac{CI}{\text{CN}} \tag{9.6}$$

9.3.2.6 Average Service Availability Index (ASAI)

This index gives the percentage of time that a customer has received power during the defined reporting period.

$$\text{ASAI} = \frac{\text{Customer Hours of Service Availability}}{\text{Customer Hours of Service Demand}}$$

$$= \frac{N_T \times \text{Number of Hours} - \sum_i \frac{r_i N_i}{60}}{N_T \times \text{Number of Hours}} \tag{9.7}$$

If the reporting period is one year, the number of hours are 8760 for normal years and 8784 for a leap year.

9.3.2.7 Customers Experiencing Multiple Interruptions (CEMI$_n$)

This index gives the fraction of individual customers experiencing more than n sustained interruptions.

$$\begin{aligned}\text{CEMI}_n &= \frac{\text{Total Number of Customers that Experienced } n \text{ or More Sustained Interruptions}}{\text{Total Number of Customers Served}} \\ &= \frac{\text{CN}_{(k \geq n)}}{N_T}\end{aligned} \tag{9.8}$$

9.3.2.8 Customers Experiencing Long Interruption Durations (CELID)

This index gives the fraction of individual customers who experience interruptions with durations longer than or equal to a given time. The time is either the duration of a single interruption (S) or the total amount of time (T) that a customer has been interrupted during the reporting period. For the single interruption duration, we get

$$\begin{aligned}\text{CELCELID}_S &= \frac{\text{Total Number of Customers that Experienced an Interruption of } S \text{ or More Hours}}{\text{Total Number of Customers Served}} \\ &= \frac{\text{CN}_{(t \geq S)}}{N_T}\end{aligned} \tag{9.9}$$

And for the total interruption duration, we get

$$\begin{aligned}\text{CELID}_T &= \frac{\text{Total Number of Customers that Experienced Total Interruption Duration of } T \text{ or More Hours}}{\text{Total Number of Customers Served}} \\ &= \frac{\text{CN}_{(t \geq T)}}{N_T}\end{aligned} \tag{9.10}$$

9.3.3 Load-based Indices

These indices use load interrupted instead of customers affected. They are useful for measuring system performance in areas that serve relatively few customers and have large concentrations of load, such as industrial and commercial customers.

9.3.3.1 Average System Interruption Frequency Index (ASIFI)

This index gives the average number of times the system experienced sustained interruptions over a predefined period.

$$\begin{aligned}\text{ASIFI} &= \frac{\text{Total Connected kVA of Load Interrupted}}{\text{Total Connected kVA Served}} \\ &= \frac{\sum_i L_i}{L_T}\end{aligned} \tag{9.11}$$

9.3.3.2 Average System Interruption Duration Index (ASIDI)

The index gives the average duration of system load interruption based on connected load.

$$\text{ASIDI} = \frac{\text{Total kVA duration of Load Interrupted}}{\text{Total Connected kVA Served}}$$

$$= \frac{\sum_i r_i L_i}{L_T} \tag{9.12}$$

9.3.4 Momentary Interruption Indices

Momentary interruption indices are based on momentary interruptions experienced by customers.

9.3.4.1 Momentary Average Interruption Frequency Index (MAIFI)

This index gives the average frequency of momentary interruptions experienced by customers over a duration.

$$\text{MAIFI} = \frac{\text{Total Number of Customer Momentary Interuptions}}{\text{Total Number of Customers Served}}$$

$$= \frac{\sum_i \text{IM}_i N_{mi}}{N_T} \tag{9.13}$$

9.3.4.2 The Momentary Average Interruption Event Frequency Index (MAIFI_E)

This index gives the average frequency of momentary interruption events. It does not include the events immediately preceding a sustained interruption.

$$\text{MAIFI}_E = \frac{\text{Total Number of Customer Momentary Interuptions Events}}{\text{Total Number of Customers Served}}$$

$$= \frac{\sum_i \text{IM}_E N_{mi}}{N_T} \tag{9.14}$$

9.3.4.3 Customers Experiencing Multiple Sustained Interruption and Momentary Interruption Events Index (CEMSMI_n)

This index is the ratio of individual customers experiencing n or more of both sustained interruptions and momentary interruption events to the total customers served. It is useful in identifying customer issues that are hidden in averages.

$$\text{CEMSMI}_n = \frac{\begin{array}{c}\text{Total Number of Customers Experiening } n\\ \text{or more Interruptions}\end{array}}{\text{Total Number of Customers Served}}$$

$$= \frac{\text{CNT}_{(k \geq n)}}{N_T} \tag{9.15}$$

9.3.5 Sustained Interruption Example

Consider a feeder serving 1000 customers as shown in Figure 9.1. The feeder also has several single-phase laterals connected to it through fuses. We are considering the fuse protecting scheme for the recloser and breaker. In this scheme, the recloser opens whenever there is a fault downstream. The recloser closes after a short delay and opens again if the fault is not cleared. It will reclose multiple times based on the selected settings for number of reclosing operations and will lock out after that if the fault is still there. Following that, the fuse opens if the fault is downstream of the fuse, or the recloser opens if the fault is on the main feeder. Similarly, the breaker opens if there is a fault anywhere downstream or it but upstream of the recloser. It follows an operation procedure similar to that of the recloser. Note that opening of the breaker causes an interruption to all the customers served from this feeder.

Table 9.1 gives a part of the outages logged in the system. Note that the outage on 12 May has a total duration of 3 minutes and 11 seconds, which is a momentary interruption according to the cutoff time of 5 minutes. Since this outage interrupted all the customers, it must have been due to operation of the breaker or could be due to an event in the transmission systems. Note that 400 customers

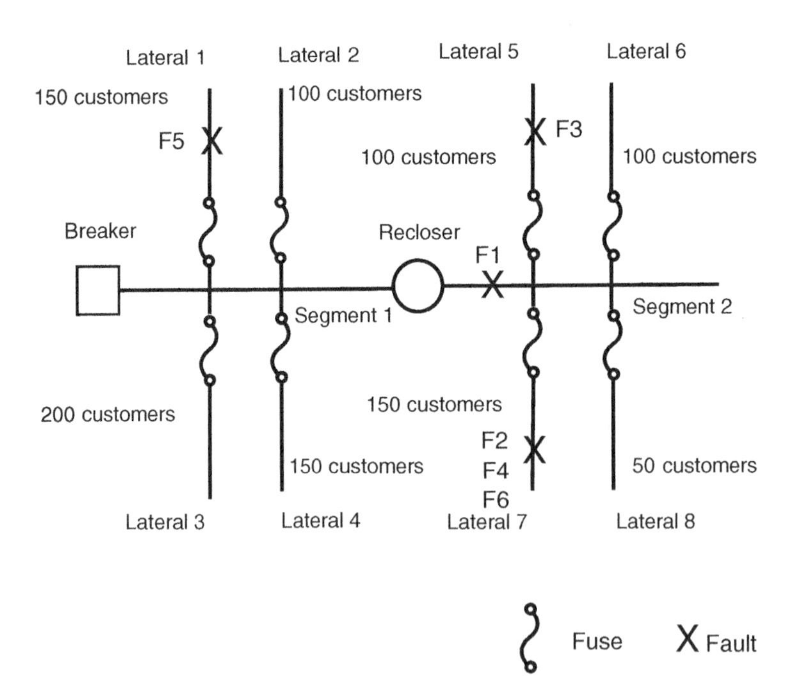

Figure 9.1 An example distribution feeder.

Table 9.1 Log of interruptions in the system of Figure 9.1.

Date	Time off	Time on	No. of customers	Load (kVA)	Interruption type	Location
20 March	20:18:30	22:20:00	400	800	S	F1
9 April	15:15:20	15:28:40	150	300	S	F2
12 May	08:25:25	08:28:36	1000	2200	M	Unknown
11 June	04:39:40	04:50:10	100	100	S	F3
8 July	23:45:15	00:15:00	150	350	S	F4
19 Aug	15:20:45	15:45:00	100	250	S	F5
22 September	18:22:23	18:42:53	150	320	S	F6

experienced at least one sustained interruption, and 150 customers (connected to lateral with F2, F4, and F6 faults) experienced four sustained interruptions and total interruption of longer than three hours.

We compute various indices described previously based on the data in this table.

$$\text{SAIFI} = \frac{400 + 150 + 100 + 150 + 100 + 150}{1000} = 1.05$$

$$\begin{aligned}\text{SAIDI} &= \frac{\begin{array}{c}(400 \times 121.5) + (150 \times 13.33) + (100 \times 10.5) + (150 \times 44.75)\\ + (100 \times 24.25) + (150 \times 20.5)\end{array}}{1000}\\ &= 63.86 \text{ minutes}\end{aligned}$$

$$\text{CAIDI} = \frac{\text{SAIDI}}{\text{SAIFI}} = \frac{63.86}{1.05} = 60.82 \text{ minutes}$$

$$\begin{aligned}\text{CTAIDI} &= \frac{\begin{array}{c}(400 \times 121.5) + (150 \times 13.33) + (100 \times 10.5) + (150 \times 44.75)\\ + (100 \times 24.25) + (150 \times 20.5)\end{array}}{500}\\ &= 127.72 \text{ minutes}\end{aligned}$$

$$\text{CAIFI} = \frac{400 + 150 + 100 + 150 + 100 + 150}{500} = 2.1$$

$$\begin{aligned}\text{ASAI} &= 1 - \frac{\begin{array}{c}((400 \times 121.5) + (150 \times 13.33) + (100 \times 10.5)\\ + (150 \times 44.75) + (100 \times 24.25) + (150 \times 20.5))\end{array}}{8760 \times 1000 \times 60}\\ &= 0.999878\end{aligned}$$

$$\text{ASIFI} = \frac{800 + 300 + 220 + 300 + 200 + 300}{2200} = 1.06$$

$$\text{ASIDI} = \frac{\begin{array}{r}(800 \times 121.5) + (300 \times 13.33) + (220 \times 10.5) + (300 \times 44.75) \\ + (200 \times 24.25) + (300 \times 20.5)\end{array}}{2200}$$

$$= 58.15 \text{ minutes}$$

Note that 150 customers experienced four sustained interruptions with a total duration of 200 minutes and 5 seconds, which gives $\text{CN}_{(k \geq 4)}$ and $\text{CN}_{(t \geq 3)}$ equal to 150. Also, the longest interruption of 121 minutes and 30 seconds due to F1 affected 400 customers, which gives $\text{CN}_{(s \geq 2)}$ equal to 400. If we count the total interruption including momentary and sustained, 250 customers experienced three or more interruptions, which gives $\text{CNT}_{(k \geq 3)}$. We can use this information to compute the additional indices.

$$\text{CEMI}_4 = \frac{150}{1000} = 0.15$$

$$\text{CELID}_{\text{S}(2)} = \frac{400}{1000} = 0.4$$

$$\text{CELID}_{\text{T}(3)} = \frac{150}{1000} = 0.15$$

$$\text{CEMSMI}_5 = \frac{150}{1000} = 0.15$$

$$\text{CEMSMI}_3 = \frac{250}{1000} = 0.25$$

9.3.6 Momentary Interruption Example

Table 9.2 shows a sample of momentary interruptions recorded for the system of Figure 9.1. The recorded data shows a total of 6600 momentary interruptions (IM_i) and 3600 momentary events (IM_E).

Table 9.2 Log of momentary interruptions in the system of Figure 9.1.

Date	Time	Device	No. of operations	No. of operations to lockout	No. of customers interrupted
4 October	08:18:22	Breaker	1	3	1000
15 October	14:13:21	Recloser	3	4	400
16 November	05:21:40	Breaker	2	3	1000
25 November	18:45:37	Recloser	2	4	400
8 December	02:18:45	Recloser	1	4	400
18 December	19:10:06	Recloser	3	4	400

The momentary interruption indices are computed below.

$$\text{MAIFI} = \frac{1000 + 400 \times 3 + 1000 \times 2 + 400 \times 2 + 400 + 400 \times 3}{1000} = 6.6$$

$$\text{MAIFI}_E = \frac{1000 + 400 + 1000 + 400 + 400 + 1200}{1000} = 3.6$$

9.4 Major Event Day Classification

Although the distribution systems are designed to handle outages that happen under normal operation, certain unforeseen events, mainly due to extreme weather, can push the system to the limit by causing numerous outages [3]. Such events skew the reliability performance of the system and thus are excluded from the calculations of reliability indices. Hence, by definition, a MED is a day in which the daily system SAIDI exceeds a threshold value, T_{MED}. The SAIDI index is used for this purpose because it has led to consistent results regardless of the utility size. Also, SAIDI is a good indicator of operational and design stress on the system. For calculating the daily system SAIDI, any interruption that spans multiple days is accrued to the day on which the interruption begins. The T_{MED} value is calculated at the end of each reporting period (typically one year) for use during the next reporting period. The process known as the "Beta Method" [3] is used to identify MEDs. Its purpose is to allow major events to be studied separately from daily operation and in the process to better reveal trends in daily operation that would be hidden by the large statistical effect of major events. Specific steps of the Beta Method are:

1. Gather values of daily SAIDI for five sequential years until the last day of the full reporting period. If less than five years of historical data are available, use all the available historical data. Less than five years of data may not yield accurate results. Similarly, more than five years of data may distort the results.
2. Use only those days that had interruptions and have a SAIDI value and exclude days with no recorded interruptions. In the originally proposed Beta Method [3], the lowest nonzero SAIDI per day value in the data set was used as the SAIDI value for days with zero SAIDI. However, IEEE Std 1366 [2] suggests excluding days with zero SAIDI.
3. Compute the natural logarithm (ln) of SAIDI for all the days in the data set.
4. Calculate the average (α) of the logarithms of the data set.
5. Calculate the standard deviation (β) of the logarithms of the data set.
6. Compute k for which

$$\ln(T_{\text{MED}}) = \alpha + k\beta \tag{9.16}$$

or

$$T_{\text{MED}} = e^{(\alpha + k\beta)} \tag{9.17}$$

Table 9.3 Probability of exceeding T_{MED} as a function of k.

k	Probability	MEDs/yr
1	0.15866	57.9
2	0.02275	8.3
2.4	0.00822	3.0
2.5	0.00621	2.3
3	0.00135	0.5
6	9.9×10^{-10}	3.6×10^{-7}

7. Table 9.3 shows probabilities and the expected number of MEDs for various values of k. If we know the allowed number of MEDs per year, we can compute a value for k. However, there is no analytical method of choosing an allowed number of MEDs/year. Therefore, $k = 2.5$ is the recommended value of k, which was determined based on the consensus reached among IEEE Power and Energy Society's Distribution Reliability Working Group members.
8. Any day with daily SAIDI greater than the threshold value T_{MED} that occurs during the subsequent reporting period is classified as a MED.

Activities that occur on days classified as MEDs should be separately analyzed and reported. The readers are referred to [3] for examples on the use of the Beta Method.

9.5 Causes of Outages

Each piece of equipment in a distribution system has a certain probability of failing. When first installed, a piece of equipment can fail due to poor manufacturing, damage during shipping, or improper installation. Healthy equipment can fail due to very high currents, extreme voltages, animals, and severe weather conditions such as tornadoes and lightning storms. Sometimes equipment may fail due to chronological age, chemical decomposition, contamination, and mechanical wear [4, 5]. The factors that cause failures in distribution systems can be divided into three main groups.

1. **Intrinsic Factors**: age of equipment, manufacturing defects in equipment, and the size of conductors.
2. **External Factors:** trees, birds/animals, wind, lightning, and icing.
3. **Human Factors**: vehicular accidents, accidents by utility or contractor work crew, and vandalism.

Since distribution system overhead lines are highly exposed to the atmosphere, external factors are the major causes of damages or failures. Intrinsic factors, on their own, generally do not endanger the reliability of such lines. Their effect can be seen only when they are combined with some other factors. For example, a very old small conductor would not break down or burn down by itself, but if lightning strikes lines with such conductors, probability of its break down or burn down is much higher than that of a new conductor under the same situation. A brief description of various failure causes and the possible preventive actions to mitigate their effect are given below.

9.5.1 Trees

Trees are among the major factors that affect the reliability of an overhead distribution line. Trees can cause failure of such lines in the following ways:

- Overhead conductors can be damaged when struck by a falling branch of tree.
- Wind can blow a tree branch into overhead conductors, resulting in two wires contacting each other.
- A growing branch of a nearby tree can push two phase conductors together resulting in a two-phase fault.
- During regular tree trimming, a tree branch can be accidentally dropped on the overhead line.
- Ice accumulation on tree branches can cause limbs to break off and fall on the conductors.

Tree trimming (periodically pruning vegetation adjacent to power lines) is the best possible solution to avoid overhead line failures caused by trees. Most distribution systems have trimming on a two- to six-year cycle. Operating and maintenance cost can also be reduced by selective trimming, which means trimming of only those trees that cause more customer interruptions. Some utilities perform trimming only on main trunks of the feeders and not on lateral branches. Tree trimming should always be performed by a trained crew to ensure safety and direct regrowth away from the conductor location. For additional information on predicting vegetation-related failure rates and optimal vegetation maintenance scheduling, the readers are referred to [6, 7].

9.5.2 Lightning

Lightning can be defined as a transient, high-current electric discharge. It occurs when some region of the atmosphere attains an electric charge sufficiently large that the electric fields associated with the charge cause electric breakdown of the air. Lightning can occur in four ways: it can travel between points within a cloud, from a cloud to the surrounding air, from a cloud to an adjacent cloud, and from a cloud to the ground. These flashes are referred to as intracloud, cloud-to-air, cloud-to-cloud, and cloud-to-ground, respectively. Although the most frequently occurring form of lightning is the intracloud discharge, we are more concerned about cloud-to-ground lightning, which is a big threat to power systems. A detailed discussion on lightning phenomenon is given in [8]. For information on models for effects of lightning storms on distribution system reliability, the readers are referred to [9, 10].

Lightning can affect power systems in two ways:

- **Direct Strokes**: Lightning directly strikes the power system. Although the incidents of direct strokes are very few, they are very dangerous for the system.
- **Indirect Strokes**: Most of the lightning strikes are of this type. They do not strike the power system directly, instead they strike some nearby objects such as a tall building or a tree. In this case, a traveling voltage wave is induced, which is less severe than the direct strokes.

Lightning can cause severe damages to an overhead line, which cannot be fully avoided but can be reduced by careful application of shield wires and surge arresters. Surge arresters should be inspected for any manufacturing defects, and also, very old arresters should be replaced by new ones to prevent any damage caused by lightning. The level of damage caused by lightning depends on some other factors also. For example, lightning could be more destructive for a very small and very old conductor compared to a big and new conductor. Some suggestions to protect the distribution lines from lightning are given in [11, 12].

9.5.3 Wind

As mentioned in [4], the probability of equipment failure increases rapidly with increasing wind speed because the pressure exerted on trees and poles is proportional to the square of the wind speed. Wind can cause supply interruptions in the following ways:

- Wind can cause a tree branch to touch two phase conductors together, resulting in a fault.
- Wind induces several types of conductor motions – swinging, galloping, and aeolian vibrations. If the swing amplitude is high, phase conductors could touch each other. Conductor galloping is the phenomenon when conductor starts moving up and down harmonically due to wind. Aeolian vibrations are generated by the air turbulence on the downwind side of the conductor. All these conductor motions are not good from the reliability point of view.

To reduce the possibility of interruptions caused by wind, spacing between two conductors should be kept large in highly windy areas. Also, the use of twisted pair conductor is proposed in [13] to avoid the various mentioned conductor motions.

9.5.4 Icing

Ice storms occur when supercooled rain freezes on contact with tree branches and overhead conductors and forms a layer of ice, which can cause outages in multiple ways:

- Heavy accumulation of ice on tree branches can cause them to break off and fall on the conductors.
- Ice places heavy physical load on conductors and support structures.
- Combination of ice and wind can result in sagging of conductor. The possibility of sagging is more when the conductor is of very small size. When ice breaks off, it can cause the conductor to jump into the conductor located above it.

To avoid the failure caused by ice, the strength of the overhead conductors and the supporting structure should be quite high. Also, bigger conductors should be used in the areas with high possibility of ice storms.

9.5.5 Animals/Birds

Animals and birds can cause harm to an overhead distribution line in several ways [4, 14]. Following are a few possibilities that may result in customer interruptions:

- Squirrels cause faults by bridging grounded equipment with the phase conductor.
- Raptors and roosting birds cause faults by bridging conductors with their wings.
- Woodpeckers cause damage to utility poles by pecking holes in them.
- Large animals, such as cattle, horse, and bear, can also do physical damage to utility poles, making the system more prone to future outages.

To avoid failures caused by various animals and birds, several remedies can be implemented [4, 14]. To prevent squirrels from simultaneously touching the tank and phase conductors, plastic animal guards on bushings and insulators should be installed. Anti-roosting devices should be used on attractive perches to prevent birds from roosting. Problems caused by woodpeckers can be mitigated using steel or concrete poles instead of wooden poles. Use of barricades near wooden poles can reduce the problems caused by large animals.

9.5.6 Vehicular Traffic

Collision of a fast speeding vehicle with a distribution pole may result in the damage of the pole itself, sagging or swinging of the phase conductors, and damage of other equipment. Pad-mounted equipment are also vulnerable to vehicular accidents. Use of concrete barriers and concrete poles can reduce the frequency of automobile collisions. Pad-mounted equipment should also be protected by concrete barriers.

9.5.7 Age of Components

Every component of a distribution feeder has its own probability of failure. It is a common practice to assume that performance of distribution system components deteriorates when they approach 30 years of age. However, "age" alone does not create any reliability problem. Its effects can be seen when it is combined with some other factors such as conductor size, wind velocity, and lightning intensity in the area under study. Very old components (around 30 years old) should be replaced by new ones on the feeders having more reliability problems.

9.5.8 Conductor Size

Conductor size, when considered alone, does not cause any reliability problem. Similar to "age," its effects can also be seen when combined with some other factors. For example, in high wind velocity areas, the swinging amplitude of small conductors will be high compared to big conductors.

9.6 Outage Recording

Utilities record daily outages in the service territory. The recorded data include time, duration, location, number of customers affected, and the possible cause of the outage. Table 9.4 shows the number of outages and their causes recorded for a period of two years from January 2003 to December 2004 for 66 feeders with a total length of approximately 1000 km in the distribution system of a city in Kansas. The operating voltage of most of the lines is 12.47 kV, but few are operated at 4 and 34.5 kV.

The data show that trees/vegetation and animals/wildlife caused 53.47% of outages followed by equipment failure and unknown. Environmental factors, which include lightning, extreme wind, trees, animals, ice storm, and debris, caused a total of 1290 outages, which is 61% of all the outages. A significant number of outages are reported as unknown or other causes. The cause of the outage is recorded

Table 9.4 Causes and number of outages in a service territory in Kansas in 2003 and 2004.

Cause	Number of outages	Percentage
Customer request	0	0.00
Equipment failed	251	11.80
Overload	41	1.93
Trees/vegetation	634	29.79
Public damage	25	1.17
Customer problem	0	0.00
Animals/wildlife	504	23.68
Other	141	6.63
Lightning	131	6.16
Extreme wind	8	0.38
Ice storm	13	0.61
Trees outside right of way	0	0.00
Debris, nature/weather	0	0.00
Unknown	208	9.77
Company damaged	0	0.00
Procedural error	1	0.05
Load transfer	0	0.00
Safety/hazard	0	0.00
Load shed	0	0.00
Maintenance	171	8.04
Total	2128	100.00

by the field crew based on the inspection of the failed equipment and any circumstantial evidence at the site of the outage. In some cases, the crew is not able to determine the cause, and they declare it as unknown.

It is also important to determine seasonal variation in the number of outages. A plot showing outages occurring in each month during the study period is shown in Figure 9.2. The number of outages is higher during the summer months (June, July, August, and September). The main reason for this is due to thunderstorms and windy conditions during the summer months. The graph also shows a large number of outages for January 2004. Most of these outages were due to trees, which fell on the feeders during icy conditions caused by winter storms. Although spring and fall that have quieter weather will lower the probability of outages, most of

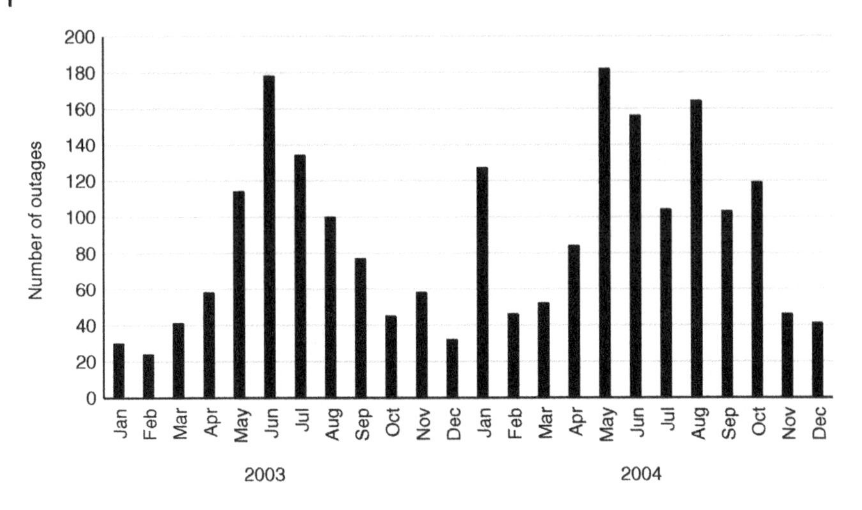

Figure 9.2 Monthly outages in a service territory in Kansas.

the outages during these seasons are caused by squirrels. Nice weather promotes higher animal activity. Also, the end of winter and the end of summer coincide with the birth of new litter of squirrels, which increases the probability of them causing outages.

9.7 Predictive Reliability Assessment

The data-based approaches for reliability evaluation are able to provide an assessment of the system performance during a period in the past. They are not able to predict the expected reliability in the future. Component failure models combined with network topology and reduction provide techniques for such prediction. In this section, we examine some aspects of predictive reliability assessment.

9.7.1 Component Failure Models

Every component in distribution is designed to function without failure for its lifetime. However, components fail due to intrinsic defects or external causes. Typically, the failure rate is higher in early stages of deployment due to manufacturing defects, damage in shipping, or incorrect installation [4]. After the break-in period, the component is expected to perform well during its expected life. However, toward the end of its life, higher failures are possible due to aging. A hazard function or failure rate is typically used to model failure of components. A bathtub curve with high failure rate in early and late stages and constant failure rate in the useful life is a good way to represent component reliability. We focus our analysis based on constant failure rate during the useful life of components.

Hazard rate or failure rate of a component is defined as the probability of a component failing at time t, given that the component has been functioning until time t. Considering the lifetime of the component as a random variable, we can represent the failures by a probability density function (PDF) $f(t)$ and the corresponding cumulative probability distribution function (CDF) $F(t)$. An expression for the hazard function $h(t)$ is obtained as given below [15]:

$$h(t) = \frac{f(t)}{1 - F(t)} \tag{9.18}$$

A common function used for constant failure rate modeling is the exponential function. With exponential function, the probability that a component will fail at time t, given that it is working at t, is independent of t. For exponential probability distribution function, we get expressions for $f(t)$ and $F(t)$:

$$f(t) = \lambda e^{-\lambda t} \tag{9.19}$$

$$F(t) = 1 - e^{-\lambda t} \tag{9.20}$$

Substituting (9.19) and (9.20) in (9.18) gives

$$h(t) = \lambda \tag{9.21}$$

9.7.2 Network Reduction

In power distributions systems, power flows through a set of components for delivery to the customers. These components are connected either in series or in parallel. For a series connection, all the components must be functional to deliver electricity. However, for a parallel connection, at least one of the components must be working for delivery of electricity, or all them must not be available for disruption of service.

Consider that two components with probability of availability P_1 and P_2 are connected in series as shown in Figure 9.3. P_1 and P_2 can be computed from the failure rate and mean time to repair (MTTR). So, for a duration of one year, with λ_1 as the annual failure rate, and R_1 as the MTTR of component 1, P_1 for a year is

$$P_1 = \frac{8760 - \lambda_1 R_1}{8760} \tag{9.22}$$

and

$$Q_1 = 1 - P_1 \tag{9.23}$$

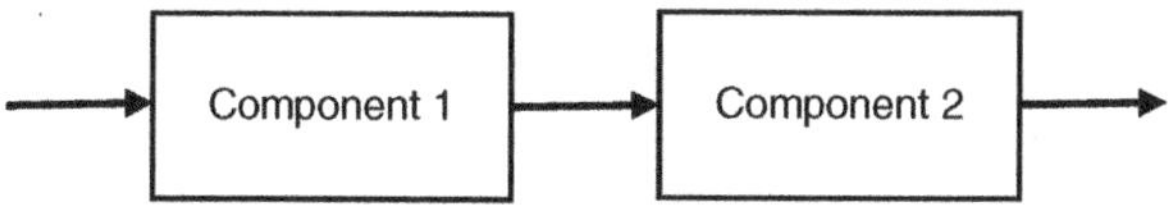

Component 1 → Component 2

Figure 9.3 Two components connected in series.

where Q_1 is the annual probability of unavailability of component 1. Hence, the combined probability of availability of both components 1 and 2 is

$$P_{\text{series}} = P_1 P_2 \tag{9.24}$$

If there are n components in series, we can generalize the equation to find probability of all of them being available, or

$$P_{\text{series}} = \prod_{i=1}^{n} P_i \tag{9.25}$$

$$Q_{\text{series}} = 1 - P_{\text{series}} \tag{9.26}$$

Now, let us consider two components connected in parallel as shown in Figure 9.4. In this case, the connection will fail if both components are unavailable. The probability of unavailability of both 1 and 2 can be computed, which is

$$Q_{\text{parallel}} = Q_1 Q_2 \tag{9.27}$$

and

$$P_{\text{parallel}} = 1 - Q_{\text{parallel}} \tag{9.28}$$

If there are n components in parallel, the equation can be generalized to determine the probability of unavailability of all of them, or

$$Q_{\text{parallel}} = \prod_{i=1}^{n} Q_i \tag{9.29}$$

Example

Consider a network of six components connected as shown in Figure 9.5. Various steps required to reduce this network for reliability evaluation are shown in Figure 9.6.

Solution

The given probabilities of availability of these components are $P_1 = 0.9$, $P_2 = 0.8$, $P_3 = 0.7$, $P_4 = 0.6$, $P_5 = 0.5$, and $P_6 = 0.4$. We can determine the probability of series

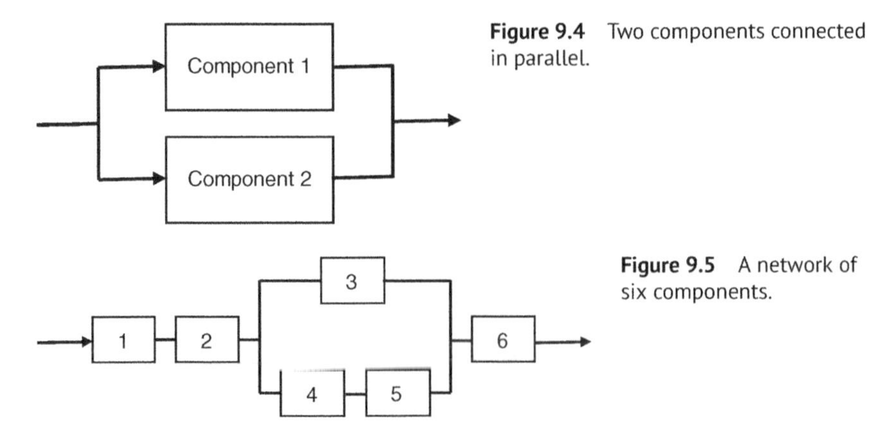

Figure 9.4 Two components connected in parallel.

Figure 9.5 A network of six components.

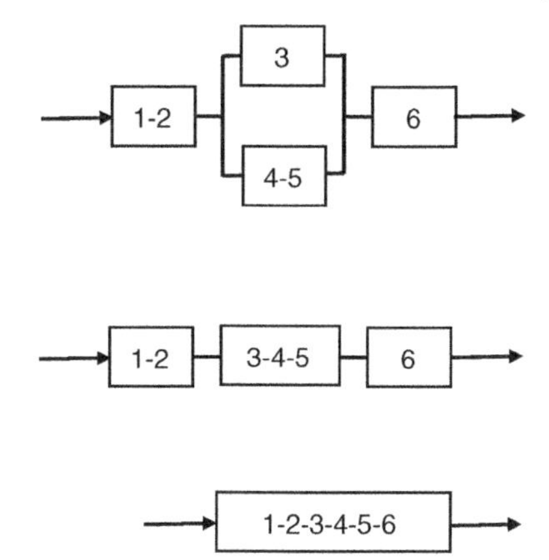

Figure 9.6 Steps for network reduction.

connected components 1–2 and 4–5.

$$P_{1-2} = P_1 \times P_2 = 0.9 \times 0.8 = 0.72$$

$$P_{4-5} = P_4 \times P_5 = 0.6 \times 0.5 = 0.30$$

and

$$Q_{4-5} = 1 - P_{4-5} = 1 - 0.30 = 0.70$$

Also,

$$Q_3 = 1 - P_3 = 1 - 0.70 = 0.30$$

Therefore,

$$Q_{3-4-5} = Q_3 \times Q_{4-5} = 0.30 \times 0.70 = 0.21$$

or

$$P_{3-4-5} = 1 - Q_{3-4-5} = 1 - 0.21 = 0.79$$

Now, we find the probability of continuity of the whole network, which is

$$\begin{aligned} P_{\text{Network}} &= P_{1-2} \times P_{3-4-5} \times P_6 \\ &= 0.72 \times 0.79 \times 0.40 \\ &= 0.22752 \end{aligned}$$

Network reduction works well for simple systems but becomes very tedious for large systems. Therefore, it is not suitable for large complex systems with many components.

9.7.3 Markov Modeling

Markov modeling, which is based on the Markov process, is a popular approach for reliability assessment of power systems. In this approach, various states of the system are defined that represent different operating states of the system. The states include a fully operational system and states with the failure of a component. Transition between states from one time to the next is a random process defined by probability of component failing and the probability of repair of a failed component. Both the failure and repair probabilities are considered to have exponential distribution, which give constant failure rate (λ) and constant repair rate (μ). An important property of the Markov process is that the probabilities in the future given the present state do not depend on the past. Hence, the Markov process is called a memoryless process.

Now, if we consider that the system is in state i at time t, we can write an expression for transition probabilities at time $(t + \Delta t)$ for a small Δt [15].

$$P\left[X(t+\Delta t) = j \,\middle|\, X(t) = i\right] = p_{ij}(\Delta t) \tag{9.30}$$

and

$$P\left[X(t+\Delta t) = i \,\middle|\, X(t) = i\right] = p_{ii}(\Delta t) \tag{9.31}$$

where $p_{ij}(\Delta t)$ is the probability that the state will change to j, and $p_{ii}(\Delta t)$ is the probability that the state will remain i at time $(t + \Delta t)$. We further define transition intensities, which are

$$q_{ij} = \lim_{\Delta t \to 0} \frac{p_{ij}(\Delta t)}{\Delta t}, i \neq j \tag{9.32}$$

and

$$q_{ii} = \lim_{\Delta t \to 0} \frac{1 - p_{ii}(\Delta t)}{\Delta t} \tag{9.33}$$

Since the sum of probabilities of being in any state is 1, we can write

$$p_{ii}(\Delta t) + \sum_{j \neq i} p_{ij}(\Delta t) = 1 \tag{9.34}$$

which gives

$$q_{ii} = \sum_{j \neq i} q_{ij} \tag{9.35}$$

Further, we can define a transition intensity matrix $\boldsymbol{A}$

$$\boldsymbol{A} = \begin{bmatrix} -q_{11} & q_{12} & q_{13} & \cdots & \cdots & \cdots \\ q_{21} & -q_{22} & q_{23} & \cdots & \cdots & \cdots \\ \vdots & \vdots & \vdots & \vdots & \vdots & \vdots \\ q_{i1} & q_{i2} & q_{i3} & \cdots & -q_{ii} & \cdots \\ \vdots & \vdots & \vdots & \vdots & \vdots & \vdots \end{bmatrix} \tag{9.36}$$

Let $P[X(t) = i]$ be $p_i(t)$, which is the unconditional probability of being in state i at time t. If we define $\dot{p}_i(t)$ as the time derivative of $p_i(t)$, we can write an equation in matrix form for all the states of the system. The readers are referred to [15] for additional details and proof.

$$\dot{\boldsymbol{p}}(t) = \boldsymbol{p}(t)\boldsymbol{A} \tag{9.37}$$

Note that $\dot{\boldsymbol{p}}(t)$ and $\boldsymbol{p}(t)$ are row vectors, or

$$\dot{\boldsymbol{p}}(t) = \begin{bmatrix} \dot{p}_1(t) & \dot{p}_2(t) & \dot{p}_3(t) & \cdots & \dot{p}_i(t) & \cdots \end{bmatrix} \tag{9.38}$$

and

$$\boldsymbol{p}(\boldsymbol{t}) = \begin{bmatrix} p_1(t) & p_2(t) & p_3(t) & \cdots & p_i(t) & \cdots \end{bmatrix} \tag{9.39}$$

Differential equation (9.37) can be solved to find probabilities of being in different states as a function of time, given the initial probabilities or $\boldsymbol{p}(0)$. For state probabilities in the long run, we let $t \to \infty$ and $\dot{\boldsymbol{p}}(t) = \boldsymbol{0}$. Equation (9.37) reduces to N ordinary linear equations for a system with N states. However, these equations are dependent, and thus, the determinant of $\boldsymbol{A}$ is zero. Thus, to solve this equation, we remove one of the equations and replace it by

$$\sum_{i=1}^{N} p_i = 1 \tag{9.40}$$

or the probabilitiöes of states must sum to 1.

Example

In a distribution system, we can use the Markov process by defining the states representing failure of a component, such as the section of a feeder. The failures and repairs of components are represented by exponential probability distribution functions giving constant failure and repair rates. Again, consider the system shown in Figure 9.1. For simplicity of explaining the concept, we consider only the failures of Segment 1 and Segment 2. Thus, as shown in Figure 9.7, the system will have three states, which are system fully operational (State 1), Segment 1 failed (State 2), and Segment 2 failed (State 3). It is assumed that once a segment has failed, no additional failures will take place in the system until repairs on the failed segment are completed. Segment 1 is 1-mile long and Segment 2 is 2-miles long. The failure rates are 0.1 faults per mile per year, and the MTTR is three hours.

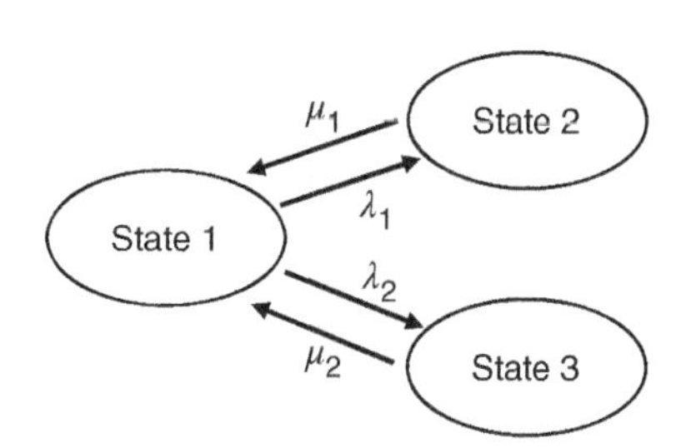

Figure 9.7 State transition diagram for the system of Figure 9.1 for outages on Segment 1 and Segment 2.

Solution

From the given data, we can determine the following values for the Markov model:

$$\lambda_1 = 0.1 \times 1 \text{ mile} = 0.1/\text{year or } 0.00001142/\text{hour}$$

$$\lambda_2 = 0.1 \times 2 \text{ miles} = 0.2/\text{year or } 0.00002284/\text{hour}$$

$$\mu_1 = \frac{1}{3} \text{ or } 0.3333/\text{hour (MTTR} = 3 \text{ hours)}$$

$$\mu_2 = \frac{1}{3} \text{ or } 0.3333/\text{hour (MTTR} = 3 \text{ hours)}$$

The $\boldsymbol{A}$ matrix for the for the system is

$$\begin{bmatrix} -(\lambda_1 + \lambda_2) & \lambda_1 & \lambda_2 \\ \mu_1 & -\mu_1 & 0 \\ \mu_2 & 0 & -\mu_2 \end{bmatrix} \tag{9.41}$$

Now, using Eq. (9.37) for steady state, we get

$$\begin{bmatrix} 0 & 0 & 0 \end{bmatrix} = \begin{bmatrix} p_1 & p_2 & p_3 \end{bmatrix} \begin{bmatrix} -(\lambda_1 + \lambda_2) & \lambda_1 & \lambda_2 \\ \mu_1 & -\mu_1 & 0 \\ \mu_2 & 0 & -\mu_2 \end{bmatrix} \tag{9.42}$$

Delete the first column of the matrix and replace it by 1s to represent $p_1 + p_2 + p_3 = 1$

$$\begin{bmatrix} 1 & 0 & 0 \end{bmatrix} = \begin{bmatrix} p_1 & p_2 & p_3 \end{bmatrix} \begin{bmatrix} 1 & \lambda_1 & \lambda_2 \\ 1 & -\mu_1 & 0 \\ 1 & 0 & -\mu_2 \end{bmatrix} \tag{9.43}$$

or

$$\begin{bmatrix} 1 & 0 & 0 \end{bmatrix} = \begin{bmatrix} p_1 & p_2 & p_3 \end{bmatrix} \begin{bmatrix} 1 & 0.00001142 & 0.00002284 \\ 1 & -0.3333 & 0 \\ 1 & 0 & -0.3333 \end{bmatrix} \tag{9.44}$$

We can solve Eq. (9.44) to find $p_1 = 0.99989722$, $p_2 = 0.0000342599$, and $p_3 = 0.0000685198$. As expected, the system spends most of the time in fully functional state. In the next step, we compute the total time the system would be in states of partial failure, which are $0.0000342599 \times 8760 = 0.3$ hour in State 2 and $0.0000685198 \times 8760 = 0.6$ hour in State 3. Further, we compute the customer minutes of interruptions in these two states as follows:

State 2: $0.3 \times 1000 \times 60 = 18\,000$

and

State 3: $0.6 \times 400 \times 60 = 14\,400$

Thus, we get a total CMI of $18\,000 + 14\,400 = 32\,400$.

Therefore, $\text{SAIDI} = \dfrac{32\,400}{1000} = 32.4$ minutes.

If we expand the problem to include failures of all the laterals in addition to the main feeder, the system will have 11 states. Since distribution systems are typically much larger than the example we have considered, implementation of this method becomes very tedious.

9.7.4 Failure Modes and Effects Analysis (FMEA)

FMEA is a method that is very effective for reliability analysis of radial distribution systems. The method uses system topology, models for devices including protective devices and switches, and models for system restoration. Capabilities include models for temporary and permanent faults, protection and switching including backup protection, isolation through protective device operation and sectionalizers, and full restoration through repair and partial restoration through switching.

In the FMEA method, failure on every segment of the system is considered as a *Failure Mode*. Since each failure mode causes interruption of service to a part of the system, these interruptions are identified as *Effects* on the system reliability.

9.7.4.1 FMEA Method Assumptions

Temporary and permanent faults are considered to be independent and mutually exclusive. Faults on each segment are considered to have a constant failure rate, or they follow exponential probability distribution. Similarly, repair rate is considered to have a constant value, which implies that the probability of a repair at time t after it has failed has exponential probability distribution. The required data include the system topology, the line segment failure rate or mean time between failures (MTBF), and the repair rate or MTTR.

9.7.4.2 FMEA Procedure

The FEMA procedure starts by identifying all the failure modes and the associated system-wide effects. The effects in terms of the number of customer outages and the duration of customer outages are determined. Following that, we sum the effects of all the failure modes to get the cumulative number of customer outages and the duration of customer outages over a period such as a year. The next step is to compute system-wide indices such as SAIFI and SAIDI.

Example

Consider the distribution system shown in Figure 9.1. Additional data for the system are given in Table 9.5. Segment 1 of the main feeder is from the breaker to the recloser, and Segment 2 is downstream of the recloser.

Solution

We start the computations by considering failures on each component one by one and recording their effects. So, for Segment 1, the expected number of faults is 0.1, which is obtained by multiplying the failure rate per mile by the length in miles. Note that the failure rates are fractional numbers. These numbers are determined from the historical data for the specific utility. For example, if the utility has 2000 miles of laterals in its service territory, and 400 faults were recorded on these

Table 9.5 Data for the distribution system shown in Figure 9.1.

Component	Length (miles)	Failure rate (faults per mile per year)	Repair time (h)
Segment 1	1	0.1	3
Segment 2	2	0.1	3
Lateral 1	0.5	0.2	2
Lateral 2	1	0.2	2
Lateral 3	1	0.2	2
Lateral 4	1.5	0.2	2
Lateral 5	0.5	0.2	2
Lateral 6	1	0.2	2
Lateral 7	1	0.2	2
Lateral 8	0.5	0.2	2

laterals, the failure rate comes out to be 0.2 faults per mile per year. This implies that not every section of the line will see a fault. However, for calculations, we use the expected number of faults. A fault on Segment 1 will interrupt service to all the customers, and they will have an expected interruption of 3 hours, which gives CMI of 100×180 or 18,000 minutes of interruption. Similarly, we can account for effects of faults on each component as shown in Table 9.6.

Further, we can compute the expected SAIFI and SAIDI for the system as shown below:

$$\text{SAIFI} = \frac{365}{1000} = 0.365$$

Table 9.6 Computation of CMI.

Component	No. of faults per year	No. of customers affected	No. of customer interruptions	CMI
Segment 1	0.1	1000	100	18,000
Segment 2	0.2	400	80	14,400
Lateral 1	0.1	150	15	1800
Lateral 2	0.2	100	20	2400
Lateral 3	0.2	200	40	4800
Lateral 4	0.3	150	45	5400
Lateral 5	0.1	100	10	1200
Lateral 6	0.2	100	20	2400
Lateral 7	0.2	150	30	3600
Lateral 8	0.1	50	5	600
Total			365	54,600

and

$$\text{SAIDI} = \frac{54{,}600}{1000} = 54.6 \text{ minutes}$$

Now, consider the option of using the tie switch at the end of the feeder for restoration of power to the customers connected to Segment 2 whenever Segment 1 has a fault. This is done by opening the recloser and closing the tie switch. Consider that this process takes four minutes. Hence, the 400 customers connected to Segment 2 will experience an interruption of only four minutes whenever a fault takes place in Segment 1. Since this interruption is less than five minutes, which is the cutoff between the momentary and sustained interruptions, the number of interruptions as well as the CMI for these customers will be removed from SAIFI and SAIDI calculations. Therefore, the total customer interruption reduces to 325, and CMI reduces to 47 400, which gives SAIFI of 0.325 and SAIDI of 47.4 minutes. Note that we are able to get information only on expected values of SAIFI and SAIDI, which are useful for comparison of different system topologies during the planning stages or for decisions relating to system upgrades.

9.7.5 Monte Carlo Simulation

The FMEA described in the previous section provides an assessment of system reliability only in terms of estimated mean values of reliability indices. For detailed assessment, we have to rely on the Monte Carlo simulation, which is typically very time consuming. In addition to knowing the mean value of failure rate (or time between failures) and time to repair (or repair rate), probability distribution of these two parameters is needed. While modeling failures with a fixed average rate and exponential probability distribution function is a good assumption, considering fixed average repair rate with exponential probability distribution function is strictly not true. Although other probability distribution functions can be used to model repairs, prior research [15] shows that exponential probability distribution function provides good approximation. Also, the average failure rates and the associated probability distribution functions can change due to external conditions, such as storms. Since analysis with variable failure rates becomes very complex, we consider fixed failure rates.

The simulation begins by building failure scenarios for the system. Using a random generator, we determine the failure rate and the repair rate of each component from their respective probability distributions. In the next step, we find the effects of each component's failure on the numbers of customers affected and the customer minutes of interruptions. The cumulative count of customer interruptions and customer minutes of interruptions provides the values of SAIFI and SAIDI. The process is repeated multiple times to get a distribution of SAIFI and SAIDI. The number of simulations is decided based on the desired results. For example, simulations can be continued until the mean of all simulations converges to a stable value. The results provide probability distribution of system indices from which the average and standard deviation as well as confidence levels can be computed.

The size of the system has significant impacts on the results. Specifically, the standard deviation of SAIFI and SAIDI increases as the system size decreases. The spread of SAIDI is usually higher than that of SAIFI. Also, systems with higher failure rates or more faults have less standard deviation. As we zoom into the system, the spread of the reliability indices increases. The results provide annual performance standards for individual feeders or parts of the system. They do not accurately forecast the system performance in the future. So, we need to guard against the apparent sense of precision in these results. Nonetheless, these results are valuable in relative comparison of feeders or systems against one another for planning improvements.

9.8 Regulation of Reliability

Maintaining adequate level of reliability requires investments in system upgrades. However, there are trade-offs between cost and reliability. While the reliability of the bulk system, which includes transmission and generation, is regulated at the federal level in the United States, electric distribution is regulated at the state level. According to a report published in 2012 [16], 35 states out of 50 states and District of Columbia actively regulate distribution system reliability. Typically, frequencies and durations are used to measure the overall system performance. Some regulators require utilities to report reliabilities in a smaller region, such as a geographic area or feeder, and may require identification of circuits with worst reliability. If a utility serves both urban and rural customers, the regulators may ask for separate reports for urban and rural regions. Some regulators have included reliability performance in their revenue regulations. These regulations may include penalties only for not meeting the standards or also rewards for exceeding the performance [17]. Other examples include setting quality of service target without any penalty or rewards and reporting only without targets. While the targets set by regulators and utilities vary widely. some examples of targets specified in [17] are listed below:

- SAIDI of the worst performing feeder exceeding system SAIDI by 300%.
- SAIDI of a feeder greater than four times the system SAIDI or in the top 10% for two consecutive years.
- More than 90% of the interrupted system restored in 36 hours for all events except extreme events, and more than 90% restored in 60 hours for extreme events.
- Customers experiencing more than six outages per year for three consecutive years or outages with total duration of more than 18 hours per year for three consecutive years.

Problems

9.1 Table 9.7 shows a log of outages due to faults in the distribution system shown in Figure 9.8.

(a) Find the number of customers interrupted by each outage.
(b) Find the CMI for each outage.
(c) Find SAIFI, SAIDI, CAIFI, CAIDI, and ASAI with an assumption that the log of outages covers the whole year.

Table 9.7 Log of outages for Problem 9.1.

Date	Time off	Time on	Location
20 March	15:15:20	15:25:40	F1
9 April	20:18:30	21:20:00	F2
11 June	04:39:40	05:15:10	F3
8 July	23:45:15	00:25:00	F4
19 Aug	15:20:45	15:35:00	F5
22 September	18:22:23	18:52:53	F6

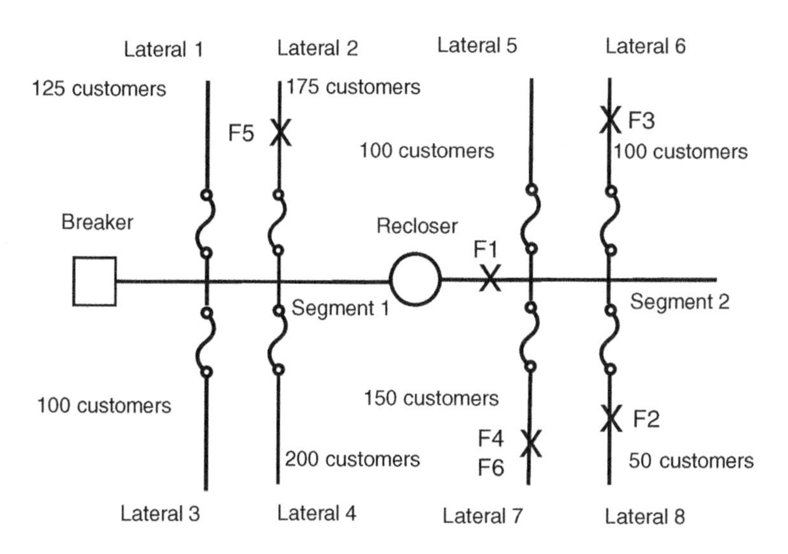

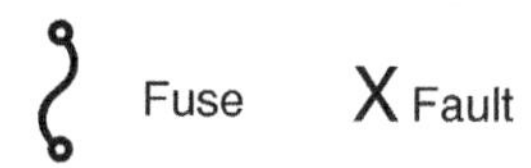

Figure 9.8 Distribution system for Problems 9.1 and 9.5.

(d) What is the maximum number of interruptions experienced by customers? How many customers experience them? Find the associated CEMI_k value.

(e) How many customers experience the longest outage? What is the corresponding CELID_S? Consider S = 1 hr if the longest duration is less than but close to it.

(f) Find the number of customers who experience the highest total interruption duration. What is the corresponding CELID_T?

9.2 A network of components is connected in series as shown in Figure 9.9. The given probabilities of availability of these components are $P_1 = 0.8, P_2 = 0.9$, $P_3 = 0.6, P_4 = 0.7, P_5 = 0.9$, and $P_6 = 0.75$.

Using the network reduction method, compute the probability of continuity in the given network.

9.3 Figure 9.10 shows a two-segment feeder of a distribution system. Breaker 2 is normally open (NO) under normal operation. Segment 1 is 2-miles long, and Segment 2 is 1-mile long. The failure rates of both the segments are 0.1 faults per mile per year, and the MTTR is two hours. Determine the states of the system. Find the transition intensity matrix (***A***) for the system. Determine the steady-state probabilities of being in different states. Compute the long-term SAIDI of the system if Segment 1 serves 300 customers and Segment 2 serves 500 customers.

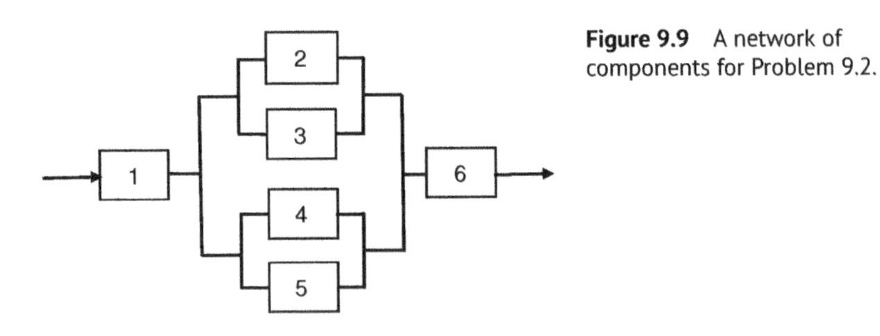

Figure 9.9 A network of components for Problem 9.2.

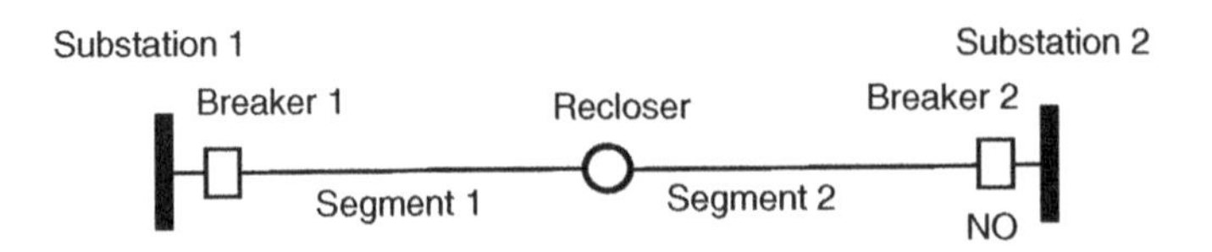

Figure 9.10 System for Problem 9.3.

9.4 Consider that following a fault on Segment 1, Recloser is opened and Breaker 2 is closed to restore power to customers on Segment 2 while repairs are being conducted on Segment 1. The mean time to switching (MTTS) of opening Recloser and closing Breaker 2 in such events is one hour. All the other parameters of the system are same as previously defined in Problem 9.3. Determine the states of the system. Find the transition intensity matrix ($\boldsymbol{A}$) for the system. Determine the steady-state probabilities of being in different states. Compute the long-term SAIDI of the system.

Now, consider that with automation, Recloser can be opened and Breaker 2 can be switched with MTTS of 10 minutes following a fault on the segment. Rework the problem to determine the steady-state probabilities of being in different states. Compute the long-term SAIDI of the system.

9.5 Table 9.8 provides information about the system shown in Figure 9.8.

(a) Use the FMEA method to estimate the expected number of customer interruptions and the associated CMI for faults on each component of the system.

(b) Find the expected SAIFI and SAIDI for the system.

Table 9.8 Failure data for Problem 9.5.

Component	Length (miles)	Failure rate (faults per mile per year)	Repair time (h)
Segment 1	2	0.15	2.5
Segment 2	1	0.15	2.5
Lateral 1	1.0	0.25	1.5
Lateral 2	0.5	0.25	1.5
Lateral 3	1.5	0.25	1.5
Lateral 4	1.2	0.25	1.5
Lateral 5	0.8	0.25	1.5
Lateral 6	1.2	0.25	1.5
Lateral 7	0.8	0.25	1.5
Lateral 8	0.6	0.25	1.5

References

1 IEEE 100: *The authoritative dictionary of ieee standards terms. IEEE Std 100-2000*, 7, 2000, 1–1362. https://doi.org/10.1109/IEEESTD.2000.322230.

2 IEEE Standard 1366 (2012). *IEEE Guide for Electric Power Distribution Reliability Indices*. New York: IEEE.
3 Warren, C.A., Bouford, J.D., Christie, R.D. et al. (2003). Classification of major event days. IEEE PES General Meeting in Toronto, Canada (July 2003).
4 Brown, R.E. (2008). *Electric Power Distribution Reliability*, 2e. Boca Raton, FL: CRC Press.
5 Gupta, S. (2002). An adaptive-fuzzy model to predict failure rates of overhead distribution system lines. Master's thesis. Kansas State University.
6 Radmer, D.T., Kuntz, P.A., Christie, R.D. et al. (2002). Predicting vegetation-related failure rates in electric power distribution systems. *IEEE Transaction on Power Delivery* 17 (4): 1170–1175.
7 Kuntz, P.A., Christie, R.D., and Venkata, S.S. (2002). Optimal vegetation maintenance scheduling of overhead electric power distribution systems. *IEEE Transaction on Power Delivery* 17 (4): 1164–1169.
8 Uman, M.A. (1969). *Lightning*, Advanced Physics Monograph Series. McGraw Hill, Inc.
9 Brown, R.E., Gupta, S., Christie, R.D. et al. (1997). Distribution system reliability assessment: momentary interruptions and storms. *IEEE Transactions on Power Delivery* 12 (4): 1569–1575.
10 Balijepalli, N., Venkata, S.S., Richter, C.W. et al. (2005). Distribution system reliability assessment due to lightning storms. *IEEE Transactions on Power Delivery* 20 (4): 2153–2159.
11 McDermott, T.E., Short, T.A., and Anderson, J.G. (1994). Lightning protection of distribution lines. *IEEE Transactions on Power Delivery* 9 (1): 138–152.
12 IEEE Standard 1410-1997 (1997). *IEEE Guide for Improving the Lightning Performance of Electric Power Overhead Distribution Lines*. New York: IEEE.
13 Baker, G.C. (2000). ACSR twisted pair overhead conductors. IEEE Rural Electric Power Conference, pp. B4/1–B4/4.
14 Frazier, S.D. and Bonham, C. (1996). Suggested practices for reducing animal: caused outages. *IEEE Industry Applications Magazine* (July/August), pp. 25–31.
15 Endrenyi, J. (1978). *Reliability Modeling in Electric Power Systems*. Wiley.
16 Hesmondhalgh, S., Zarakas, W., and Brown, T. (2012). *Approaches to Setting Electric Distribution Reliability Standards and Outcomes*. London, UK: The Brattle Group.
17 (2005). *State of Reliability Distribution Regulation in the United States*. Washington DC: Edison Electric Institute.

10

Distribution System Grounding

10.1 Basics of Grounding

10.1.1 Need for Grounding

Grounding is a mechanism to protect distribution equipment and people under normal operating conditions, abnormal operational (overcurrents and overvoltages) responses, and hazardous conditions such as shocks [1]. Grounding is necessary to assure correct operation of electrical devices, to assure safety during normal or fault conditions, to stabilize voltages during transient conditions, and to dissipate energy associated with lightning strokes. Good system grounding provides the path for normal load and fault currents while maintaining load and controls temporary overvoltages. Good equipment grounding ensures personnel safety.

10.1.2 Approaches for Grounding

Most North American distribution systems have a neutral that acts as a return conductor and as an equipment safety ground. It is recommended to ground the neutral at various strategic locations in distribution substations, overhead lines and underground cables, distribution transformers, and all loads. Four-wire with multigrounded neutral system, shown in Figure 10.1, is the most common type followed in the United States for grounding in distribution systems. Other approaches for grounding include three-wire with unigrounded neutral system, three-wire with ungrounded system, and three-wire ungrounded delta-connected system. The three-wire unigrounded system is shown in Figure 10.2. In the three-wire unigrounded system, neutral is grounded only at the substation and the distribution transformers. Safety is the major advantage of the multiground approach. If the neutral-to-ground connection at the transformer gets disconnected, the system can still operate safely with multiground, but safety would be compromised with uniground.

Electric Power and Energy Distribution Systems: Models, Methods, and Applications, First Edition.
Subrahmanyam S. Venkata and Anil Pahwa.

Companion website: www.wiley.com/go/Pahwa/ElectricPowerDistributionSystems

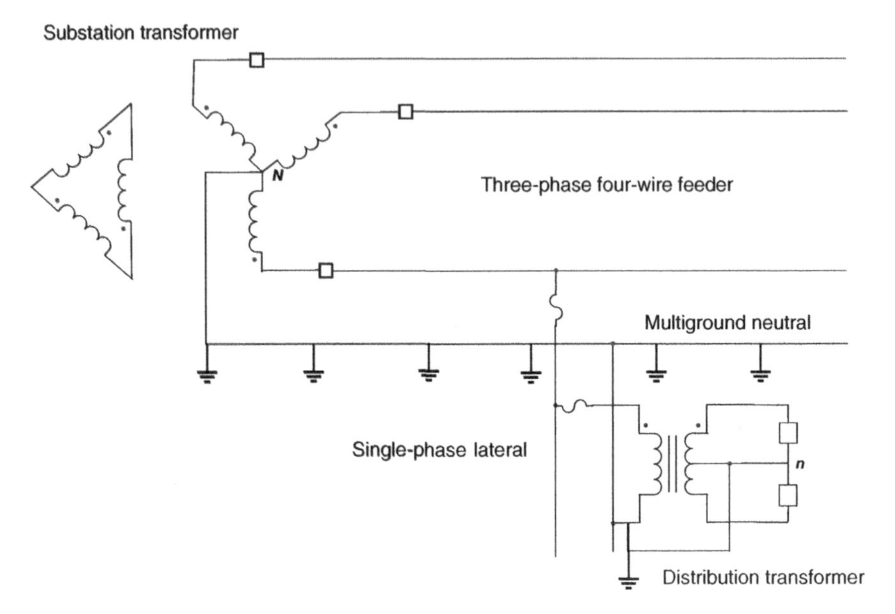

Figure 10.1 Four-wire multigrounded distribution system.

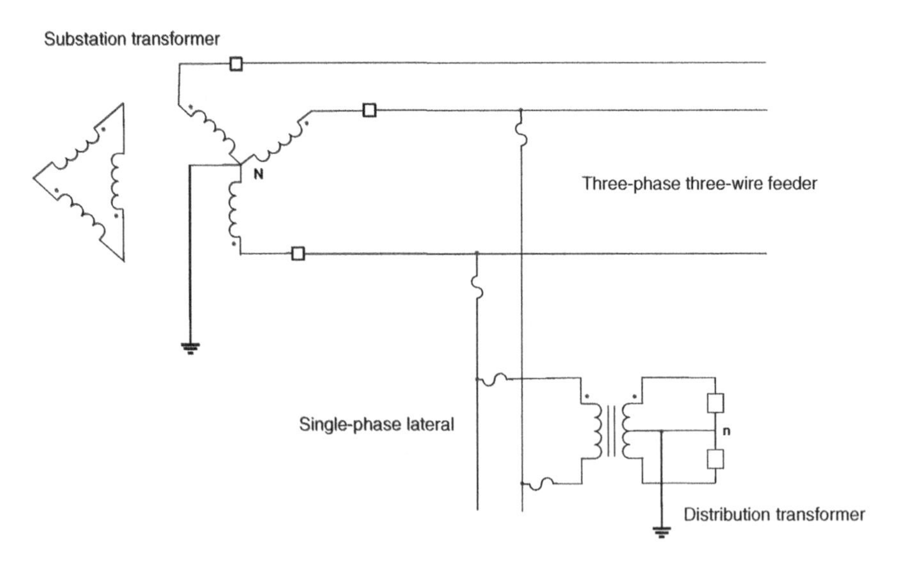

Figure 10.2 Three-wire unigrounded distribution system.

Four-wire systems are superior to three-wire systems for serving single-phase loads, and are predominant in North America [1]. In addition to safety, it is cheaper to build the system because a single cable can be used for underground single-phase load, and single-phase overhead lines are less costly. The distribution transformers with this configuration need only one bushing, one surge arrester, and one fuse on the live side.

10.1.3 Effects of Grounding on System Models

Neutral grounding, the system frequency and soil resistivity impact modeling of the distribution system components. Specifically, frequency and soil resistivity have a profound effect on online parameters. For example, in Chapter 3, while developing the models for overhead distribution feeders using Carson's equations, we assumed that the neutral is grounded at multiple locations, which resulted in voltage drop across the neutral conductor to be zero because grounding made a significant portion of the current to flow through the ground. In addition, the impedance values depend on the frequency and the soil resistivity.

10.2 Neutral Grounding

10.2.1 Neutral Shift Due to Ground Faults

A single-line-to-ground fault in distribution systems causes a shift in the potential of the ground at the fault location. The level of the shift is a function of grounding used in the system [1]. In ungrounded systems, the line-to-ground voltage on unfaulted phases increases to the line-to-line voltage or 1.73 per unit. In a system with perfectly grounded neutral, there is no shift in the neutral voltage. However, in a system with multiground neutral, the voltage can rise to 1.3 per unit. This phenomenon is quantified by two factors, which are coefficient of grounding (COG) and earth fault factor (EFF). COG is the ratio of the highest line-to-ground power-frequency voltage of a sound phase, at a selected location, during a fault to ground affecting one or more phases to the line-to-line power-frequency voltage which would be obtained, at that location, with the fault removed, or

$$\text{Coefficient of Grounding (COG)} = V'_{LN}/V_{LL} \tag{10.1}$$

EFF is defined as the ratio of the highest line-to-ground power-frequency voltage on a sound phase at a selected location due to a line-to-ground fault to the phase-to-ground power-frequency voltage at that location without the fault, or

$$\text{Earth Fault Factor (EFF)} = V'_{LN}/V_{LN} \tag{10.2}$$

where V'_{LN} is the maximum line-to-ground voltage on the unfaulted phases for a fault from one or more phases to ground, and V_{LL} and V_{LN} are the nominal line-to-line and line-to-neutral voltages.

The system is considered effectively grounded if COG is less than or equal to 80% [1]. This also results in EFF equal to 138% or less.

10.2.2 Types of Neutral Grounding

A resistance or reactance is used between the neutral of the substation transformer and the ground to reduce the level of ground fault currents. Two seminal publications [2, 3] provide excellent details on different grounding types and their impacts. In the United States, typically reactance is used to ground the neutral on the low-voltage side of the delta-wye-connected distribution substation transformers. However, this also reduces the effectiveness of grounding, which results in higher voltages on the unfaulted phases. Therefore, selection of the reactor for neutral grounding must be carefully evaluated by considering the trade-off between the decrease in fault current and increase in voltage on the unfaulted phases [1].

10.2.3 Standards for Neutral Grounding

There are several standards for neutral grounding

- ANSI/IEEE Standards 80 “IEEE Guide for Safety in AC Substations Grounding” (Equivalent to IEC 479-1).
- ANSI/IEEE Std 487-2000: “IEEE Recommended Practice for the Protection of Wire-Line Communication Facilities Serving Electric Supply Locations – Description.”
- IEEE Std 1100-1992, IEEE Recommended Practice for Powering and Grounding Sensitive Electronic Equipment (The IEEE Emerald Book).
- IEEE C62.92.5 Guide for the Application of Neutral Grounding in Electrical Utility Systems, Part IV – Distribution. The guide deals with the neutral grounding of single- and three-phase ac utility primary distribution systems with nominal voltages of 2.4–34.5 kV.
- IEEE 32 Standard Requirements, Terminology, and Test Procedure for Neutral Grounding Devices.

10.3 Substation Safety

Unbalance in three phases of the distribution system under normal operation results in flow of some current from ground to the substation transformer neutral.

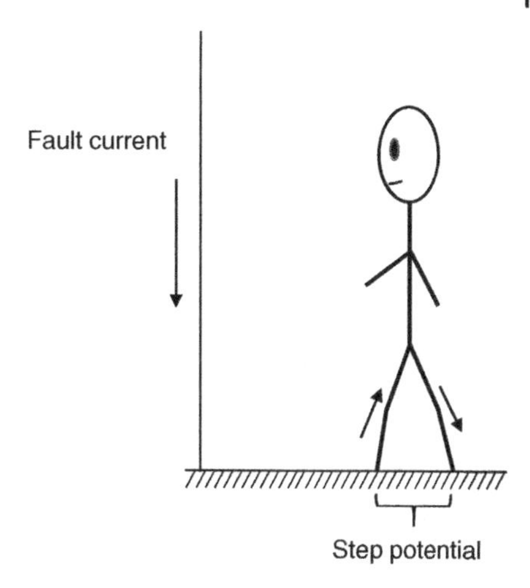

Figure 10.3 Current through human body due to step potential.

Although this current does not create high potential on earth around the transformer, ground faults can create large return current and subsequently large ground potential. The ground potential can be harmful to people working in the substation. Of specific concern are step potential and touch potential. Step potential is defined as the potential difference on the ground between two feet when a person walks. This potential difference sends a current through the legs of a person as shown in Figure 10.3. Touch potential is the potential difference between the hand and feet of a person whenever the person touches a conducting element. In this situation, current flows through the arm, body, and legs of the person to ground as shown in Figure 10.4. Multiground neutral substantially helps in reducing both the step and touch potential by distributing the flow of current from ground to neutral. Also, low impedance path for current to flow back to the substation allows the protective devices to react quickly to faults to isolate

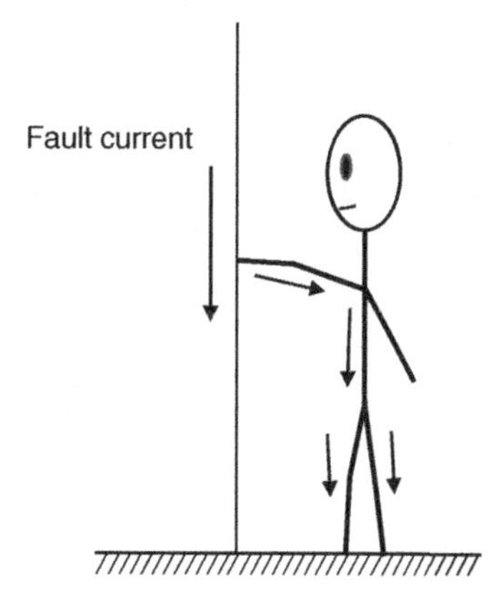

Figure 10.4 Current through human body due to touch potential.

them. A good ground mat in the substation with several grounding electrodes tied together reduces these potentials and increases safety for operating personnel.

10.4 National Electric Safety Code (NESC)

NESC [4] is designed for primary part of the distribution system and has been adopted by law by most states and Public Service Commissions across the United States. It is the authoritative source on good electrical engineering practice for over 90 years. The NESC is the single most important document for safeguarding of persons from hazards arising from the installation, operation, or maintenance of conductors and equipment in electric supply stations and overhead and underground electric supply and communication lines. Further, it contains extensive updates and critical revisions that directly impact the power utility industry. It also includes work rules for construction, maintenance, and operation of electric supply and communication lines and equipment. The standard is applicable to the systems and equipment operated by utilities, or similar systems and equipment, of an industrial establishment or complex under the control of qualified persons. While NESC governs the rules for the system under a utility's authority, safety issues related to customer's premises are governed by the National Electric Code (NEC) [5].

10.5 National Electric Code (NEC)

NEC, which has been adopted in all the 50 states of the United States, addresses safety consideration for the secondary part of the distribution systems beyond the distribution transformer. It is a benchmark for safe electrical design, installation, and inspection to protect people and property from electrical hazards. According to NEC, all grounding systems must be carefully coordinated. It provides guidance on grounding electrode systems, lightning protection, and communications grounding and serves as a reference guide for computer room signal. Grounding for communications systems must follow the requirements in EIA/TIA Standard 607: Commercial Building Grounding (Earthing) and Bonding Requirements for Telecommunications (and related bulletins).

Figure 10.5 shows the circuit diagram for safety ground for homes where the ground rod provides connection to ground at the service entrance. The green ground wire connected to the ground rod goes to all the lights, fans, and other loads as well as all receptacles in the house or building. Electricians wiring the building must ensure that there is no discontinuity between any of the load points and the ground rod. Also, the neutral wire must not be connected to the ground

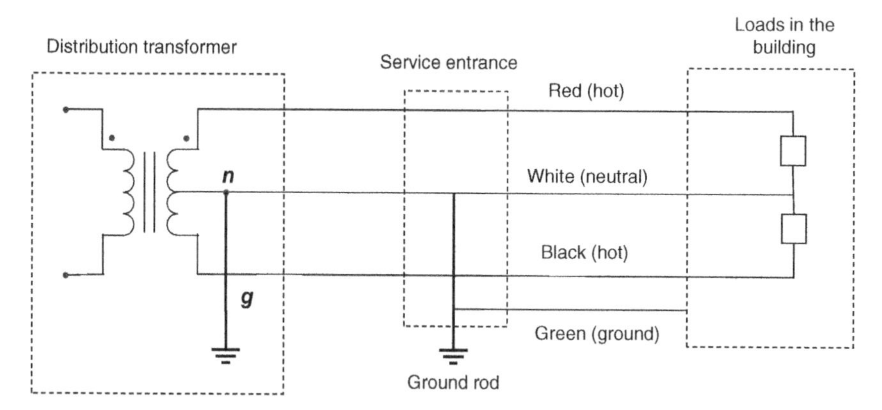

Figure 10.5 Safety ground for homes and buildings.

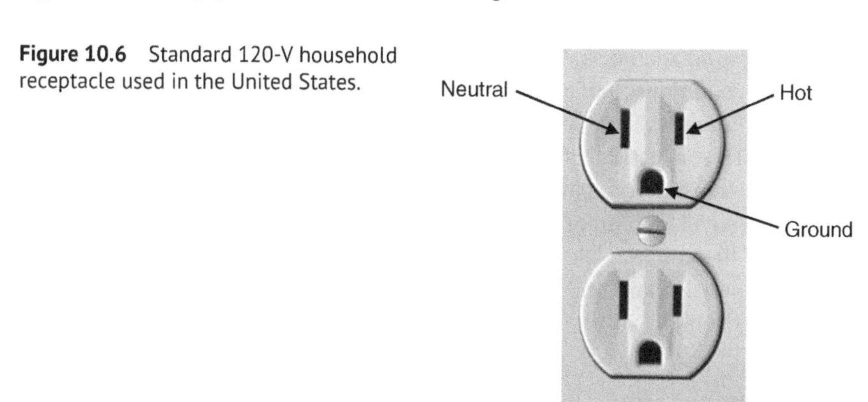

Figure 10.6 Standard 120-V household receptacle used in the United States.

wire except at the ground rod. NEC has standardized the 120-V household receptacles to be polarized and grounded. Figure 10.6 shows the standard household plug with center rounded pin at the bottom connected to the ground wire. The narrow blade on the right is connected to the hot wire, and the wide flat blade on the left is connected to the neutral. NEC requires that all receptacles must have ground connection with a minimum wire size of 14 AWG (copper) and 12 AWG (aluminum) for 15-A circuits and 12 AWG (copper) and 10 AWG (aluminum) for 20-A circuits.

Improper grounding in secondary systems can cause safety issues including fire and failure of equipment in homes. Most common problems are open secondary neutral, load incorrectly connected to the ground wire instead of neutral, and connection of the ground wire to neutral at wrong locations [1]. Open neutral of secondary systems causes the unbalanced current in the two circuits in a house to flow to the ground. If the two circuits have large difference in their loads, the neutral to

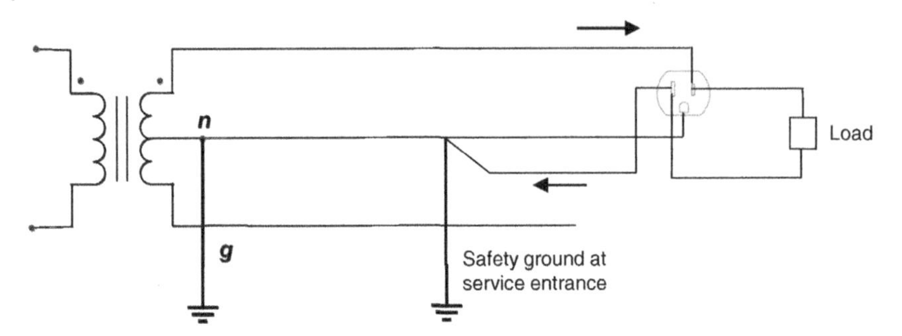

Figure 10.7 Hazard due to reversal of ground and neutral wires at load.

earth potential can become high due to large current flowing through the ground electrode, which can have high resistance. Switching on and off loads can cause fluctuations of the neutral-to-earth voltage, which in turn causes flicker. Reversal of safety ground and neutral and improper connection of safety ground and neutral create conditions that increase shock hazards. For example, consider that, as shown in Figure 10.7, the return conductor of a load is connected to the ground, and neutral is connected to the body of the load, such as an iron. Thus, touching the body of the iron creates a path in parallel to the path for load to the ground, which will send a small amount of current through the person touching the conducting element of the iron.

For additional details on NEC, the readers are referred to an excellent reference on the subject [6].

References

1 Short, T.A. (2004). *Electric Power Distribution Handbook*. Boca Raton: Taylor & Francis (CRC Press).

2 Blackburn, J.L. (1979). *Applied Protective Relaying*. Westinghouse Electric Corporation, Second Printing.

3 Blackburn, J.L. and Domin, T.J. (2014). *Protective Relaying: Principles and Applications*, 4e. Boca Raton: Taylor & Francis (CRC Press).

4 (2016). *Draft National Electric Safety Code (NESC) 2017*. New York: IEEE.

5 (2020). *NFPA 70 – National Electric Code (NEC)*. Quincy, MA: National Fire Protection Association (NFPA).

6 Keller, K. (2010). *Electrical Safety Code Manual, A Plain Language Guide to National Electrical Code, OSHA, and NFPA 70E*. Elsevier.

11

Distribution System Protection

11.1 Overview and Philosophy

The famous protection engineer Mason wisely mentioned in his classic book [1] that power system protection is more an "art" than a "science." It is one of the most complex and difficult topics in power system engineering. Though scientific principles provide the needed guidance to design a proper protection system, one can only master it through practical experience and through the lessons learned. The first step is to learn the basic principles and understand them thoroughly. To protect the same system, each protection engineer could arrive at a different solution than his/her counterparts, and all could be valid solutions. This is unlike other topics such as power flow studies covered in Chapter 4 for which only unique solutions exist.

The primary philosophy of protection is to preserve sensitivity, selectivity, minimum time of operation, and reliability. It is a local protection philosophy covering about two to three buses (or nodes of all phases) beyond any protective device. Though dependable and robust, this philosophy does not cover a wider area for defense against catastrophic failures to provide higher reliability and resiliency. Quite often, these considerations could be conflicting, and protection engineers must make adequate compromise and trade-off to achieve proper protection of a distribution system with cost-effectiveness. Safety of personnel and equipment is the paramount consideration of power system protection. The National Electric Reliability Council (NERC) has reported that 70% of outages in electric power systems are due to protection-related issues.

Distribution systems need protection against overcurrent and overvoltage. In this chapter, protection will be limited to overcurrent considerations only. Normally, overvoltage protection against electromagnetic transients is covered in separate courses.

Electric Power and Energy Distribution Systems: Models, Methods, and Applications, First Edition.
Subrahmanyam S. Venkata and Anil Pahwa.

Companion website: www.wiley.com/go/Pahwa/ElectricPowerDistributionSystems

During the past 25 years, distribution systems have experienced unprecedented transformation under the "Smart Grid" rubric with the infusion of innovative technologies in every automation domain: sensors, generation, control, communications, and computing. With the introduction of computer-based protection devices, the existing protection systems are changing gradually. System resiliency, as a new performance metric for systems, needs particularly important consideration related to protection. Further, advanced architecture and adaptive protection ideas need to be explored. With these changes in mind, protection of both classical and emerging distribution systems will be covered in this chapter, addressing the basic principles, design, and coordination. The detailed treatment of this subject for classical systems is covered in *Electrical Distribution Protection Manual* developed by Cooper Power Systems [2] (now part of Eaton Corporation). This is a particularly useful resource for both students and practicing engineers. Other resources and references are included at the end of this chapter [3–7].

11.2 Role of Protection Studies

There are many reasons for conducting protection studies. Since faults or abnormal conditions result in voltages and currents outside the operating limits, the primary reason is to prevent damage to equipment and circuits. In addition, it is important to prevent hazards to the public and utility personnel. Further, the utilities depend on protection to maintain highest service reliability, safety, and resiliency by preventing unnecessary power interruptions. Although with proper protection equipment damage can be minimized, it is impossible to completely avoid them. The protection system minimizes the effects of damage when an interruption occurs. Additionally, it minimizes the duration of service interruptions to customers due to a fault or short circuit and the number of customers affected with proper coordination and operation of the protective devices.

The primary objectives of performing protection studies as a part of comprehensive distribution planning and/or design studies of a given system are:

- Basic addition or expansion of a distribution system
- Manual and automatic sectionalizing of portions of a system
- Decision on proper phase spacing between conductors and selection of insulation
- Vegetation management to assure the highest level of system reliability
- Inspection for other potential problems such as salt deposition on conductors and dust accumulation on insulators
- Preventive equipment maintenance

11.3 Protection of Power-carrying Devices

Adequate protection must be provided for all types of power-carrying equipment such as:

- Lines, feeders, and laterals
- Distribution substation transformers and distribution transformers
- Capacitors
- Voltage regulators
- Segments of the system itself
- Conventional and distributed energy sources (DERs)
- Loads

11.4 Classification of Protective and Switching Devices

Protective devices are weak links intentionally created to save expensive power-carrying assets such as lines (feeders and laterals) and transformers (both substation and distribution). The most basic protective devices available for overcurrent protection in a distribution system are designed to burn and open to clear overcurrent and thus protect equipment from overloads and short circuits. Details on various devices used to protect various parts of the distribution systems are presented in this section.

11.4.1 Single-action Fuses

Fuses are overcurrent protective devices with a circuit-opening fusible part that is heated and severed by the passage of current through it. They can also be used for sectionalizing feeder segments to form zones. A single-action fuse must carry the expected load of a distribution line such as a feeder and a lateral. A fault occurring at the end of the line should be cleared by the fuse. The fuse performs both sensing and fault-interrupting functions. The real drawback with this device is that it must be replaced after one operation. Although fuses are inexpensive, the labor associated with changing fuses is not, from the operational point of view.

Fuses are available in variety of types:

- Expulsion fuses
- Vacuum fuses
- Current-limiting fuses.

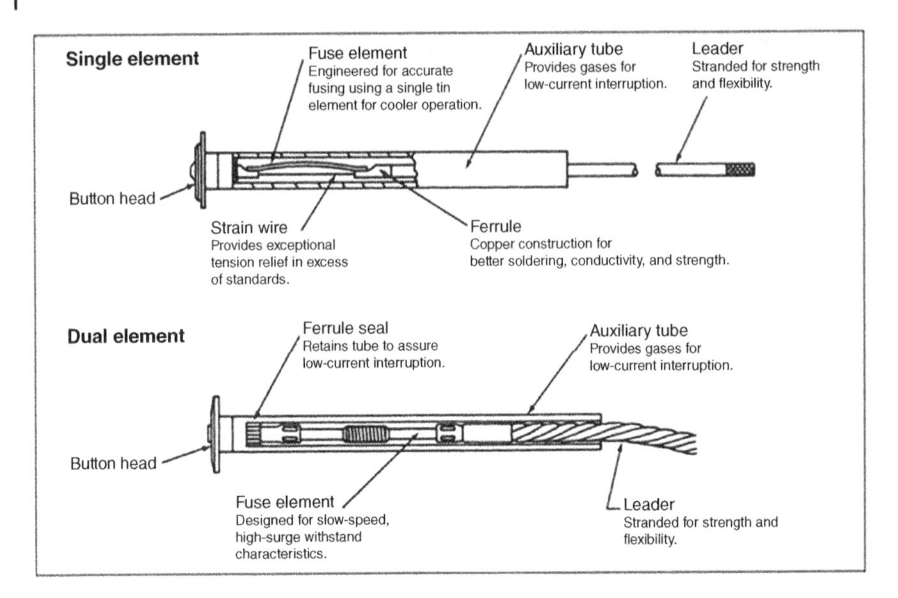

Figure 11.1 Fuse-link construction. Source: Courtesy of Eaton Corporation.

11.4.1.1 Expulsion Fuses

The principal component of a fuse link is a fusible element. It is made of various materials, including silver. It is housed inside a fuse cutout as shown in Figure 11.1. The time–current characteristic is the most common and classical model used to determine its time of operation for a specific fault current. The fuses are available with single or dual elements. The latter reduce the long-time minimum melt currents without reducing the short-time melt currents.

There are distinct types of expulsion fuses, which are designed to carry 100% of their rated current continuously. The operating characteristics of fuse links that are shown in Figure 11.2 are distinguished by the speed of operation defined by the speed ratio. The speed ratio of fuse links of 100 A and below is the ratio of the current that melts the fuse link in 0.1 second to the current that melts the fuse in 300 seconds. For fuse links rated greater than 100 A, it is defined as the ratio of current that melts the fuse link in 0.1 and 600 seconds.

- K link – “fast type” with speed ratio of 6–8.1. These are commonly used for urban systems.
- N link – This is also “fast type” with speed ratio of 6–11.
- T link – “slow type” with speed ratio of 10–13. These are suited for suburban systems.
- S link – These are “very slow” with speed ration of 15–20.

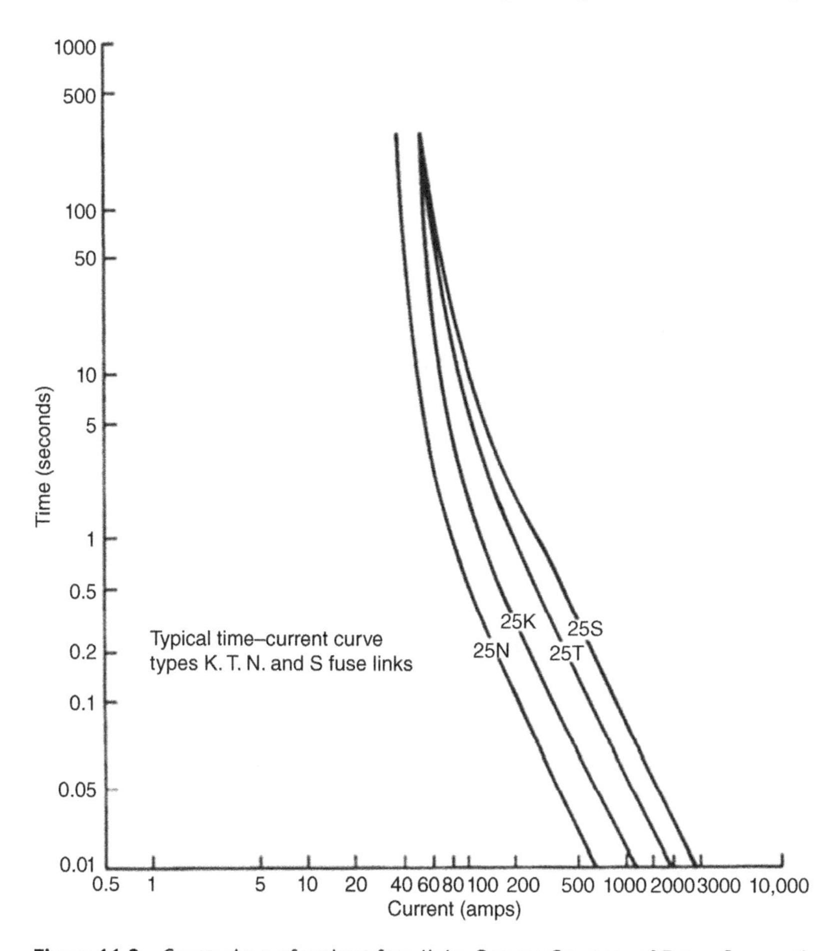

Figure 11.2 Comparison of various fuse links. Source: Courtesy of Eaton Corporation.

11.4.1.2 Vacuum Fuses

In vacuum fuses, the fusible element is enclosed in a vacuum medium. Other internal features include arc runners, shield, and ceramic insulation. Several cycles may be required for low fault currents to achieve burn-back of the fusible element. These fuses can be applied indoors and under oil.

11.4.1.3 Current-limiting Fuses

These nonexpulsion fuses limit the energy to the protective device. They are intended to reduce the possibility of catastrophic failure to the protective device. Their operation is dependent on the type of medium in which they operate. Similar to any other fuse, high-current clearing is basically the same. The key factors that determine their operation are let-through current, melt I^2t, let-through I^2t, and peak-arc voltage.

Basic Types of Current-limiting Fuses

- **Backup or Partial-range Fuse:** It must be used in conjunction with an expulsion fuse or some other device. It is capable of properly interrupting current only above a specified level.
- **General-purpose Current-limiting Fuse:** It is designed to interrupt all fault currents from its rated interrupting current down to the current that causes element melting in one hour.
- **Full-range Current-limiting Fuse:** It interrupts any continuous current (up to rated interrupting current) that will cause the element to melt.

11.4.1.4 Distribution Fuse Cutouts

Figure 11.3 shows an example of fuse cutout, which is a housing for connecting fuse link. This arrangement assists the field crews to replace a burned fuse link with a new one.

11.4.2 Automatic Circuit Reclosers

A recloser is a multifunction protective device which has both fault-sensing and fault-clearing capabilities. Unlike the fuse, it is a self-contained intelligent device with the ability to sense overcurrent and interrupt the current flow, depending on the value of the current. Further, it is designed to reclose automatically

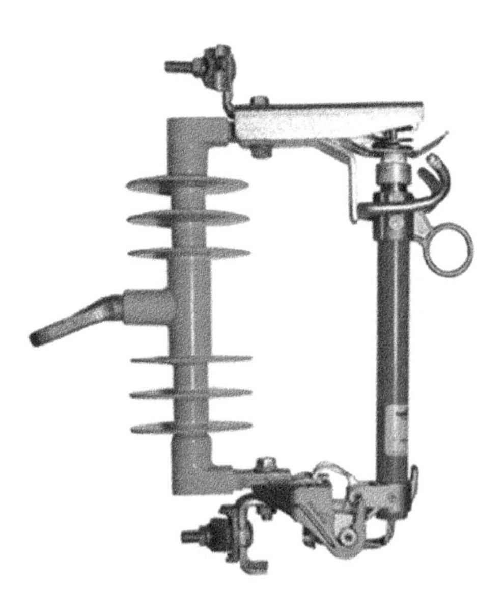

Figure 11.3 Example of a distribution system fuse cutout. Source: Courtesy of Eaton Corporation.

(a)

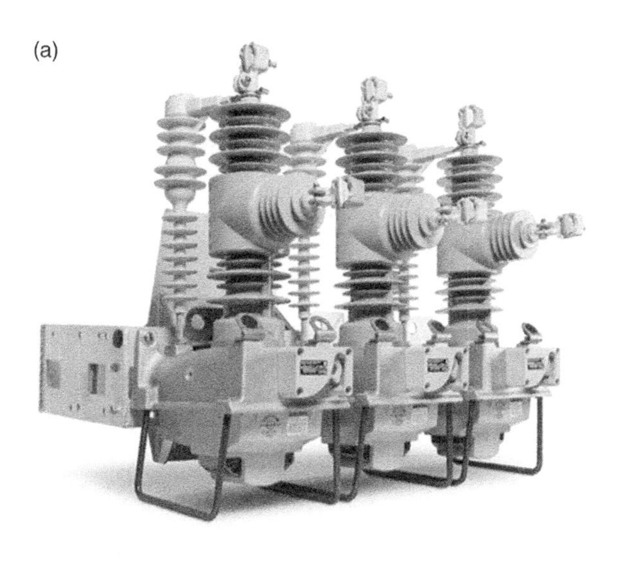

(b)

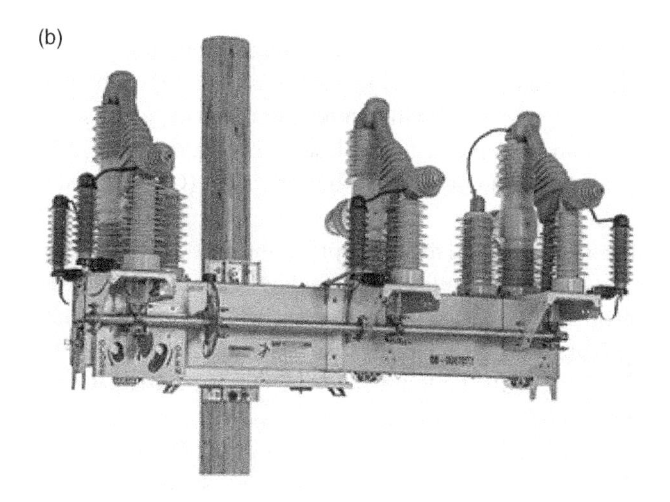

Figure 11.4 Examples of modern reclosers: (a) Nova NXT and (b) IntelliRupter® PulseCloser®. Source: (a) Courtesy of Eaton Corporation. (b) Courtesy of S&C Electric.

to reenergize the line. Some old reclosers are not as intelligent as modern microprocessor-protective relays. However, many modern reclosers now have fully capable microprocessor relays. Figure 11.4 shows examples of modern reclosers. They are much lighter than circuit breakers and are mounted on poles in overhead distribution systems. Reclosers with advanced microprocessor

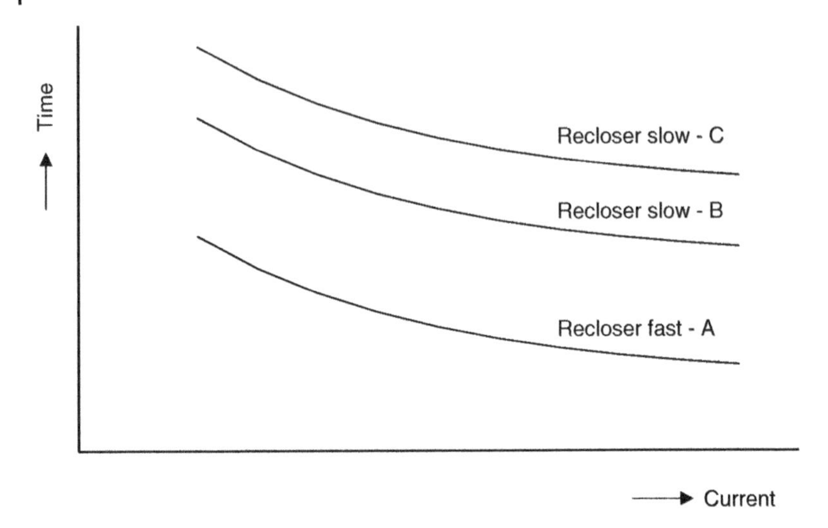

Figure 11.5 Example of recloser characteristics with one fast (A) and two slow curves (B and C).

protective relays are also commonly used at point of common coupling (PCC) to microgrids. Reclosers have considerable ability to distinguish between temporary and permanent faults, unlike fuse links, which interrupt either type indiscriminately. Figure 11.5 shows an example of the recloser's time–current operating characteristics. They typically have one fast (A) and one slow (C) or two slow (B and C) characteristics.

An automatic circuit recloser trips and recloses a preset number of times to clear temporary faults or isolate permanent faults. After a fault is detected, reclosers trip and automatically reenergize to "test" the line by successive "reclose" operations while giving temporary faults repeated chances to clear or be cleared by downstream protective devices. Should the fault not clear, the recloser recognizes it as a permanent fault and locks open or "locks out." A drawback of many reclosers is limited fault interruption capability. Reclosers must be coordinated with upstream protective relay-controlling circuit breakers in a substation. These circuit breakers are designed for interrupting fault currents.

Figure 11.6 shows two fast and two slow operations of a recloser. If Curve C is used for slow operations, the operating sequence is called 2A2C. Similarly, if Curve B is used for slow operations, the operating sequence is called 2A2B. The second two operations are deliberately slowed to allow the fault to clear if it is temporary or a downstream fuse to clear it if it is permanent. In this example, the recloser opened and locked out after the third reclosing action because the fault was either not cleared by the downstream fuse or it was upstream of the fuse.

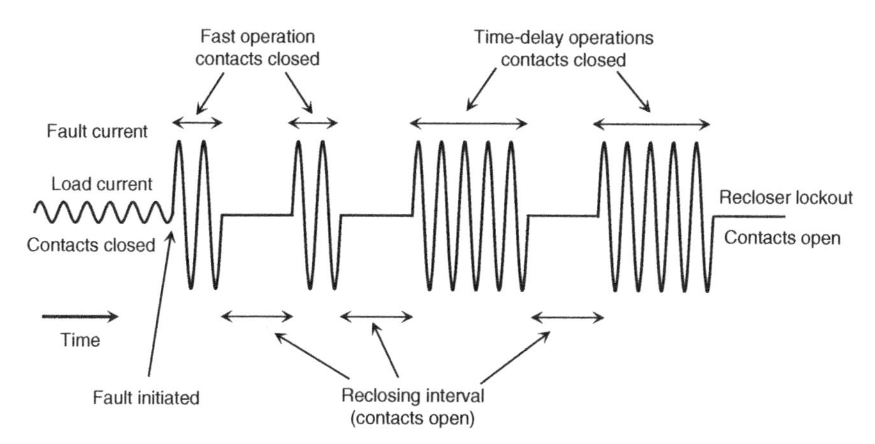

Figure 11.6 Typical recloser operating sequence to lockout. Source: Courtesy of Eaton Corporation.

11.4.2.1 Recloser Classifications

Reclosers are classified as single phase for single-phase lateral applications and three phase for three-phase feeders. They can be hydraulically or electronically controlled. The interrupting media could be oil or vacuum. Modern reclosers are usually electronically controlled. Figure 11.7 shows a control box for the recloser. Most reclosers interrupt fault current in oil-filled chambers, but recent designs have been built around vacuum circuit interruption.

11.4.3 Sectionalizers

Sectionalizers are circuit-interrupting devices and resemble reclosers but can be less expensive if they do not have fault-interrupting mechanism. While the legacy sectionalizers did not have fault-interrupting capability, most of the modern sectionalizers are reclosers programmed to operate as sectionalizers.

A sectionalizer applied in conjunction with a recloser or circuit breaker has the memory of counting the number of operations of the upstream device, but it does not have any fault-interrupting capability of its own. It counts the number of operations of the backup device (recloser or circuit breaker) during fault conditions, and after a preselected number of current-interrupting operations (reclose attempts), the sectionalizer opens and isolates the faulted section of line. If the fault is temporary, both the sectionalizer and the recloser reset to normal state. If the fault is persistent, however, the recloser operates on its sequence, but the sectionalizer isolates the fault before the recloser starts it final reclose operation; thus, recloser lockout is avoided, and only that portion of the circuit beyond the sectionalizer is interrupted. Figure 11.8 shows the operating sequence of a sectionalizer, where

Figure 11.7 Control box for a recloser. Source: Courtesy of Eaton Corporation.

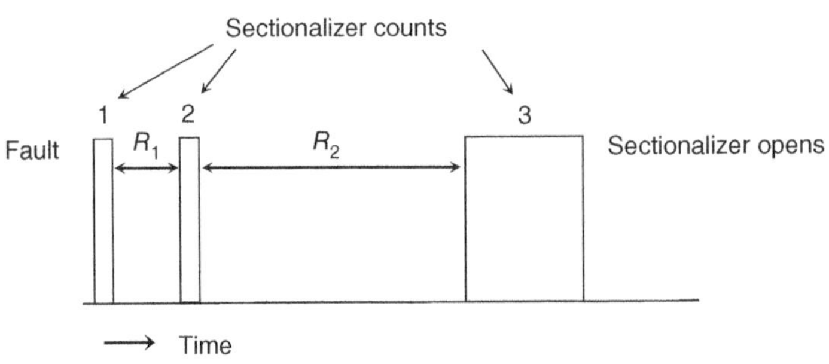

Figure 11.8 Operational sequence of a sectionalizer. Source: Courtesy of Eaton Corporation.

the sectionalizer is set for three counts to work with a four-sequence operation of a recloser on the upstream side.

Sectionalizers can be used between two protective devices with operating curves that are close together. They can also be used on close-in taps where high fault magnitude prevents coordination of fuses with the backup recloser or breaker. They are also ideal at locations where temporary faults could frequently occur. They are designed to automatically reset with the mechanism provided in them.

11.4.4 Circuit Breakers

A circuit breaker is normally employed at the substation level for overcurrent protection of the feeders connected to them. It is a mechanical switching device capable of making, carrying, and breaking currents under short-circuit or normal operating conditions. A circuit breaker is an expensive and bulky protective device which can only be cost justified at the substation level.

Circuit breakers are classified by the interrupting medium and the method of storing energy. These are:

- Oil interruption
- Vacuum interruption
- Air-blast interruption
- SF_6 (gas) interruption
- Air-magnetic interruption

The medium in which the circuit interruption is performed may be designated by a suitable prefix, e.g. air-blast circuit breaker and gas circuit breaker. Within the distribution systems, feeder breakers normally utilize oil, vacuum, or air magnetic as the interrupting medium and energy storage. Figure 11.9 shows an example of 15-kV and 38-kV class circuit breakers, which are air insulated and interruption is done in vacuum.

In general, relay-controlled circuit breakers are preferred to reclosers due to their better accuracy. Thus, opening and closing of substation circuit breakers are always controlled by protective relays. An automatic circuit breaker is equipped with a trip coil connected to a relay (which senses an abnormal condition) or other means, designated to open the breaker automatically under abnormal conditions, such as fault and overcurrent. It is frequently used to restore service quickly after a line trips out owing to lightning or a temporary fault. The stored-energy mechanisms in a circuit breaker are designed to close its contacts several times. These operations use (i) motor-compressed spring for one closing/opening

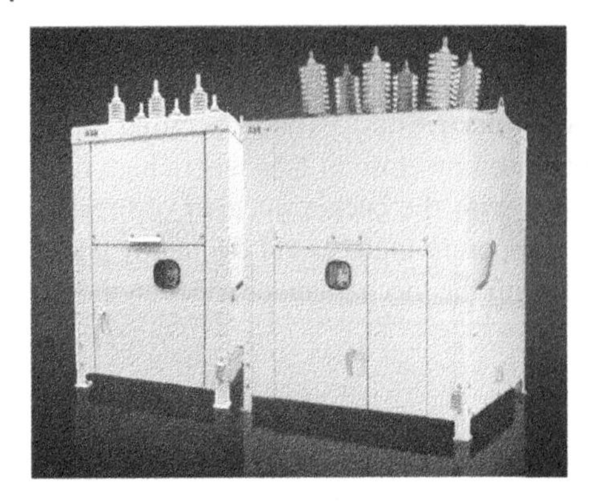

Figure 11.9 Examples of 15-kV and 38-kV circuit breaker. Source: Courtesy of ABB.

operation with spring reset within 10 seconds, (ii) compressed air or other gas for two closing/opening operations, and (iii) pneumatic or hydraulic breakers for higher numbers such as five closing/opening operations.

11.4.5 Time Overcurrent Relays

Historically, the classical electromechanical protective relays have been in existence since the dawn of electricity in the early 1880s. Electromechanical relay designs evolved through the 1950s. The image of the inside of a vintage Westinghouse electromechanical overcurrent relay shot by the authors is shown in Figure 11.10. Most of the world still uses these devices and they will continue to be utilized for the foreseeable future. Electromechanical relays are reliable; however, they cannot perform complex or adaptive protection and cannot advise operations about their own failure.

Relays have the intelligence to detect an abnormal condition and send proper signals to circuit breakers to achieve automatic tripping and closing of the circuit breaker contacts. For primary distribution or medium voltage systems, sensing function requires instrument transformers to step down both voltages and currents to standard 120 V and 5 A, respectively. For example, for a 10-MVA, 115-kV/12.47-kV three-phase substation transformer, a voltage transformer (VT) with turns ratio of 60 : 1 will be required to step down 7.2 kV (L–N voltage on the low-voltage side) to 120 V for the main feeder connected to the substation. Similarly, a current transformer (CT) with turns ratio of 500 : 5 will be needed to step down the full load current of 463 to a lower value suitable for relays.

The time-inverse overcurrent relay is the most used relay for overcurrent protection. This relay has plugs to select tap setting (TS), which is the minimum current at which the relay starts operating. Typical TSs range from 1 to 12 A. The other setting on these relays is the time dial (TD) setting, which delays the operation

Figure 11.10 A vintage electromechanical overcurrent relay.

of the relay. TD values range from 0.5 to 11. Figure 11.11 shows the characteristics of a commonly used time-inverse overcurrent relay (CO-8). The x-axis shows multiples of tap value current or multiples of tap setting (MTS), which is obtained by dividing the current flowing in the relay by the selected TS.

To understand the use of Figure 11.11, consider a fault current of 960 A for CT of 300 : 5 and TS of 4 A for the relay. The current seen by the relay can be obtained by dividing the fault current by the CT ratio, or

$$I_{\text{relay}} = 960 \times 5/300 = 16\ \text{A}$$

Now, we divide the relay current by TS to get MTS, or

$$\text{MTS} = I_{\text{relay}}/\text{TS} = 16/4 = 4$$

If we consider that a TD of 5 is selected, we can find the operating time of the relay under these conditions by noting the time on the graph with TD = 5 at MTS of 4, which is about 0.4 second.

Although the present generation of overcurrent relays use digital technology to process the input current, these relays continue to mimic the time-inverse overcurrent characteristics provided by the classical electromechanical relays. The standard characteristics of relays used in the United States are given by an equation in terms of TD and MTS, which is

$$t = \frac{\text{TD}}{7}\left[\frac{\beta}{(\text{MTS})^{\alpha} - 1} + K\right] \tag{11.1}$$

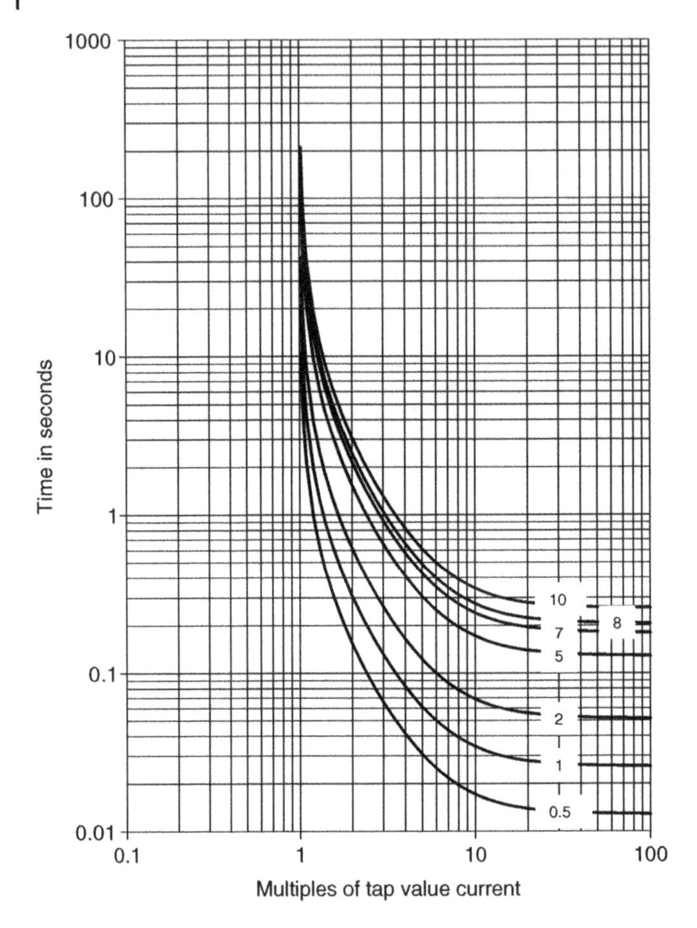

Figure 11.11 Time-current characteristics of CO-8 time-inverse overcurrent relay at different time dial values.

where t is the operating time of the relay, and α, β, and K are constants, which have specified values for different relay types as shown in Table 11.1. The corresponding graphs for these relays are shown in Figure 11.12. The equations and constants for International Electrotechnical Commission (IEC) relays used in Europe are available in [8].

In distribution systems, predominantly overcurrent and reclosing relays are utilized. The control mechanisms are beyond the scope of this book and are not covered in this chapter. The readers are encouraged to refer to standard distribution protection books, handbooks, and manuals listed at the end of this chapter.

Table 11.1 Constants for different time-inverse overcurrent relays.

Characteristics	α	β	K
IEEE extremely inverse	2	28.2	0.1217
IEEE very inverse	2	19.61	0.491
CO8 inverse	2	5.95	0.18
IEEE moderately inverse	0.02	0.0515	0.114
CO2 short time inverse	0.02	0.02394	0.01694

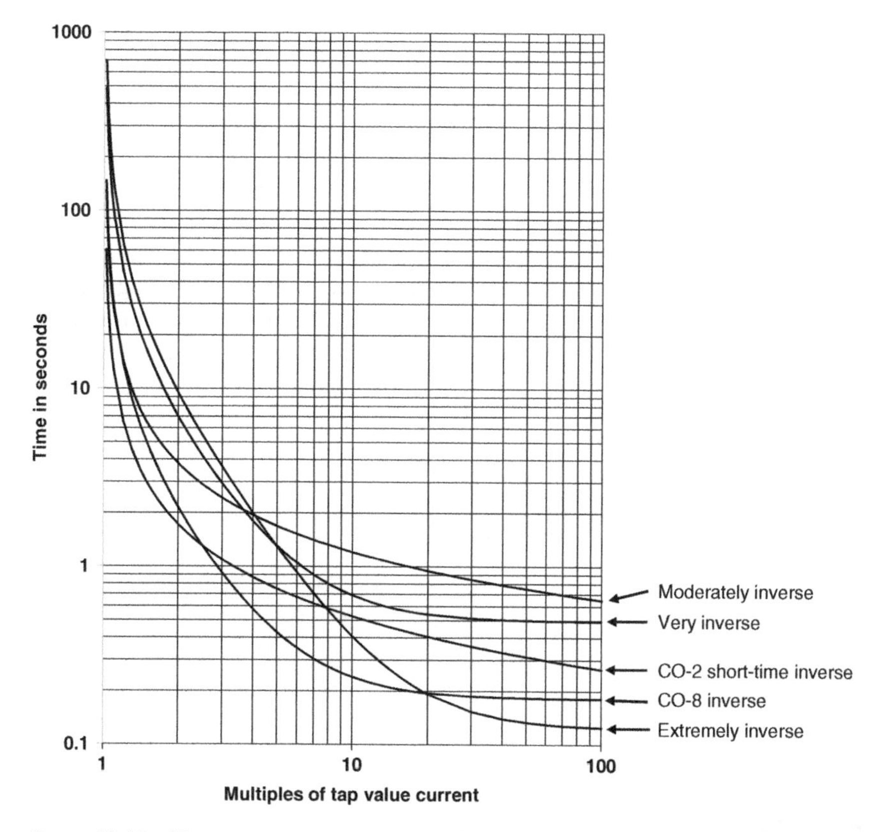

Figure 11.12 Time–current characteristics of different time-inverse overcurrent relays at time dial of 7.

11.4.6 Static or Solid-state Relays

Unlike electromechanical relays, these do not have any moving parts. The invention of transistors made the realization of the static relays a reality. Though they are less reliable than their classical counterparts, they are more accurate, and the response times are faster. This class of relays required very high-quality direct current (DC) power supplies, which is not so practical in a substation environment. Static or solid-state relays came into existence in the early 1960s; however, these devices were short lived due to their low reliability.

11.4.7 Digital or Numerical Relays

Digital or numeric relays were introduced in the mid-1980s due to the availability of low-cost microprocessors. These have become the preferred choice because of their multifunction capabilities and accuracy. Figure 11.13 shows a modern digital relay for overcurrent protection of distribution system feeders. Digital relays are becoming increasingly reliable and are now the preferred choice for protection of circuits from 480 V to 765 kV. In addition to performing the most complicated protection and control functions, digital relays have self-diagnosis capabilities and advise operations if they are failed. These relays are now commonly used as both protective relays and microgrid controllers. The digital relays have revolutionized protection with functions never possible including breaker failure, digital communications, adaptive protection, subcycle fast protection, and harmonic restraints, to name a few. These multifunction devices represent a significant advancement in parts, cost, and maintenance in a substation.

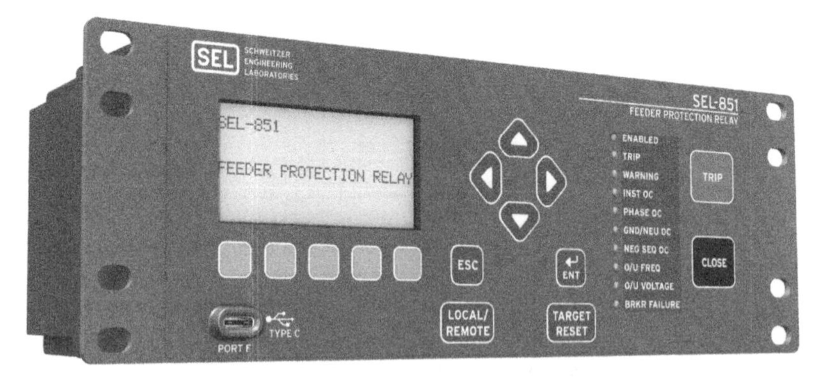

Figure 11.13 A modern digital relay for protection of distribution system feeders. Source: Courtesy of Schweitzer Engineering Laboratories.

Substations before the digital relay were always commonly manned. These new relays have provided better data collection for continuous monitoring and event root cause analysis than ever before possible.

11.4.8 Load Break Switch

This is a circuit disconnect device designed to make or break a circuit at specified currents. Load break switches have some auxiliary equipment to increase the speed of the disconnect switch blade and alter the arcing phenomenon to allow safe interruption.

11.4.9 Circuit Interrupter

This is a device designed to open and close a circuit by nonautomatic means and to open the circuit automatically at a predetermined overcurrent value without damage to the device when operated within its rating.

11.4.10 Disconnecting Switch

This is a mechanical device having a movable member adapted to connect or disconnect the contact members to which conductors are securely bolted. Disconnecting switches are usually operated on dead circuits only but are sometimes operated on energized low-capacity circuits where the arc cannot sustain and extinguishes by itself. These switches were used commonly for the earliest low-voltage circuits. As currents and voltages increased, it was found that the arc burns while opening the switch damaged or destroyed the contacts.

11.4.11 Sectionalizing Switch

Sectionalizing switch is a general term used to describe any switch that allows breaking the feeder into multiple sections but does not have any other function. Thus, both a disconnecting switch and load break switch are sectionalizing switches. However, sectionalizers and reclosers are not sectionalizing switches.

11.4.12 Example Distribution System

Figure 11.14 shows the layout of a distribution system in which several sectionalizing switches are deployed. The solid circles are used for sectionalizing switches

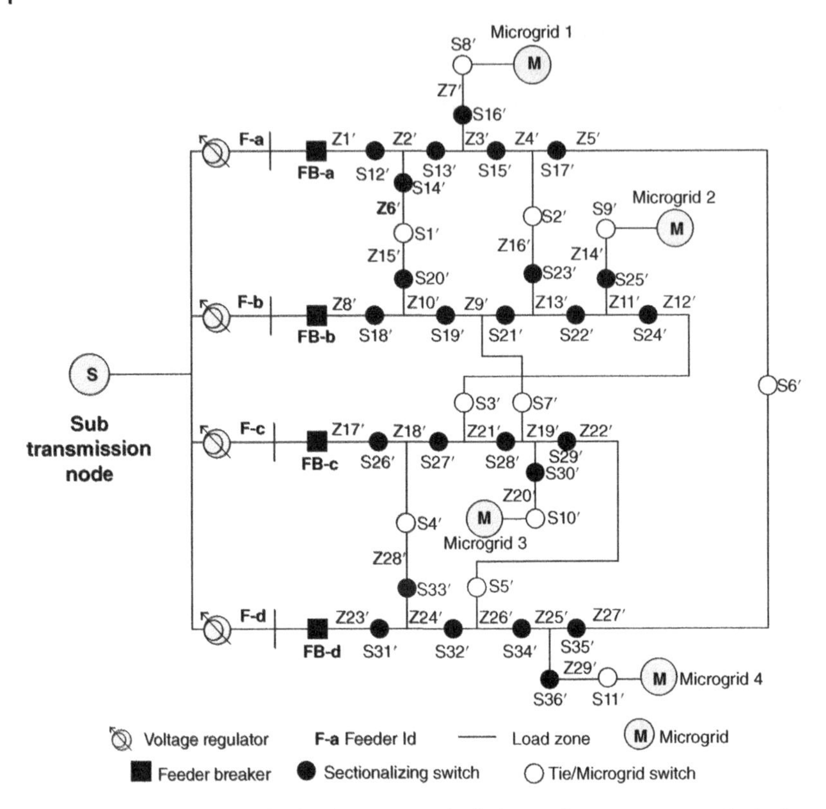

Figure 11.14 A typical distribution system depicting various components including tie and sectionalizing switches. Source: Courtesy of IEEE P2030.12 WG.

on the feeders (normally closed), and open circles are used for end-of-the-feeder tie switches (normally open).

11.5 New Generation of Devices

11.5.1 Smart Switching Devices

Recent advances in technology have resulted in a new class of smart switching devices. The concept of these devices is still in several stages of development. A lot more research needs to be done before one could realize an automated fuse or a similar device. A brief description of some of these devices under development follows.

11.5.1.1 Smart Fuses

A smart fuse device is a combination of a conventional fuse with intelligent sensor that simulates conventional current-limiting characteristics during high current faults and has the inherent ability to self-monitor and the capability to be triggered from an external source. Still the fuse needs to be replaced manually after it melts when it blows following a fault. The smart fuse can be used for both substation transformer and secondary conductor overcurrent protection. The device allows the medium voltage system to be grounded with a low resistance to minimize ground fault currents while still allowing coordination between upstream and downstream devices. It provides protection against single phasing without sacrificing the current-limiting features.

11.5.1.2 Smart Reclosers (Interrupters)

A smart recloser is a combination of a recloser with some form of intelligence and control incorporated to achieve automation. One such device was developed by S&C Electric in 2008. This Wi-Fi-enabled electrical equipment is a kind of "smart switch" that utilities use to more quickly detect and correct outages along their distribution systems.

11.5.1.3 Smart Circuit Breakers

Most of the ac circuit breakers deployed in the field are simple, electromechanical devices that sit idle most of the time. But the latest versions are coming with features such as wireless connectivity and computing power that are meant to turn them into something more like a smart meter or a smartphone.

11.6 Basic Rules of Classical Distribution Protection

All faults must be given a chance to be temporary by providing a reclosing operation for a fault anywhere on the system where momentary outages are acceptable. In responding to faults found to be permanent after the designated number of reclosing operations have been performed, the protection devices must remove from service only the smallest possible portion of the system necessary for isolation of the faulted segment. This assures that minimum number of customers are affected and thus assures higher reliability for the system under study. Data on overcurrent relays, reclosers (courtesy of Eaton Corporation), and selected fuse links (courtesy of S&C Electric) are too large for inclusion in the book, but they are available on the book's companion website.

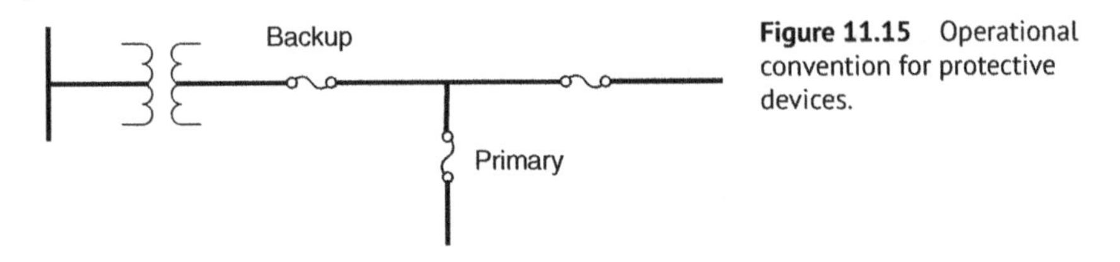

Figure 11.15 Operational convention for protective devices.

11.6.1 Operational Convention for Protective Devices

By conventional definition, when two or more protective devices are applied to a system, the device nearest to the fault on the supply side is the "primary" device. The other ones toward the upstream are called the "backup" devices, as shown in Figure 11.15.

11.6.2 Protecting Feeder Segments and Taps

To minimize the effects of faults on the main feeder, sectionalizing devices can be used to divide the feeder into smaller segments using devices such as reclosers, sectionalizers, and/or a combination of both. All taps branching off the feeder should have a protective device where it connects to the main feeder. Fuses are normally used for taps serving single-phase loads for short distances, say less than a mile, while reclosers and sectionalizers are utilized for large taps serving larger loads for longer distances.

The fast-trip curve of the recloser is used to clear all temporary faults on the main feeder and taps. For permanent faults on the taps, the recloser time-delay curve allows the tap fuse to clear, resulting in an outage on the tap only.

Although reclosers reduce the operation of fuses for temporary faults, they cause momentary interruptions on main feeders, which can be detrimental to certain loads. Momentary interruptions can be reduced by midpoint sectionalizing devices. Critical industrial or commercial loads can be protected by installing a recloser on the main feeder just downstream from the point of coupling of the critical load. Reclosers can also be added to longer taps off the main feeders to relieve the main feeder from momentary interruptions caused by faults on the tap.

11.7 Coordination of Protective Devices

The distribution protection coordination is a complex and complicated process. Here, the art of protection comes into picture since there are many variables and valid solutions that exist based on the philosophy of the protection engineer. The coordination of protective devices for traditional distribution systems with radial topology and one-way power flow is introduced in this section. Conducting proper coordination of protection devices provides several benefits including eliminating service interruptions due to temporary faults, minimizing the number of customers affected, and optimal service restoration.

The following data are needed for conducting proper coordination:

- Feeder configuration diagram.
- Location of protective devices.
- Proper mathematical models for protective devices (time–current characteristics).
- Expected range of normal load currents at all locations throughout the system under consideration.
- Expected range of fault currents at all locations throughout the system under consideration.

11.7.1 General Coordination Rule

The classical coordination of two local devices is performed using time–current coordination characteristics, which form the models for the devices. Essentially, two successive adjacent devices are coordinated proceeding from the load side to the source side for all device pairs and all feeder segments in the system. The objective is to keep a minimum coordination time interval between the primary device close to the point of fault and the immediate backup device(s), depending on the topology of the system. This process should be repeated for all device pairs in the system. The coordination of results tends to be very subjective because of the very nature of protection. One could conceive of multiple successful solutions for the exact fault situation.

11.7.2 Fuse–Fuse Coordination

11.7.2.1 Model for Fuses

Fuse is an overcurrent device with a circuit-opening fusible member, which is directly heated and destroyed by the passage of overcurrent in the event of an overload or short circuit. Thus, the time needed to melt the fuse link decreases with increase in current. The inverse-time melting characteristics of a fuse link are represented by the minimum-melting curve (MMC) and the total clearing curve (TCC) as depicted in Figure 11.16. The difference in two curves is the arcing time within the fuse. Typically, a new fuse will follow these curves. However, over time with multiple overloading situations, the melting time will decrease. The damaging time curve is approximately 75% of the MMC. The 25% margin considers some operating variables such as ambient temperature and loading. This margin is subjective in nature, and different values can be used, depending on the climate and

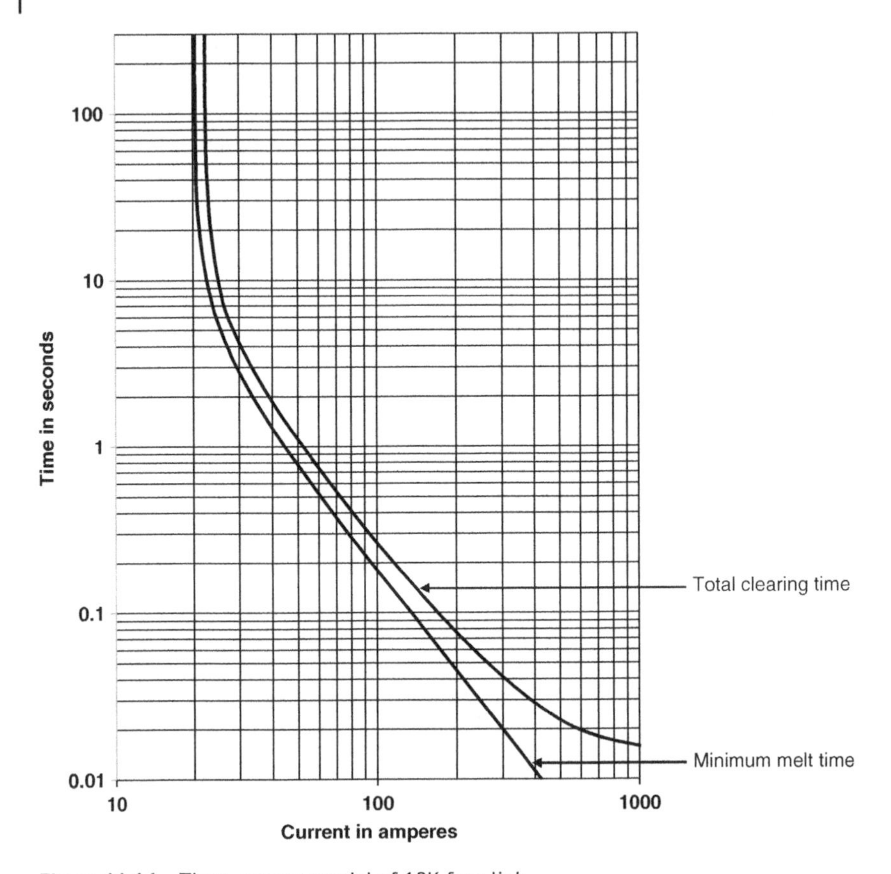

Figure 11.16 Time-current model of 10K fuse link.

operating conditions at the specific location. Note that the 10K fuse link starts to melt at 20 A, which is twice the rating of the fuse. As a general rule, fuse link rating must be greater than the full load value divided by 1.5. Therefore, 10K fuse link will be appropriate for full load less than 15 A.

Although general equations to model different fuse links are not available, some researchers have attempted to develop equations for fuse links [9]. Most often, fuse link characteristics are digitized for use in computer-based coordination software.

11.7.2.2 Rule for Fuse–Fuse Coordination

Figure 11.17 shows a simple radial feeder. For the fault at the location shown in the figure, to coordinate fuse links A and B successfully, the total clearing time

Figure 11.17 Fuse–fuse coordination of a simple radial system.

curve of B must be lower than the damaging time curve of fuse A within the desirable coordination current range. According to the damaging curve, as the name implies, the fuse deteriorates whenever the current reaches or exceeds the corresponding value. A generally acceptable method is that the total clearing time of B should not exceed 75% of the minimum melt time (MMT) of A. This coordination procedure can be further illustrated with the following example.

Example

Figure 11.18 shows part of a 12.47-kV distribution system with maximum fault, minimum fault, and full load current at the respective fuse link locations. We select fuse links Type T for locations A and B to achieve proper protection with coordination between the two fuse links.

Solution

As the first step, we select the fuse links based on the specified full load. For location B, the fuse link must have rating higher than $40/1.5 = 26.66$ A, and for location A, it must have a rating higher than $70/1.5 = 46.66$ A. A 30T fuse link for location B and a 50T for location A will work. However, we have to make sure that they follow the coordination rules. Figure 11.19 shows the MMC and TCC of the two selected fuse links. The critical current for coordination is 1794 A, which is the maximum fault current seen by both fuse links. The total clearing time of 30T is 0.037 seconds, and the MMT of 50T is 0.05 seconds at this current. The ratio of these two times is $0.037/0.05 = 0.74$, which is acceptable because this ratio must be lower than 0.75.

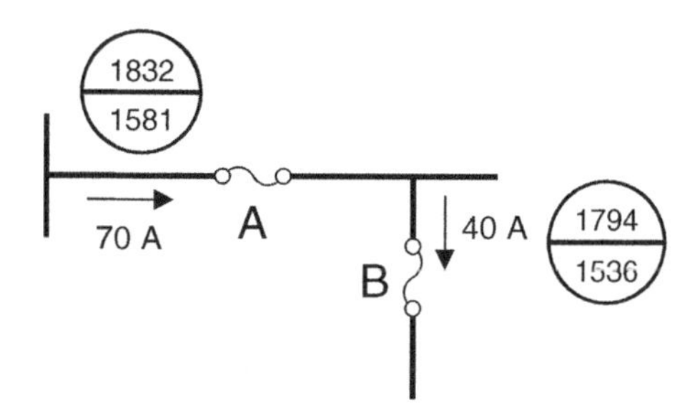

Figure 11.18 Part of a distribution system protected by fuse links. Source: Courtesy of Eaton Corporation.

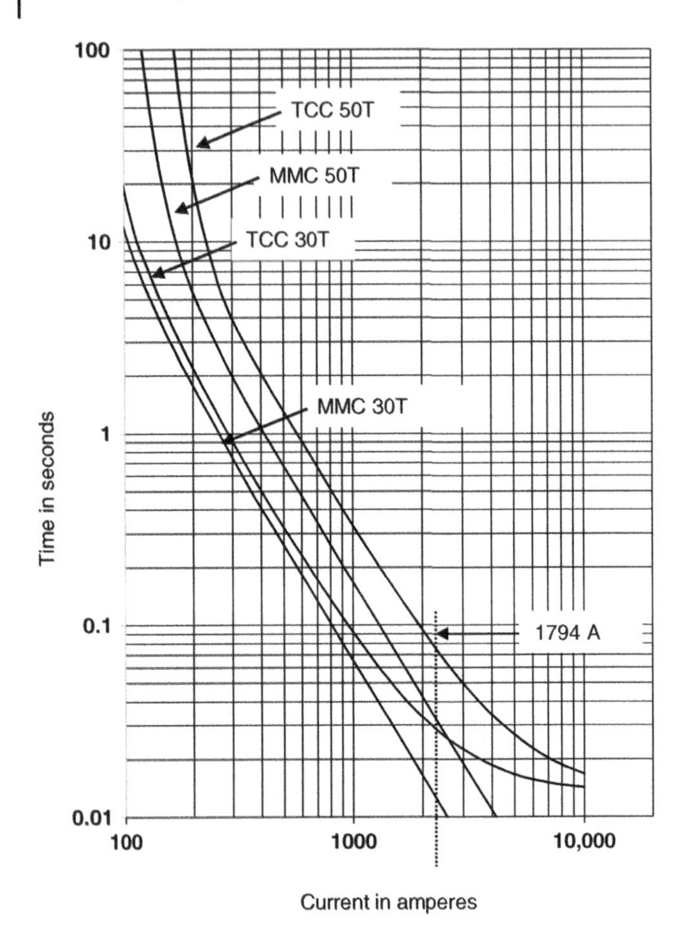

Figure 11.19 MMC and TCC curves of 30T and 50T fuse links and their coordination at the maximum fault of 1794 A. Source: Courtesy of Eaton Corporation.

11.7.3 Recloser–Fuse Coordination

Figure 11.20 shows a typical model for a recloser, which is time versus current curve. It has a fast curve A and time delay curve C. The curve in the middle is that of the downstream fuse link, shown in Figure 11.21. The fuse link curve intersects the slow and fast curves of the recloser at "*a*" and "*b*," which give the minimum fault current and the maximum fault current for which coordination between the recloser and the downstream fuse is required.

The coordination should ensure that for permanent faults downstream of the fuse, the fuse should clear the fault. Similarly, for a permanent fault upstream of the fuse, the recloser should open. But for a temporary fault, none of the devices

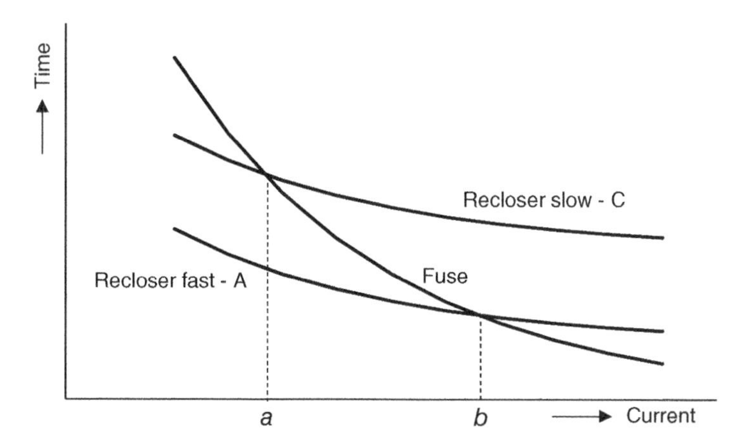

Figure 11.20 Time–current characteristic curve for a recloser and a downstream fuse link. Source: Courtesy of Eaton Corporation.

should stay open. Thus, for any fault downstream of the fuse, the recloser should operate on the fast curve first to open the recloser. This is followed by one reclosing and another fast-trip operation. If the fault is still there, the fast curve of the recloser is disabled, and the recloser follows two additional close and trip operations on the slow curve to allow the fuse to clear the fault if it is permanent and is downstream of the fuse. If the fault is permanent, but upstream of the fuse, the recloser clears it by locking out after operating twice on the slow curve.

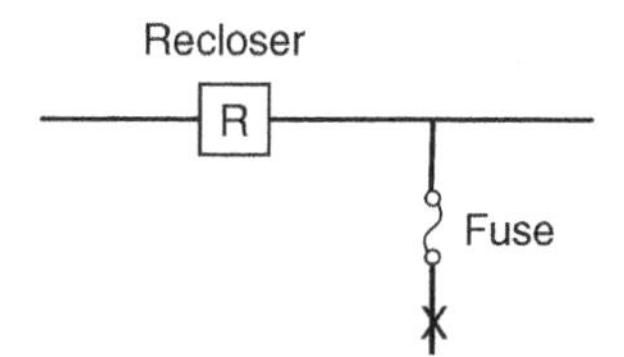

Figure 11.21 Recloser–fuse locations in a simple distribution feeder system.

Determining values of "*a*" and "*b*" is important for proper coordination between the fuse and the recloser. Since the operating characteristics of these devices are not very precise and may change due to the operating conditions, some adjustments are implemented to account for manufacturing tolerance and the cumulative temperature rise in the fuse link during the recloser operations. Hence, the recloser's curves are multiplied by a factor called *K-factor*. These factors are derived experimentally for different operating conditions, and a value of 1.35 is typically used for the fast curve of the recloser with an operating sequence of two fast and two slow operations. For an operating sequence of one fast and three slow operations, the multiplying factor reduces to 1.2 because the cumulative temperature rise is lower. Thus, for two fast and two slow operating sequences, the intersection of fuse damaging time curve (75% of minimal-melting-time curve) and fast-trip operation of the recloser scaled up to 1.35 gives the value of "*b*". The value of "*a*" is determined by the intersection

Table 11.2 Recloser K-factors for coordination with source-side fuse links.

Reclosing time in cycles	Two fast and two delayed sequences	One fast and three delayed sequences	Four delayed sequences
25	2.7	3.2	3.7
30	2.6	3.1	3.6
50	2.1	2.5	2.7
90	1.85	2.1	2.2
120	1.7	1.8	1.9
240	1.4	1.4	1.45
600	1.35	1.35	1.35

Source: Courtesy of Eaton Corporation.

Table 11.3 Recloser K-factors for coordination with load-side fuse links.

Reclosing time in cycles	Two fast operations	One fast operation
25–30	1.8	1.25
60	1.35	1.25
90	1.35	1.25
120	1.35	1.25

Source: Courtesy of Eaton Corporation.

of the maximum clearing curve of the fuse with the slow characteristics of the recloser.

Similarly, if there is a fuse upstream of the recloser, the multiplying factor must be used to scale up the recloser's slow characteristics. Hence, for the maximum fault at the recloser location, the minimum melting time of the fuse must be higher than the average clearing time of the recloser on the slow curve multiplied by the specified multiplying factor. The suggested multiplying factors range from 1.7 to 3.5 based on the recloser type, the reclosing time (30–120 cycles), the recloser operating curves, and the operating sequence. Tables 11.2 and 11.3 give recommended K-factor values for coordination with upstream and downstream fuse links.

We illustrate the local coordination philosophy with hierarchical approach for the feeder system with an example. The first step is to identify and enumerate all primary-backup device pairs in the given system. Then, the coordination procedure should be conducted using the coordination rules discussed previously.

Example

Consider the system shown in Figure 11.22. The objective is to select recloser settings, fuses F_1 and F_2, and the fuse upstream of the transformer for the given system. The maximum and the minimum fault currents at distinct locations are shown in circles drawn to the locations. The maximum load currents at distinct locations are shown next to the arrows.

Solution

Recloser Selection

Since the maximum load current at Bus 1 is 270 A, it is necessary to choose a set of reclosers with a maximum continuous current rating higher than 270 A. This is achieved by employing three single-phase reclosers type L with a trip coil rating of 280 A, which have a minimum trip rating of 280 A and interrupting capability of 4000 A. The interrupting capability is higher than the maximum fault current of 3000 A at Bus 1. Note that this selection limits the ability to increase the maximum load currents in the future without upgrading the reclosers.

Source-side Fuse and Recloser Coordination

The source-side fuse needs to be selected so that it does not melt for any fault currents on the load side of the recloser, which means that for the maximum fault current seen by the recloser (3000 A), the minimum melting time of the fuse must be greater than the clearing time of the recloser's delayed curve. Some other factors as discussed below must be considered to select this fuse.

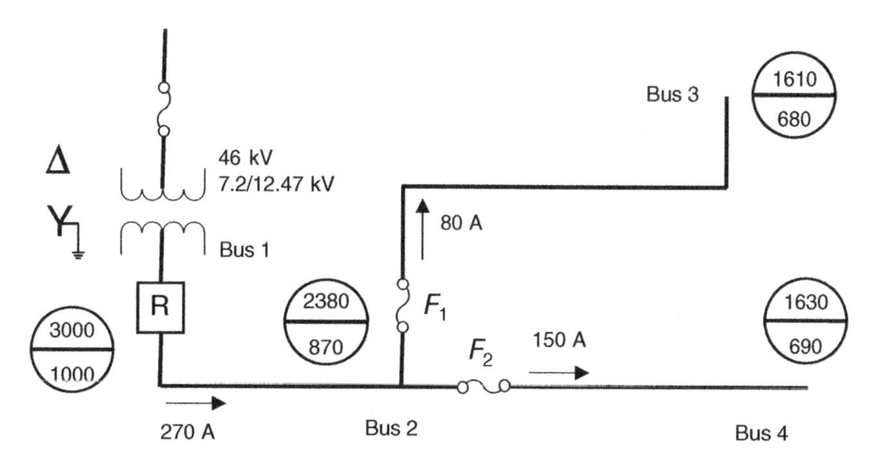

Figure 11.22 Example distribution system. Source: Courtesy of Eaton Corporation.

Transformer Turns Ratio

Since the source-side fuse is on the primary side of the transformer, the turns ratio of the transformer needs to be considered because currents from the low-voltage side have a smaller magnitude when seen from the high-voltage side since power across the transformer must be equal. Given that the transformer turns ratio is $N = 3.7$, the fuse time–current curves across the transformer have the following multiplying factors due to Δ-Y connection of the transformer:

Three-phase fault: $N = 3.7$
Phase-to-phase fault: $0.87N = 3.2$
Phase-to-ground fault: $1.73N = 6.4$

The phase-to-phase fault current has the lowest multiplying factor, and it is used as the limiting factor for the source-side fuse and recloser coordination because it results in the tightest requirement.

Continuous Load Current

The source-side fuse also needs to consider the normal operating conditions, that is the continuous peak load current. In this case, the maximum continuous load current is 270 A on the low side of the transformer, which corresponds to 73 A of current on the high side of the transformer. Generally, fuses start operating at currents that are greater than two times their rating. Thus, they can easily carry currents up to 1.5 times their rating. For instance, a 65E fuse link can carry up to 97.5 A of continuous current, and thus, it is suitable. For this example, we select a 46-kV 65E slow fuse link.

Recloser K-factor

The K-factor value for the recloser is found according to the type of the recloser sequence and the reclosing intervals. For the recloser sequence of two fast operations followed by two delay operations, a K-factor of 1.7 is used to scale the recloser's slow characteristics.

Selection

The original and the scaled versions of the fuse link and recloser curves are shown in Figure 11.23. Note that the operating time of the recloser's slow curve is multiplied by the K-factor, and the current is scaled by a factor of 3.2 for the fuse link curve. The figure also shows the fast curve of the recloser (A). From this figure, we can find the intersection of the MMC for a fuse link type 65ES and the modified C curve for the recloser at about 1800 A, which is lower than the maximum fault current value seen by the recloser. It is necessary to choose a different fuse link so that at a fault current of 3000 A, the gap between the MMC of the fuse link and the recloser operating time on the modified curve is larger than 0.5 seconds for proper coordination. Since the operating time of the recloser on the modified curve is 0.5 seconds at 3000 A, we need a fuse link with melt time greater than

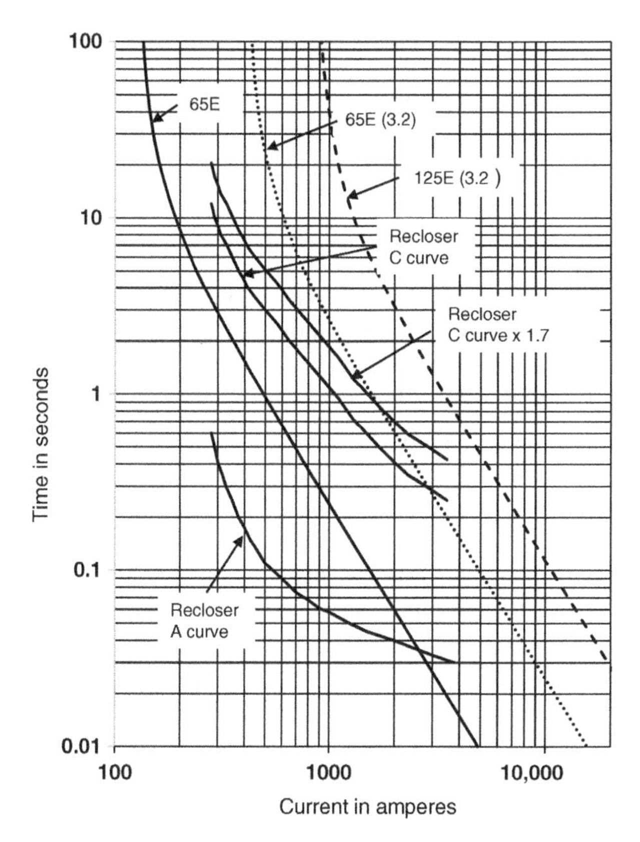

Figure 11.23 Operating time of L-type recloser and E-type slow upstream fuse links. Source: Courtesy of Eaton Corporation.

1 second (0.5 seconds + 0.5 seconds). Thus, with the adjustment factor of 3.2, we need a fuse link with MMT higher than one second for a fault current of 3000 A. The scaled MMC of slow 125E, shown in the figure, has a melt time slightly higher than one second for 3000 A. Therefore, fuse links higher than this will comply with this condition, and we choose 125E slow fuse link.

Load-side Fuse and Recloser Coordination

Coordination between fuses F_1 and F_2 and the recloser needs to be set up so that:

- In the event of a temporary fault between buses 2 and 3 or 4, neither F_1 nor F_2 burns out before the recloser's fast operation clears the fault.
- In the event of a permanent fault, fuses F_1 and F_2 clear the fault by melting before the recloser's delay operation happens.

Continuous Load Current

Since the recloser has already been selected, it is necessary to choose appropriate fuse links to achieve proper coordination. Considering the maximum continuous load currents, it is possible to find the minimum ratings of fuse links for F_1 and F_2. The load currents for F_1 and F_2 are 80 and 150 A, respectively. This means that for F_1 it is necessary to choose a rating of 65T or higher (80/1.5 = 53.333 A), and for F_2 it is necessary to choose 100T or higher (150/1.5 = 100 A). Like the slow curve, a K-factor must be applied to the recloser fast curve. The K-factor value of 1.35 is found from Table 11.3.

Minimum Melt Time (MMT)

The MMT of each of the fuses needs to be higher than the adjusted fast operation (curve A times 1.35) of the recloser between the minimum fault current (680 A for F_1 and 690 A for F_2) and the maximum fault current seen by the fuses (2380 A). This means that:

- F_1 and F_2 MMT should be higher than 0.05 seconds for a current of 2380 A. Figure 11.24 shows the A curve of the recloser times 1.35 and MMCs of 65T, 80T, and 100T fuse links. Both 80T and 100T meet the requirement, but 65T does not.
- F_1 and F_2 MMT should be higher than 0.1 seconds for currents of 680 and 690 A. All the fuse links meet the requirement.

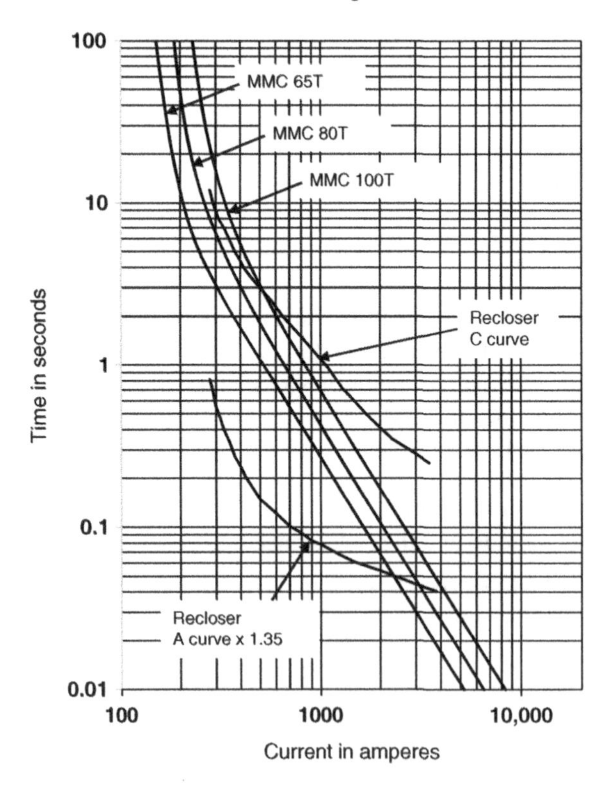

Figure 11.24 Adjusted fast curve (Curve A) of recloser and MMC of selected fuse links. Source: Courtesy of Eaton Corporation.

Maximum Clearing Time (MCT)

The maximum clearing time (MCT) of the fuses must be lower than the delayed operation (curve C) of the recloser for the maximum and minimum currents seen by the fuses. This means that:

- F_1 and F_2 MCT should be lower than 0.37 seconds approximately for a current of 2380 A. Figure 11.25 shows the C curve of the recloser and TCC of 65T, 80T, and 100T fuse links. All the fuse links meet the requirement.
- F_1 and F_2 MCT should be lower than 1.8 seconds approximately for currents of 680 and 690 A, respectively. Both 65T and 80T meet the requirement, but 100T does not.

Selection

The fuse selection should be made as the intersection of all the constraints mentioned above. The analysis shows that 80T is the only fuse that will meet all the requirements. While 65T was desired for F_1, increasing it to 80T will provide adequate protection. On the other hand, 100T is desired for F_2, but it must be reduced

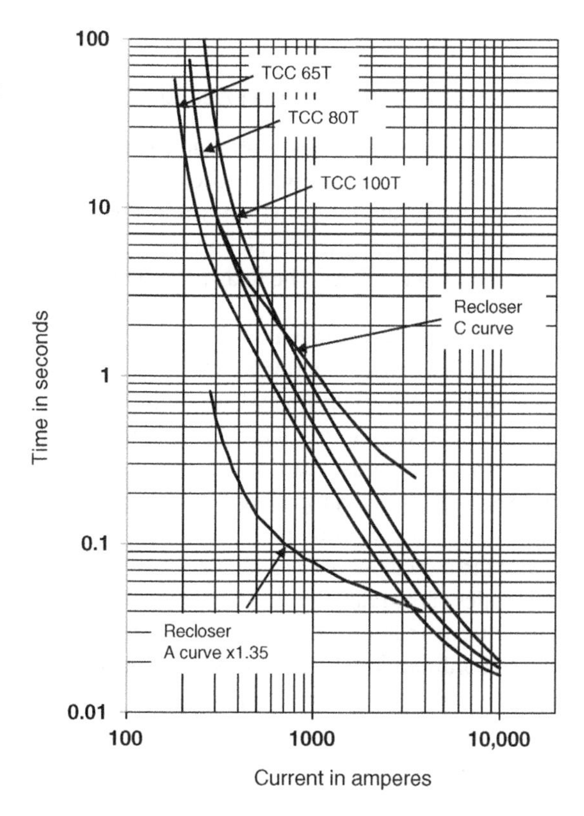

Figure 11.25 Slow curve (Curve C) of recloser and TCC of selected fuse links. Source: Courtesy of Eaton Corporation.

Figure 11.26 Basic sectionalizer–recloser coordination. Source: Courtesy of Eaton Corporation.

to 80T. 80T will coordinate with the recloser, but it could melt under maximum loading conditions.

11.7.4 Recloser–Sectionalizer Coordination

11.7.4.1 Rule for Coordination

The sectionalizer is set for one less operational count than the immediate recloser on the source side, as depicted in Figure 11.26. In this specific case, the recloser is set for four operations, and the sectionalizer should be set for three operations. The continuous current rating of the both the devices should be the same. The actuating current of the sectionalizer is set to 80% of the recloser minimum trip rating.

11.7.5 Circuit Breaker–Recloser Coordination

11.7.5.1 Models for Relay-controlled Circuit Breakers

The overcurrent relay-controlled circuit breakers have a time delay unit with characteristics similar to that shown in Figure 11.11 and an instantaneous unit. The relay characteristics, its pickup value, and TD are selected based on the given load and fault currents. The relay type CO-8 is used often.

11.7.5.2 Rule for Coordination

With a circuit breaker acting as a backup device to a recloser or any other nonreclosing device as the primary, one should make sure that the relay time–current characteristics of the circuit breaker are much above the primary device characteristics. In other words, the relay's pickup and time-dial settings must be chosen to have enough safety margin with the recloser characteristics selected. More specifically, the instantaneous relay characteristics should have proper safety margin above the fast curves of the recloser or other devices. Similarly, the time delay characteristics of the relay should allow the slow curves of the recloser to act first. Static overcurrent relays do not have issues of overtravel and coasting, which makes their coordination with reclosers easier. However, appropriate margins must be included if the relays are of mechanical type. While these are the general rules, the following examples will clarify them much better. The reader is reminded that many practical details can be found in [2, 6].

Example

Figure 11.27 shows an example of a distribution feeder protected by a recloser and a circuit breaker along with the minimum and maximum faults. Find settings of the relay for the CB to coordinate with the recloser.

Solution

The first step is to select a recloser. Since the full load current is 110 A, a recloser with 140 A trip coil setting is appropriate. Figure 11.28 shows the recloser curves.

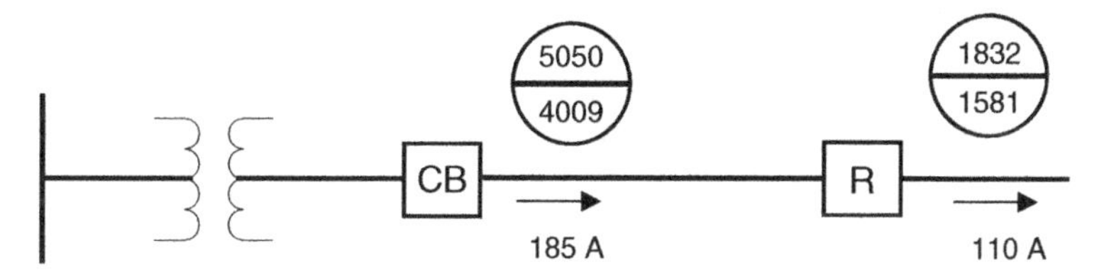

Figure 11.27 Main feeder of a distribution system. Source: Courtesy of Eaton Corporation.

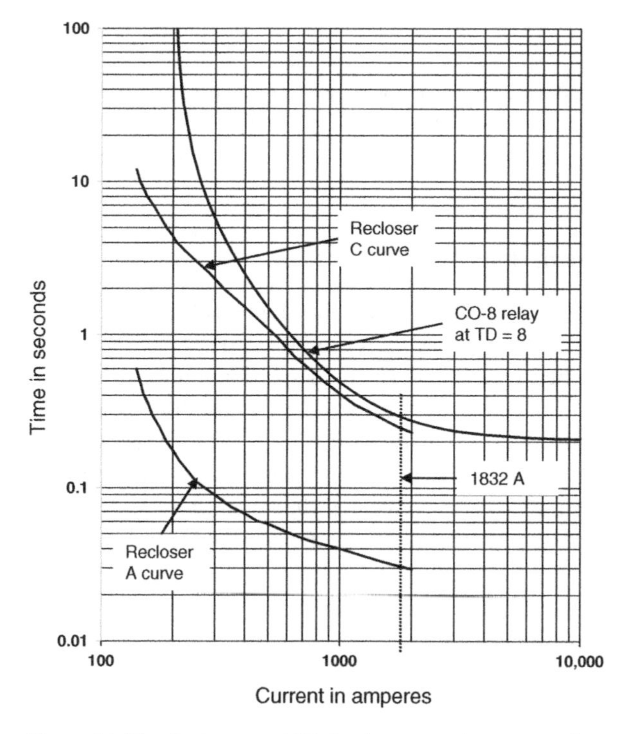

Figure 11.28 Recloser and CO-8 relay-operating curves. Source: Courtesy of Eaton Corporation.

We select a CO-8 relay for the circuit breaker. For a load current of 185 A, 200 : 5 CT will work. Further, we select 5 A as the TS for the CO-8 relay. This will allow some margin for overload on the feeder. Now, we have to select the TD setting for the relay to coordinate with the recloser. At 1832 A, which is the largest fault current seen by both the recloser and the circuit breaker, the recloser takes 0.24 seconds to operate on the slow curve (curve C). We want the relay to take higher time than that at the same current. We compute the MTS for this current, which is

$$\mathrm{MTS} = \frac{I_{\mathrm{fault}}}{(\mathrm{CT\ Ratio})(\mathrm{Tap\ Setting})} = \frac{1832}{\left(\frac{200}{5}\right)(5)} = 9.16$$

CO-8 relay with TD of 8 takes 0.3 seconds to operate, which provides sufficient margin between the operating time of the recloser and the circuit breaker.

11.8 New Digital Sensing and Measuring Devices

While standard instrument transformers namely CTs and VTs will continue to exist, all stakeholders are making a tremendous effort to capitalize on new digital technologies for improving the real-time sensing and monitoring for achieving truly smart grid systems. This is a key technology required for achieving the real-time protection and control of intelligent distribution systems. The U.S. Department of Energy is leading this development as part of the grid modernization effort.

11.8.1 Phasor Measurement Units (PMUs)

Synchrophasors are time-synchronized numbers that represent both the magnitude and phase angle of the sine waves found in ac electricity and are time synchronized for accuracy. They are measured by high-speed monitors called phasor measurement units (PMUs), shown in Figure 11.29, that are 100 times faster than supervisory control and data acquisition (SCADA). PMU measurements record grid conditions with great accuracy and offer insight into grid stability or stress. Synchrophasor technology is used for real-time operations and off-line engineering analyses to improve grid reliability and efficiency and lower operating costs.

11.8.2 Microphasor Measurement Units

PMUs installed in the power grid are currently positioned on the transmission system or in substations. A PMU creating real-time synchrophasor data from the consumer voltage level, called μPMUs, could provide new insights into modern power systems. An example of μPMUs is shown in Figure 11.30. These units can

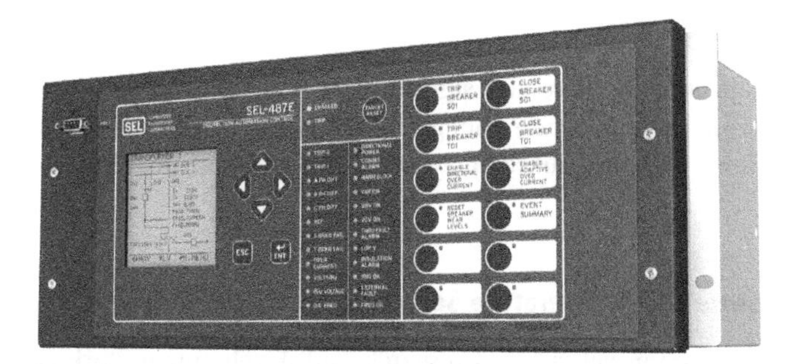

Figure 11.29 Phasor measurement unit. Source: Courtesy of Schweitzer Engineering Laboratories.

Figure 11.30 A microphasor measurement unit. Source: Courtesy of Powerside.

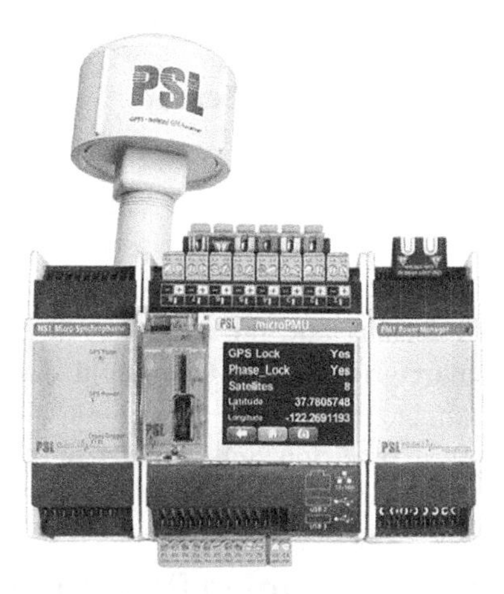

be created more cheaply, an order of magnitude less, than current commercial PMUs. For this reason, many more PMUs could be deployed and provide a much higher resolution of the distribution grid. There are many new applications for such a visible grid in postmortem event analysis and identification as well as near real-time monitoring. One such application is the fault location on a distribution system or microgrid with μPMUs, which provides accurate results.

11.8.3 Optical Line Current Sensors

These are directly hung from an overhead line to obtain direct, digital current signals in each phase. These pole-top units use Faraday effect to get accurate and precise current measurements of line currents. Thus, three sensors will be required for a three-phase system. These can be used for overhead lines from 120 V up to 34.5 kV levels. These could also facilitate power quality measurements via a power quality (PQ) meter. Soon, it is likely to be adopted for underground systems.

11.8.4 Optical Voltage Sensors

These utilize Pockels effect to derive voltage signals. Here also, three sensors will be required to seek digital and accurate voltage values from each of the three phases in overhead lines first and subsequently for underground distribution systems.

11.8.5 Digital Pressure and Temperature Sensors

Though these are not common yet, a lot of research and development work is underway in developing these newer sensors by both utilities and equipment manufacturers. It is a matter of time before they become a reality.

11.8.6 Evolving Sensors

Other developing state-of-the art-sensors such as extremely high-frequency point on wave, dynamic high-range sensors, optical PMU, and MagSense should be considered for adoption as time passes on. As nanotechnology matures, it is possible to envisage the development of newer sensors such as nanosensors for all sensing and measurement of electrical and nonelectrical quantities in the smart grid system. One could even expect genetic sensors such as bacterial nanobionics sensors to penetrate the future distribution systems to improve the speed, precision, and accuracy of the above mentioned electrical variables.

11.9 Emerging Protection System Design and Coordination

While the conventional approaches for protecting distribution systems have worked well and will continue to work, we should be aware of the changes that are taking place, which could make the present practices obsolete. The most important point is that the distribution systems have been designed to operate in a radial manner, and subsequently, all the protection schemes consider radial topology. However, over the past 20 years we are witnessing increased deployment of distributed energy resources (DER) in distribution systems. Most of these resources are either renewable resources, such as solar- or wind-based generation, or battery-based energy storage. All of them are interconnected to the system through inverters. By design, the inverters have current-limiting features, which limit the currents to small values in the event of fault. So, it is difficult to detect faults in systems powered by inverter-based resources (IBRs). Further, in addition to current flowing from substation to fault, it flows from the distributed resources toward the fault, which further complicates protection. These issues are very challenging and open to further research. While there are no standard approaches available to deal with emerging issues, reference [10] highlights some possible solutions.

With all the remarkable system changes that have occurred in the past 20 years and more expected changes in the future, the distribution system protection design and coordination deserve a more radical approach. The authors believe that future protection schemes will rely on advanced methodologies to detect faults fed by IBRs and communication between critical parts of the systems for coordination. Further, computer models for all protective devices will have to be implemented to select device settings, and the device coordination process will have to be adaptive and automated. These can be done for local coordination of any two adjacent devices or the global coordination using effective computer communication systems. This approach is valid for systems of any size that is at the feeder level, substation level, or even for several substations at a control center level. This approach can be applied for grid and microgrid systems and their transitions from one to the other. Finally, it should be an effective tool for both planning and automation of any given distribution system.

Problems

11.1 Fill in the blanks in Table 11.4 for the relay shown in Figure 11.31 with characteristics shown in Figure 11.11.

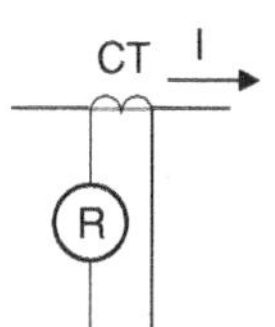

Figure 11.31 Overcurrent relay with current transformer (CT) for Problem 11.1.

Table 11.4 Data for Problem 11.1.

I (A)	CT ratio	Tap setting	TD setting	Op. time (s)
4500	300 : 5	3	3	
3840	400 : 5	8		0.5
2000	500 : 5		1	0.08
1600		4	10	3.0
	300 : 5	10	2	0.1

11.2 Figure 11.32 shows part of a 12.47-kV distribution system with maximum fault, minimum fault, and full load current at the respective fuse link locations. Select fuse link Type T for locations A, B, and C to achieve proper protection with coordination between the fuse links. Draw plots of the selected fuse links to verify coordination between them.

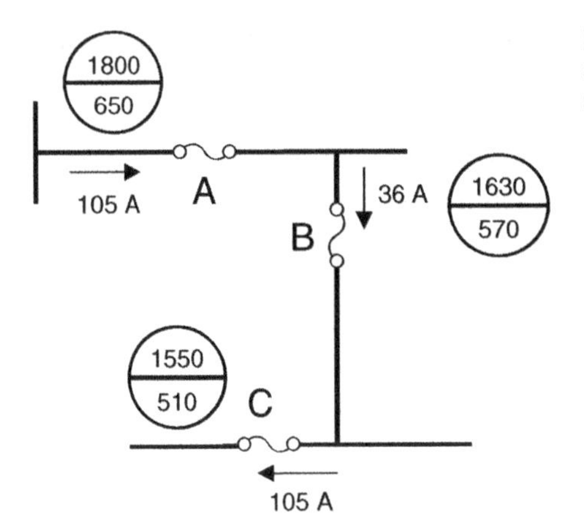

Figure 11.32 Part of a distribution system for Problem 11.2. Source: Courtesy of Eaton Corporation.

11.3 Select the pickup current of the L-type recloser for the system given in Figure 11.33. The recloser has available pickup currents of 100, 140, 200, and 280 A with symmetrical interrupting ratings of 4000 A at 14.4 kV. The operating curves A and C of the recloser are the same as those shown in Figure 11.28 but will be shifted based on the selected pickup current.

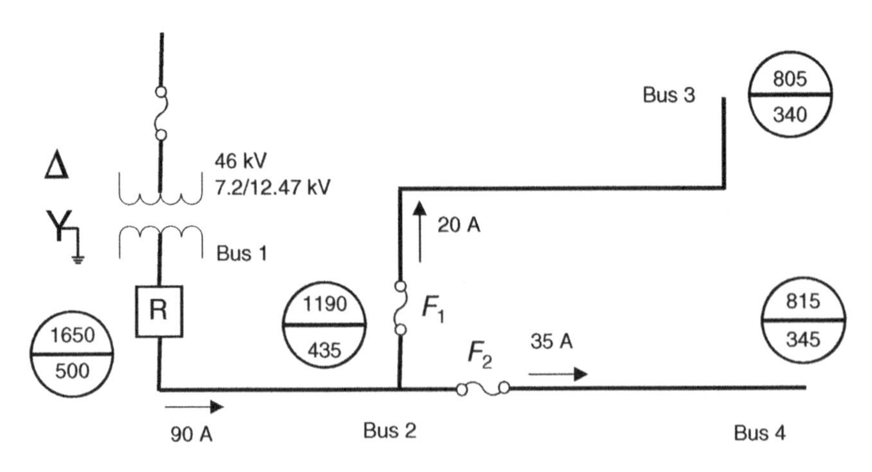

Figure 11.33 Distribution system for Problem 11.3. Source: Courtesy of Eaton Corporation.

The recloser is set for 2A2C operation with reclosing time of 90 cycles. Select the fuse link on the primary side of the transformer and fuse links F_1 and F_2 to coordinate with the recloser. Draw plots similar to that of Figures 11.23–11.25 to verify coordination.

11.4 Figure 11.34 shows part of a 12.47-kV system with maximum fault current, minimum fault current, and maximum load current for the circuit breaker and recloser. Select recloser setting and CT ratio, TS, and TD setting for the CO-8 relay for the CB. Draw plots of the recloser and CO-8 relay to verify coordination.

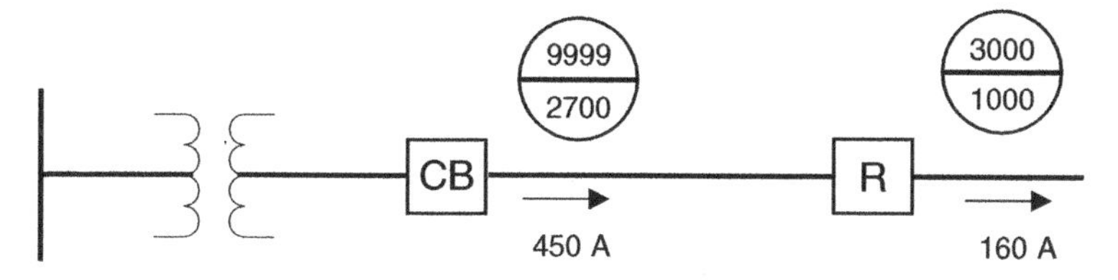

Figure 11.34 Distribution system for Problem 11.4.

References

1 Russell Mason, C. (1956). *The Art and Science of Protective Relaying*. New York: Wiley.
2 (1990). *Electrical Distribution-System Protection*, 3e. Cooper Power Systems.
3 Blackburn, J.L. (1979). *Applied Protective Relaying*. Coral Springs, FL: Westinghouse Electric Corporation.
4 IEEE (1979). Tutorial Course on *Surge Protection in Power Systems*, 79 EHO 144-6-PWR, IEEE, New York.
5 Sakis Meliopoulos, A.P. (1988). *Power System Grounding and Transients*. Marcel Dekker Inc.
6 Short, A. (2004). *Electric Power Distribution Handbook*. CRC Press.
7 Abdelmoumene, A. and Bentarzi, H. (2014). A review on protective relays' developments and trends. *Journal of Energy in Southern Africa* 25 (2): 91–95.
8 (2011). *Network Protection & Automation Guide*. Alstom Grid. ISBN: 978-0-9568678-0-3.
9 Tian, W., Lei, C., Zhang, Y. et al. (2016).Data analysis and optimal specification of fuse model for fault study in power systems. *IEEE PES General Meeting*, Boston, MA (July 2016). https://doi.org/10.1109/PESGM.2016.7741864.
10 Reno, M.J., Brahma, S., Bidram, A., and Ropp, M.E. (2021). Influence of inverter-based resources on microgrid protection: part 1: microgrids in radial distribution systems. *IEEE Power and Energy Magazine* 19 (3): 36–46.

12

Power Quality for Distribution System

12.1 Definition of Power Quality

Power quality (PQ) is related to receiving distortion-free electric supply, which is predominantly alternating current (ac). PQ, therefore, deals with the performance of signals in a distribution system or a portion of it [1–4]. While it is hard to maintain ideal signals for quantities such as voltage and current as perfect sinusoidal waves at constant frequency (60 Hz in North America) and/or perfect direct current (dc) signals at constant magnitude, all efforts must be made to keep the behavior of signals as close to the ideal conditions as possible for any given distribution system. Various issues that cause deviation from ideal conditions include harmonics, flicker, momentary events, noise, voltage fluctuations, and outages. To conduct the PQ analysis, the use of Fourier series, Fourier transforms, and other signal analysis background is of paramount importance [5, 6]. Distortion in voltage and current waveforms recorded (the highly distorted waveform is the current and less distorted waveform is the voltage) at a veneer plant is shown in Figure 12.1 to illustrate the PQ problem.

To assure the highest level of performance of a distribution system, the PQ assessment is always needed since it is hard, if not impossible, to realize the ideal conditions mentioned above. The best and practical approach is to make sure that the system performs as close to the ideal conditions as possible. For this reason, standards such as IEEE-Std 519 [7] have been developed to meet the stipulated levels of performance. Since many of the harmonic-related problems inject a periodic current signal, Fourier series is a powerful tool to determine the components of different frequencies in the signal.

Electric Power and Energy Distribution Systems: Models, Methods, and Applications, First Edition.
Subrahmanyam S. Venkata and Anil Pahwa.

Companion website: www.wiley.com/go/Pahwa/ElectricPowerDistributionSystems

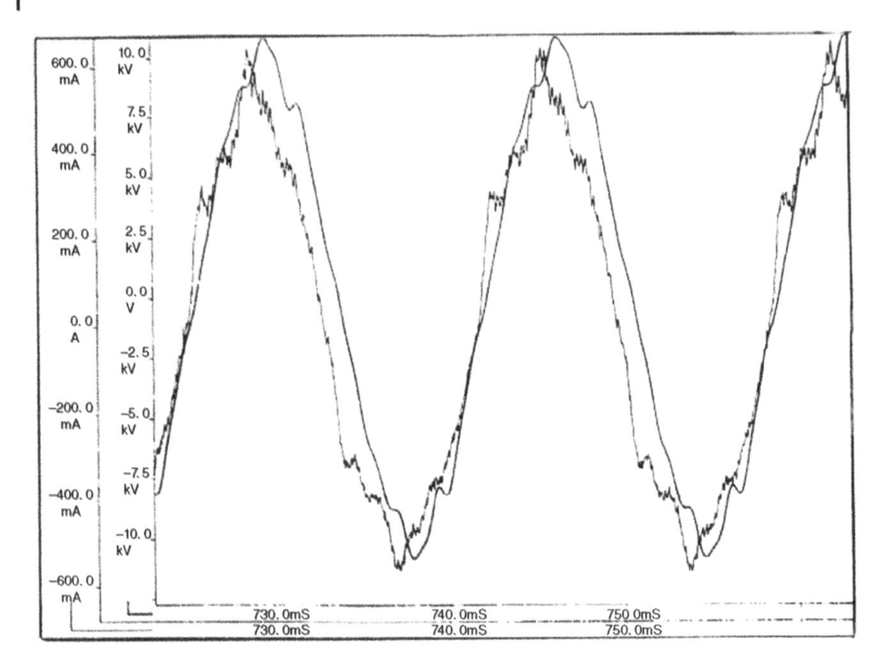

Figure 12.1 Voltage and current waveforms recorded at a veneer plant.

12.2 Impacts of Power Quality

12.2.1 The Customer Side

Computers and communication equipment are susceptible to power system disturbances, which can lead to loss of data and to erratic operation. Automated manufacturing processes such as paper-making machinery and chipmaking assembly lines can shut down even in the case of short voltage sags. Induction and synchronous motors can have excessive losses and heating due to harmonics. Home electronic equipment is vulnerable to PQ problems: for example, blinking digital clocks due to interruption of power, and flickering lights. Equipment and process control malfunction translates to high expense for replacement parts and also for downtime, with adverse effects on profitability and product quality. While passive loads, such as incandescent lights and motors, do not create any harmonics when supplied by a sinusoidal voltage source, contemporary loads in houses, such as fluorescent lights, light emitting diode (LED) lights, computers, and other electronic equipment, create current harmonics. Several of these loads depend on the conversion of ac supply voltage to dc for operation. This conversion process is the primary source of harmonics. Typically, harmonic problems are primarily associated with process-oriented industries; residential customers seldom experience any harmonic-related problems.

12.2.2 The Utility Side

Poor PQ can cause failure of power-factor correction capacitors due to the resonance condition. It also causes increased losses in cables, transformers, and conductors, especially neutral wires; errors in energy meters, which are calibrated to operate under sinusoidal conditions; and incorrect operation of protective relays, particularly in solid-state and microprocessor-controlled systems. PQ issues result not only in unhappy customers but also in malfunction and failure of system components and control systems, with adverse impact on profitability.

12.2.3 Importance of Power Quality

PQ is important because it affects both the utilities as suppliers and the customers as users in many ways. It increases losses on equipment, with increased loss of life of equipment, and higher downtime. It causes metering errors and creates electromagnetic compatibility (EMC) issues including telephone interference and computer interference. Subsequently, it degrades the quality of service to customers, increases cost, and decreases competitiveness of utilities.

12.2.4 Cost of Power Quality

The cost of poor PQ is determined by adding the cost of actions taken to improve PQ, the cost of customer losses in industrial production, and the payment to customers for improving PQ problems. In addition, there is cost to the utility due to higher energy loss, extra cost to serve higher peak load, potential loss of revenue due to metering errors, and cost associated with reduced life of equipment. According to an estimate by Electric Power Research Institute in the U. S. (EPRI) done in 2020, between $120B and $188B are spent annually for addressing the PQ problems. Hence, it is important to address PQ issues to reduce cost and improve the quality of service to customers.

12.3 Harmonics and PQ Indices

12.3.1 Total Harmonic Distortion (THD)

For periodic waves of period T $\left(\omega_0 = \frac{2\pi}{T}\ rad/s \text{ or } f_0 = \frac{1}{T}\text{Hz}\right)$, the most widely used measure of PQ in North America is the total harmonic distortion (THD). It is usually expressed in percent and can be calculated for either current or voltage. For balanced three-phase voltages, L–N voltages are used in the formula, but for unbalanced voltages, the THD will be different in each phase. For aperiodic signals, THD is not defined. However, quasi-periodic signals (i.e. signals with discrete spectra – but which may not be periodic) can be handled analogously to periodic signals.

It is mathematically defined as

$$\mathrm{THD} = \frac{\left(\sqrt{\sum_{i=2}^{\infty}(I_i)^2}\right)}{I_1} \tag{12.1}$$

where I_i is the root-mean-square (RMS) value of the ith harmonic for $i = 2$ to ∞, and I_1 is the RMS value of the fundamental component of the signal. Typically, a predefined value n is chosen to truncate the series, which is dependent on the analysis under consideration. It is a general-purpose index and is commonly used in IEEE and IEC standards. Although the general definition of THD includes all harmonics, it is theoretically impossible to have even harmonics in voltage and current in power systems because they are symmetric about the time axis.

12.3.1.1 Properties of THD

THD is zero for a perfectly sinusoidal waveform. It becomes indefinitely large as distortion increases. Also, if the fundamental term has amplitude zero, THD becomes infinity. For example, let $v(t) = \cos(5t) + \cos(7t)$. Let ω_0 be the fundamental frequency. Therefore, $5 = k\omega_0$ and $7 = m\omega_0$, and $(5/7) = (k/m)$. Subsequently, the smallest integer solution is $k = 5$ and $m = 7$, and $\omega_0 = 1$ rad/s. Hence, the Fourier series of $v(t)$ is $v(t) = 0{*}\cos(t) + \cos(5t) + \cos(7t)$. Since the amplitude of the fundamental component, which is the denominator in Eq. (12.1), is zero, THD becomes infinity. On the other hand, if a single harmonic of frequency h dominates the frequency spectrum of a signal above ω_0, the THD is given by I_h/I_1.

A commonly cited figure of 5% THD is the dividing line between the high and low harmonic distortions. One should, however, take care in using the figure of 5% distortion. It is too high for subtransmission and transmission circuits but could be readily tolerated in distribution circuits.

Example

Find THD for the voltage and current given below at a location in a power system.

$$v(t) = \cos(\omega t + 30^\circ) + 0.3\cos(3\omega t + 60^\circ) + 0.2\cos(5\omega t + 10^\circ)\ \mathrm{V}$$

and

$$i(t) = 2\cos(wt) + \cos(3\omega t + 15^\circ) + 0.5\cos(5\omega t - 50^\circ)\ \mathrm{A}$$

Solution

$$\mathrm{VTHD} = \frac{\sqrt{(0.3)^2 + (0.2)^2}}{1} = 0.3605 \text{ or } 36.05\%$$

$$\mathrm{ITHD} = \sqrt{(1)^2 + (0.5)^2}/2 = 0.559 \text{ or } 55.9\%$$

12.3.2 Total Demand Distortion (TDD)

Total demand distortion is a measure of the THD considering the circuit rating. As circuit rating increases for a fixed load current, TDD drops.

$$\text{TDD} = \text{THD}\frac{I_1}{\text{Circuit Rating}} \tag{12.2}$$

12.3.3 Power Factor (PF)

The power factor in the presence of harmonics is different from the power factor and the fundamental frequency. The power factor with harmonics is a commonly used metric for PQ and is defined as

$$\text{PF}_h = \frac{P_{\text{tot}}}{|V_{\text{rms}}|\,|\,I_{\text{rms}}\,|} \tag{12.3}$$

where P_{tot} is the total active power including all harmonics; $|V_{\text{rms}}|$ is the effective rms value of the voltage; and $|\,I_{\text{rms}}|$ is the effective rms value of the current.

Since voltage and current have signals of other frequencies in addition to the fundamental frequency, we use Parseval's theorem to determine the RMS values. According to this theorem

$$V_{\text{rms}} = \sqrt{V_{1\text{rms}}^2 + V_{2\text{rms}}^2 + V_{3\text{rms}}^2 + \dots} \tag{12.4}$$

If the voltage has multiple terms for the same frequency, they must be combined before using the above equation. The same equation can also be used for currents.

Examples

Find V_{rms} for the voltage given below.

Solution

$$v(t) = 3\cos\omega t + 4\sin(\omega t)\text{V}$$

$$v(t) = 3\cos\omega t + 4\cos(\omega t + 90^\circ)$$

$$= 5\cos(\omega t + 53^\circ)$$

$$V_{\text{rms}} = \frac{5}{\sqrt{2}}$$

Consider voltage and current given below. Find the harmonic power factor.

$$v(t) = \cos(\omega t + 30^\circ) + 0.3\cos(3\omega t + 60^\circ) + 0.2\,\cos(5\omega t + 10^\circ)\,\text{V}$$

and

$$i(t) = 2\,\cos(\omega t) + 1\cos(3\omega t + 15^\circ) + 0.5\,\cos(5\omega t - 50^\circ)\,\text{A}$$

Solution

PF at fundamental frequency = cos(30° − 0) = 0.866 lagging

Compute real powers for different frequencies:

$$P1 = (1)(2)\ \cos 30^\circ = 1.732\ \text{W}$$

$$P3 = (0.3)(1)\ \cos 45^\circ = 0.2121\ \text{W}$$

$$P5 = (0.2)(0.5)\ \cos 60^\circ = 0.05\ \text{W}$$

$$P_{\text{tot}} = \sum P = 1.732 + 0.2121 + 0.05 = 1.9941\ \text{W}$$

$$V_{\text{rms}} = \sqrt{(1)^2 + (0.3)^2 + (0.2)^2} = 1.063\ \text{V}$$

$$I_{\text{rms}} = \sqrt{(2)^2 + (1)^2 + (0.5)^2} = 2.291\ \text{A}$$

Therefore, the power factor with harmonics is

$$\text{PF}_h = \frac{P_{\text{tot}}}{|V_{\text{rms}}||I_{\text{rms}}|} = \frac{1.9941}{(1.063)(2.291)} = 0.8188$$

Note that $\text{PF}_h \leq \text{PF}$. Although customers prefer to use PF, the losses are more closely associated with the PF_h. Therefore, the utilities prefer to use PF_h.

12.3.4 Standards for Harmonic Control

Various international organizations have developed PQ standards. IEEE has several standards including IEEE Std 519-2014 [7], which is dedicated to harmonic control in electric power systems. The philosophy of this standard is that the utility is responsible for maintaining the quality of voltage waveform, and the customer is responsible for limiting harmonic currents injected onto the power system. Accordingly, the standard specifies voltage and current distortion limits at the point of common coupling.

The recommended distortion for bus voltage below 1 kV is 5% for individual harmonic and 8% for THD. For buses above 1 kV but below 69 kV, the limits are 3% for individual harmonic and 5% for THD. These values reduce for higher voltages. For systems of voltages rated 120 V to 69 kV with the ratio of short-circuit current to maximum load demand current (short-circuit ratio or SCR) at fundamental frequency less than 20, the limits as percentages of maximum demand load current are specified in Table 12.1 for different odd harmonics. The maximum demand load current is defined as the sum of the currents corresponding to the maximum demand during each of the 12 previous months divided by 12. For systems with higher SCR ratios, higher current distortion values are acceptable. This is due to the fact that higher short-circuit current implies lower equivalent source impedance, which will result in lower voltage drop and thus lower voltage distortion at the point of common coupling (PCC).

Table 12.1 Current distortion limits for systems rated 120V through 69 kV for SCR ratio less than 20.

Harmonic order (odd harmonics)	Maximum % distortion
3–9	4.0
11–15	2.0
17–21	1.5
23–33	0.6
35–50	0.3
TDD	5.0

Source: Adapted from Ref. [7].

Even harmonics are limited to 25% of the odd harmonic limits as specified in Table 12.1. Current distortions that result in a dc offset, such as half-wave rectifiers, are not allowed. IEEE Std 519-2014 also provides values for maximum allowed distortion for systems with higher short-circuit current to maximum load demand current. The readers are referred to this specific standard for additional information.

Examples

1. A distribution system has a short-circuit impedance $0 + j\,0.05$ per unit on a 0.333-MVA, 7800-V base (this is a single-phase system). If a load effectively injects 1 A at the fifth harmonic into the bus, estimate the fifth harmonic voltage that results in the system.

Solution

$$I_{\text{base}} = \frac{S_{\text{base}}}{V_{\text{base}}} = \frac{0.333 \times 10^6}{7800} = 42.69\ \text{A}$$

$$1\ \text{A} = \frac{1}{42.69} = 0.023\ \text{pu}$$

$$|V| = |Z||I| = 5(|j0.05|)(0.023) = 0.0058\ \text{pu}$$

Note that we have used a multiplier of 5 to scale the impedance because the frequency is five times that of the fundamental.

$$|V| = 0.0058 \times 7800 = 44.85\ \text{V}$$

2. A residence has a dedicated single-phase distribution transformer rated 27.5 kVA, 12.47 kV/120 V, and 8.1% reactance. Estimate the maximum third harmonic current that can be taken from the transformer by a nonlinear load in this residence. Consider the impedance of the feeder to be $0.0508 + j49.230\ \Omega$

and maximum load demand current to be 60% of the current at rated value and use reasonable assumption for the SCR.

$$Z_{base} = \frac{(120)^2}{27.5 \times 10^3} = 0.5236\ \Omega$$

For the transformer, $X_{actual} = 0.081 \times 0.5236 = 0.0424\ \Omega$.

Convert the impedance of the feeder from the high side to the low side, or

$$Z' = \frac{0.0508 + j49.23}{(12.47 \times 10^3/120)^2} = 4.7 \times 10^{-6} + j0.0045\ \Omega$$

Neglect the resistance because it is too small. Therefore,

$$Z'_{tot} = j0.0045 + j0.0424 = j0.0469\ \Omega$$

or $$Z'_{tot} = j\frac{0.0469}{0.5236} = j0.0895\ \text{pu}$$

Therefore, $I_{SC} = \frac{1}{0.0895} = 11.164$ pu

and $I_L = 0.6$ pu as specified

This gives $\frac{I_{SC}}{I_L} = \frac{11.164}{0.6} = 18.6$, which is less than 20. Therefore, from Table 12.1, the limit for the third harmonic current is 4%.

$$I_L = 0.6\frac{27.5 \times 10^3}{120} = 137.5\ \text{A}$$

Therefore, the third harmonic current must be less than $0.04 \times 137.5 = 5.5$ A.

12.4 Momentary Interruptions

Momentary interruptions occur in distributions systems due to the operation of reclosers or reclosing circuit breakers while trying to clear temporary faults and fuse-saving operation. While momentary interruptions are mainly considered a reliability issue, they are also a PQ issue because they cause inconvenience to customers by requiring them to reset clocks, switch off computers, and reset industrial process equipment. For reliability consideration, the interruption duration for a momentary interruption is defined as five minutes, but for PQ, the duration is much smaller. IEEE Standard 519-1995 provides a PQ definition of momentary interruptions as

> A type of short duration variation. The complete loss of voltage (<0.1 pu) on one or more phases for a time period between 0.5 cycles and 3 seconds.

12.5 Voltage Sag and Swell

12.5.1 Definition

Voltage sags are defined as reduction in rms value of voltage for a duration up to a few seconds. Figure 12.2 illustrates an example of voltage sag. Sags are usually

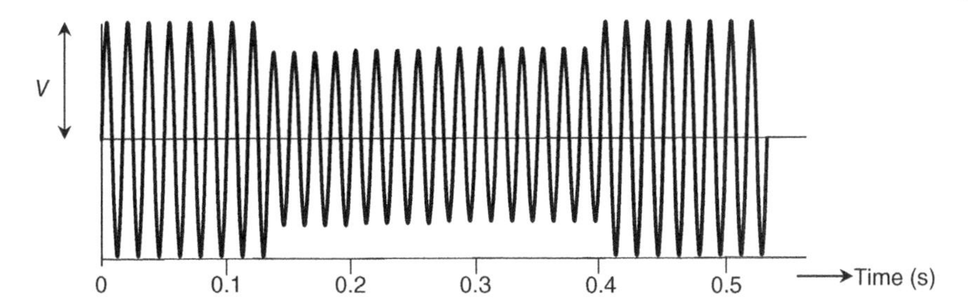

Figure 12.2 Example of voltage sag.

caused by starting of heavy loads or faults in the system. The duration of a voltage sag is the time taken by the system to return to a normal state after the switched load has stabilized or by the protective device to clear the fault. While most of the equipment owned by customers can ride through short duration sags, long duration sags can be detrimental to equipment. Sags can also have a transient in the beginning in some cases. These transients are not shown in Figure 12.2.

Swell could happen when a large load is removed from the system. It could also happen at locations where a distributed energy resources (DER) is connected to the system. In addition, lightning strikes on the system and capacitor switching cause voltage swells.

A PQ index called SARFI (system average RMS (variation) frequency index) is typically used to account for sag and swell events. According to an EPRI report [8], it is defined as

> SARFI_x represents the average number of specified rms variation measurement events that occurred over the assessment period per customer served, where the specified disturbances are those with a magnitude less than X for sags or a magnitude than X for swells.

$$\text{SARFI}_X = \frac{\sum N_i}{N_T} \tag{12.5}$$

where X, RMS voltage threshold value in % of rated value; N_i, Number of customers experiencing short-duration voltage deviations with magnitudes below X% or above X%; N_T, Number of customers served from the part of the system to be assessed.

Determining this index would require monitoring voltage at selected locations. While this index is similar to reliability indices, it is difficult to extend it to find system wide values because that would require a large number of locations for voltage monitoring [4].

12.5.2 ITI (CBEMA) Curve

While most of the equipment can tolerate sags and swells, Information Technology Equipment (ITE) is sensitive to them. To address this issue, Information

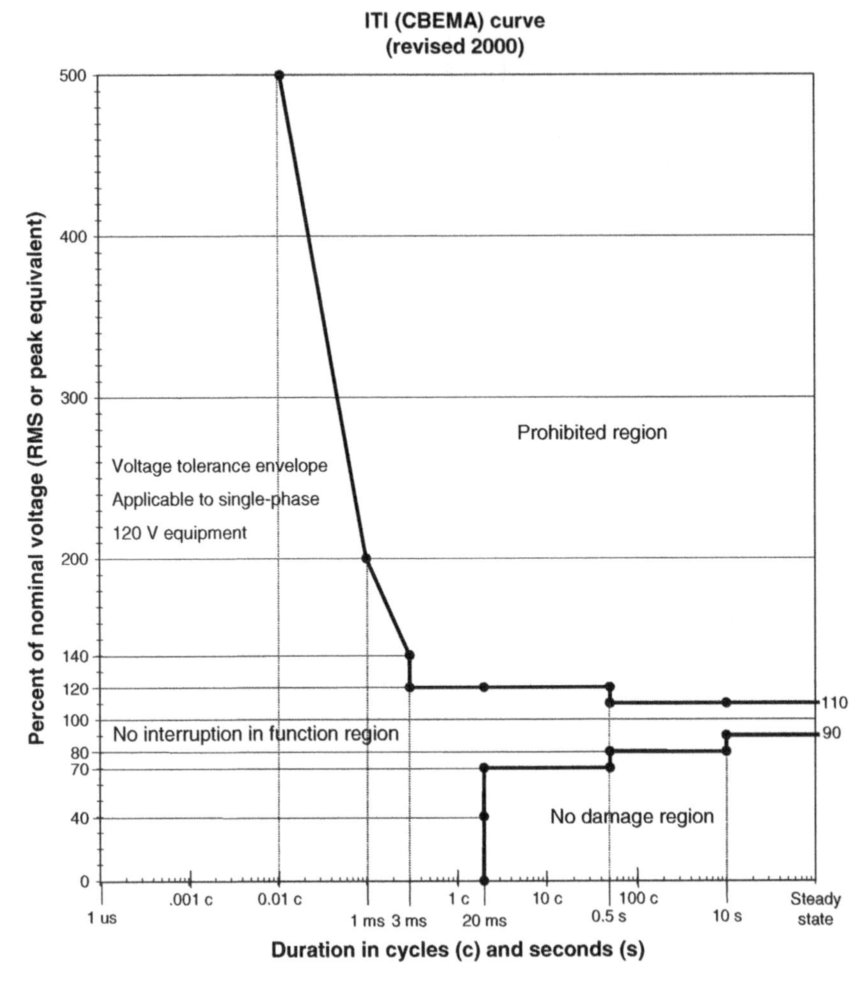

Figure 12.3 Voltage sag and swell limits for information technology equipment. Source: Reprinted with permission of ITIC.

Technology Industry Council (ITI, formerly known as the Computer & Business Equipment Manufacturer's Association) has set voltage ride through design specifications for ITE and distribution systems. The ITI (CBEMA) curve shown in Figure 12.3 applies to nominal voltage of 120 V RMS at 60 Hz. The curve has three regions along with the voltage tolerance envelope. Accordingly, large swell and sag can take place if they are for extremely short durations. The range of voltages shrinks as the duration increases. Nominal voltage between 90% and 110% is permitted for an indefinite duration. Events in the No Damage Region will not permit normal operation of the equipment, but no damage to the equipment is expected. However, events with swell above the upper limit (Prohibited Region) of the curve will damage the equipment.

12.6 Flicker

Voltage flicker is fluctuation of voltage amplitude at a frequency much lower than the power frequency. Periodic switching on of large loads such as sawmills, irrigation pumps, welding machines, and elevators can cause flicker. Figure 12.4 shows an example where the amplitude of voltage drops by ΔV periodically. In this example, there are two dips in the voltage in 0.5 second or four dips in 1 second. This will cause eight fluctuations per second. This example shows the fundamental voltage modulated by a square wave, but if the fluctuations are sinusoidal of frequency ω_f radians/second, the nominal instantaneous voltage $V_m \cos(\omega_0 t)$ gets modulated by the flicker signal $V_f \cos(\omega_f t)$ [1]. Therefore, we get the flicker component of the voltage

$$v_f(t) = V_f \cos(\omega_f t) V_m \cos(\omega_0 t) \tag{12.6}$$

and the total voltage is

$$\begin{aligned} v(t) &= V_m \cos(\omega_0 t) + v_f(t) \\ &= V_m \cos(\omega_0 t) + V_f \cos(\omega_f t) V_m \cos(\omega_0 t) \\ &= (1 + V_f \cos(\omega_f t)) V_m \cos(\omega_0 t) \end{aligned} \tag{12.7}$$

Usually, the flicker effect is defined by a flicker factor F, which is the ratio of the flicker voltage amplitude and the amplitude of the nominal voltage, or

$$F = \frac{V_f}{V_m} \tag{12.8}$$

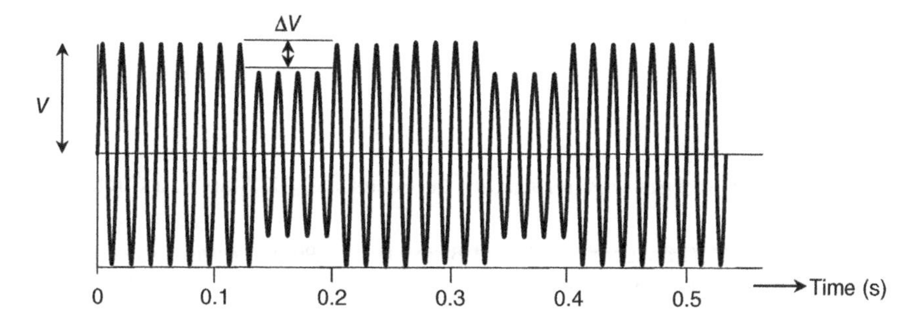

Figure 12.4 Example of periodic voltage change that causes flicker.

Cases where motors start only a few times in a day are considered special cases of flicker. Although there is no periodic flicker signal, motor start draws current that is five to six times larger than the normal current. Voltage drops suddenly cause lights to dim and return to normal in a few seconds. Utilities usually have a criterion for voltage drop during motor starts. Most utilities use a limit of 3% for voltage drop.

Problems

12.1 The voltage and current at a single-phase 120-V load connected to a 20-kVA transformer rated 7.2 kV/120 V with 12% impedance are

$$v(t) = 110\cos(\omega t + 45^\circ) + 20\cos(3\omega t + 60^\circ) + 10\ \cos(5\omega t)\ \text{V}$$

$$i(t) = 83.33\ \cos(wt) + 2.08\ \cos(3\omega t + 30^\circ) + 1.25\ \cos(5\omega t - 60^\circ)\ \text{A}$$

(a) Compute the RMS value of the voltage and current.
(b) Find the THD in the voltage and current.
(c) Compute PF and PF_h for the given voltage and current.

12.2 Consider the voltage in Problem 12.1 to be

$$v(t) = 110\cos(\omega t + \delta^\circ) + 20\cos(3\omega t + 60^\circ) + 10\ \cos(5\omega t)\ \text{V}$$

and the current is the same. Consider δ to be 0°, and compute PF and PF_h. Change δ in increments of 15° up to the maximum of 90°. Compute PF and PF_h for each value of δ. Draw plots of PF and PF_h as functions of δ.

12.3 The impedance of the feeder upstream from the transformer in Problem 12.1 is $0.1 + j\ 20\ \Omega$ at the fundamental frequency. Compute the voltage drops for the given load currents at the PCC. Are the currents and voltages within the limits specified by IEEE-Std 519 given in Table 12.1?

References

1 Heydt, G. (1996). *Electric Power Quality*, 2e. Scottsdale, AZ: Stars in a Circle Publications.
2 Dugan, R.C., McGranaghan, M.F., and Beaty, H.W. (1995). *Electrical Power Systems Quality*. New York, NY: McGraw Hill.
3 Arrillaga, J., Smith, B.C., Wood, A.R., and Watson, N.R. (1994). *Power System Harmonic Analysis*. London: Wiley.

4 Short, T.A. (2004). *Electric Power Distribution*. Boca Raton, FL: Francis & Taylor.
5 Porter, G.J. and VanSciver, J.A. (1998). *Power Quality Solutions: Case Studies for Troubleshooters*. Englewood Cliffs, NJ: Prentice Hall.
6 McGillem, C. and Cooper, G. (1974). *Continuous and Discrete Signal and System Analysis*. New York: Holt Rinehardt and Winston.
7 IEEE-Std 519 (2014). *IEEE Recommended Practice and Requirements for Harmonic Control in Electric Power Systems*. New York: IEEE.
8 Sabin, D.D., Grebe, T.E., and Sundaram, A. (1999). *RMS Voltage Variation Statistical Analysis for a Survey of Distribution System Power Quality Performance*. IEEE PES Winter Meeting.

13

Distributed Energy Resources and Microgrids

13.1 Introduction

Technological advances and decreasing prices are making deployment of distributed energy resources (DERs) attractive. In Chapter 4, we gave a brief introduction to DERs. In this chapter, we provide detailed information on some of the popular DER technologies. In addition, we discuss the concept of microgrid (MG) and how deployment of DERs is facilitating formation and operation of MGs. Technologies associated with different DERs and their effective deployment in distribution systems are evolving. While it is not possible to cover every aspect associated with DERs and MGs, in this chapter we present some of the most relevant topics. The readers are encouraged to explore various published articles on the subject. Some of the key references are provided at the end of this chapter.

13.2 DER Resources and Models

DERs based on wind and solar technologies, batteries, and microturbines are the most common resources at present. In this section, we present information related to them.

13.2.1 Wind Generation

Wind generation has limitations related to deployment in distribution systems. Firstly, wind generations lower than 1 MW are not very efficient, and thus, they are not viable for installation by homeowners or small businesses. Larger businesses and distribution system operators (DSOs) may find them attractive under certain conditions. Secondly, if distribution systems are in the vicinity of homes and businesses, aesthetics becomes important. So we seldom see a wind turbine

Electric Power and Energy Distribution Systems: Models, Methods, and Applications, First Edition.
Subrahmanyam S. Venkata and Anil Pahwa.

Companion website: www.wiley.com/go/Pahwa/ElectricPowerDistributionSystems

in the middle of the city. However, in rural communities, wind turbines have been deployed as DERs by the city or the local energy service provider.

The mechanical power generated by a wind generator depends on the wind speed and the size of the machine. The relationship governing them is given by:

$$P_m = \frac{1}{2}\rho A v^3 C_p \quad \text{W (Watts)} \tag{13.1}$$

where ρ is the air density in kg/m^2, A is the area swept by the blades, v is the wind speed in m/s, and C_p is the power coefficient. For blades with length l, the sweep area is πA^2 m^2.

The power captured by the wind turbine is a cubic function of wind speed as given by Eq. (13.1). However, this relationship is valid between the cut-in speed and the cut-out speed as shown in Figure 13.1. The cut-in speed is the lowest speed at which the turbine creates enough motion to produce output power. The cut-out speed is the speed at which the turbine must be stopped to avoid damage due to high wind speeds. The rated speed is in between the two speeds, and at this speed, the turbine produces the rated output. The cut-in speed is 6 to 9 miles per hour, the rated speed is about 30 miles per hour, and the cut-out speed is about 55 miles per hour of steady wind.

The wind turbine is connected to a gear box, which is connected to the generator that produces electrical power. The initial designs of wind generators were based on squirrel cage induction generator, and they operated at fixed speed. Doubly fed induction generators (DFIGs) are the most used for wind generators presently. These generators operate in a range of speeds around the synchronous speed. Synchronous generators can also be used for generating electrical power. The details are beyond the scope of this book, but the readers are suggested to other look at published books on the subject [1].

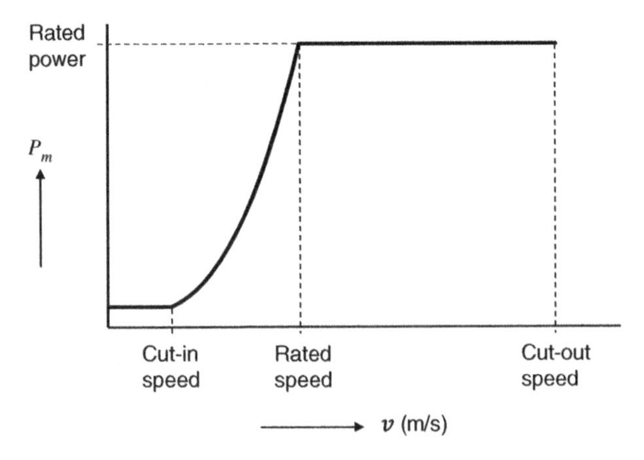

Figure 13.1 Mechanical power output of wind turbines as a function of wind speed.

13.2.2 Solar Generation

Solar photovoltaic (PV) and concentrated solar power (CSP) are the two common approaches used for producing electricity. CSP requires a large infrastructure, and it is only suitable for large-scale deployment in bulk power systems. In contrast, solar PV has a range from 200 W to hundreds of MW. Therefore, it can be installed on rooftops of homes and businesses and as a stand-alone on-ground facility within a city. Solar PV is also attractive from aesthetics because it blends in with the roof lines of structures, and on-ground installations look attractive if properly done. According to Solar Energy Industry Association (SEIA), the total energy generated from solar plants in the United States was 104 billion kWh in 2019, of which approximately 33% was from rooftop distributed PV systems with less than 1 MW generation capacity. Various countries, and states and utilities within the U.S. have used policies and incentives to promote distributed solar energy, which has resulted in different levels of deployment of solar energy around the world. Deploying solar PV as DER has many challenges, which span technical, economic, social, and policy domains. On the technical side, sizing, siting and aggregate capacity of the PV systems, impacts on the local system as well as adjoining utility grid, and coordination with the existing generation are some of the issues. Economic issues include cost of installation and maintenance of the PV system; individually owned versus community-owned systems; economic impact of various energy pricing models on citizens, the service provider, and the utility. Social aspects for consideration include attitudes of the stakeholders in the community toward solar energy and social equity for the citizens. Local governments can implement policies to increase deployment of solar PV in their communities while ensuring that economic and social equity is maintained.

Solar PV systems are built by connecting several PV cells in series and in parallel. The PV cell is a p–n junction electronic device built on the principle of Schottky barrier. Illuminating the PV cell by light creates a voltage at the cell terminals. Connecting a load at the terminals creates current flow, which is determined by the characteristics of the cell as given in the equation below:

$$I = I_{PH} - I_0 \left(e^{\frac{qV}{kT}} - 1 \right) \quad \text{A} \tag{13.2}$$

where I_{PH} is the photocurrent; I_0 is the saturation current of the cell; T is the cell temperature in K; q is the charge of an electron, which is 1.6×10^{-19} C; and k is a constant equal to 1.38×10^{-23} Joules/°K. Normally, $I_{PH} \gg I_0$ with the ratio of photocurrent to saturation current in the 10^{10} range. Although the actual characteristics are slightly different, Eq. (13.2) provides an ideal version of the PV cells. A graph based on it is shown in Figure 13.2.

Multiplying V by I gives the power output of the cell, as shown in Figure 13.2. Note that there is a maximum power point on this graph. Typically, solar PV

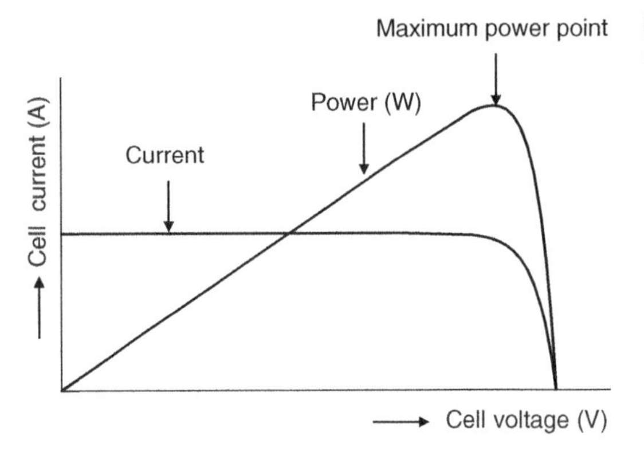

Figure 13.2 Solar PV cell characteristics.

systems are operated close to this point by maximum power point tracking (MPPT). While the maximum cell voltages remain the same, the current and the power output increase with higher illumination because the photocurrent increases. Also, note that the cell characteristics change with temperature. Although the short-circuit current remains almost constant, the open-circuit voltage increases with lower cell temperature. Consequently, the cell power increases with decrease in cell temperature. Typically, the cell power decreases by 0.5% per °C increase in temperature [2]. That is why the solar plants produce lower output during the peak summer season.

13.2.3 Battery Energy Storage System (BESS)

Batteries are used by people in many everyday applications, such as smart phones, computers, and cars. Their use for energy storage in power systems has been limited due to cost. Now, with declining cost, increased performance, and longer life, their deployment is gradually increasing. Large-scale battery storage is being deployed at the bulk power system level to support variable output of wind and solar resources and as a source of energy to enhance resiliency in the event of grid failure. However, deployment at the distribution level has been limited. With further decline in price, batteries are being deployed in distribution systems. Nonetheless, an indirect form of battery storage is appearing in distribution systems in the form of electric vehicles (EVs). Although the subject of vehicle-to-grid (V2G) interaction has been discussed for many years, individual owners will not find participation in two-way V2G interaction attractive due to degradation of batteries as well as warranty concerns. But V2G interaction is quite feasible for fleet operators in exchange for attractive rates provided by the utilities.

In the future, with further decline in prices, we can expect to see increased deployment of BESS in distribution systems as well as in homes. However, it is important to understand that any energy that is stored in batteries and used later will have a higher cost per kWh. Therefore, there must be a compelling need to store energy for use at a later period. A specific example includes storing energy when the price of electricity is low and using it when the price is high. In distribution systems with high penetration of solar PV, batteries can store energy during peak hours to prevent excessive reserve flow from the distribution system into the transmission system. The stored energy can be used during sundown to prevent excessive ramping of other generators to meet the load. In addition, the use of BESS in conjunction with solar PV plant can mitigate adverse effects of rapid fluctuations in power and voltage due to intermittent clouds. BESS also provides a good value for locations with high frequency of electricity supply interruptions and for sensitive loads that require continuous power supply.

Lead-acid batteries were the most used batteries in the past for BESS due to their maturity. However, lately, several chemistries for batteries have been tested. Lithium-ion is the most used battery chemistry at present. However, there are significant concerns about the long-term availability of rare earth elements needed to build them and environmental issues related to mining the needed elements and disposal of used batteries. The quest for more environment friendly and cheaper batteries is ongoing. In addition, recent advances in redox-flow batteries are promising. Redox-flow stores energy in the electrolyte, which is pumped through the cell to charge or discharge.

Modeling of batteries to determine the state of charge (SOC) and the state of health (SOH) for integrating them optimally into power systems is a very complex process because internal measurements are not available. Typically, only the terminal voltage, current flow, and temperature are available. A simple and effective method to model batteries is based on empirical approaches. In this approach, a mathematical function is used to fit the past performance, which is then used to predict the future. These models lack physical meaning, and they are not accurate outside of the conditions in which they were built. Equivalent circuits-based approach is another way to model batteries. They also do not effectively account for physical basis and lack accuracy for efficient operation of batteries. To achieve higher accuracy, we must use physics-based models. These models use mathematical representation of the chemical and thermal phenomena inside the battery and combine them with externally available measurements of voltage, current, and temperature to represent the overall dynamics of the batteries. We will not discuss the details of such models in this chapter. The readers can get a detailed overview of battery modeling in [3].

Since batteries produce dc voltage, they must be connected to inverters to produce ac voltage for interconnection to the power grid. The ac side will also have a transformer to step up the voltage if the output ac voltage of inverter is lower than the power grid voltage.

13.2.4 Microturbine

Microturbines are gas turbines with a radial compressor and turbine rotors. They rotate at very high speed, such as 96 000 rpm, and their capacity ranges from 30 to 250 kW. They typically use a heat recovery system to recover exhaust heat to preheat compressed inlet air, which increases efficiency. They produce ac voltage at a very high frequency, which is converted to ac at 60 Hz using a series of converters, which convert ac to dc and then dc to ac. Due to their size, they are usually suitable for industrial or commercial facilities, where high reliability of power supply is required. They are also used as a standby source of power in MGs.

13.2.5 Electric Vehicles

Although the first electric car was introduced over 100 years ago, only now they are entering the mainstream due to declining cost and performance of batteries. Several manufacturers are offering attractive models for customers to choose from. While they are still more expensive than the traditional automobiles, their cost is expected to decrease in the future. Several major automobile manufacturers have announced their plans to completely phase out the production of fossil fuel vehicles or to reduce their share in the portfolio within the next 10–20 years. Increased ownership of EVs will pose new planning and operating challenges for the utilities. They will increase the overall energy consumed as well as the power demand on the system. The increased power demand will have consequences from the distribution transformer feeding a few customers to adequacy of feeder sizes as well as other equipment in the system. If all the customers being fed power from a distribution transformer migrate to EVs, they will likely create a new peak in the night if all of them charge their vehicles simultaneously. The new peak could be higher than the capacity of the transformer in many cases. Since Level 1 chargers take several (10–12) hours to recharge the batteries, time-staggered charging is not a viable option. Time-staggered charging is possible with Level 2 chargers, which take 3–5 hours to recharge the batteries. However, Level 2 chargers are an additional expense, and how many customers will opt for it is not fully clear at present. Similarly, the feeders and the substation transformer could get overloaded if large number of EVs charge their vehicles simultaneously. While the present level of EVs has not created any issues for the DSOs, major issues can arise in the future for which the DSOs have to be prepared.

Although in most cases, EVs receive power from the grid, they can also inject power into the grid. The idea of V2G power transfer has been discussed for many years now, but it has not been implemented at a wide scale. There is no incentive for an individual EV owner to engage in V2G power transfer. The operator of a fleet of EVs could possibly engage in V2G activities if there is adequate financial incentive offered by the DSO. As far as modeling of EVs for distribution system analysis is concerned, the models of BESS will apply to EVs too because parked EVs are nothing but batteries.

13.3 Interconnection Issues

Interconnecting DERs in distribution systems could cause reverse power flows, power and voltage fluctuations, voltage rise, and protection issues. These issues become more pronounced with increasing deployment of DERs in the system. Since customer-owned rooftop solar PVs are behind the meter with no direct control, they create maximum concern for the system operator. Rapid fluctuations of power due to intermittent clouds cause voltage fluctuations, which is also a concern. Under the ideal situation, every customer in the system would have a rooftop solar PV, but practical issues limit such a scenario. Presently, 15–20% of load served by distributed rooftop solar PV is regarded as the hosting capacity limit by several utilities. However, with implementation of smart inverters and batteries, the hosting capacity can be increased in the future. IEEE Standard 1547-2018 discusses interconnection issues for inverter-based resources (IBRs) in general. We have already presented some of those issues in Chapter 4. In this section, we present some practical issues.

13.4 Variable Solar Power

Figure 13.3 shows examples of solar irradiation in Manhattan, Kansas, on three different days. The first example shows a day with full sunshine, which would result in steady electricity generation in proportion to the irradiation. The second example shows an overcast day with low solar irradiation and electricity generation. The third example shows a day with rapidly moving clouds on a sunny day causing fluctuating output from the solar panels, which result in fluctuations in power flow and voltage at several points in the system.

While several approaches are being investigated to leverage smart inverters to mitigate these fluctuations, one possible approach is to control the reactive power in response to change in real power. Using Eq. (4.56) of Chapter 4, we can write the approximate expression for voltage drop across a line, or

$$\Delta V = \frac{(R\,P + X\,Q)}{V_r} \tag{13.3}$$

Now, we consider that at the end of the feeder or bus i, the real power of load is P_L^i, and reactive power of load is Q_L^i, and the solar PV is generating the real power

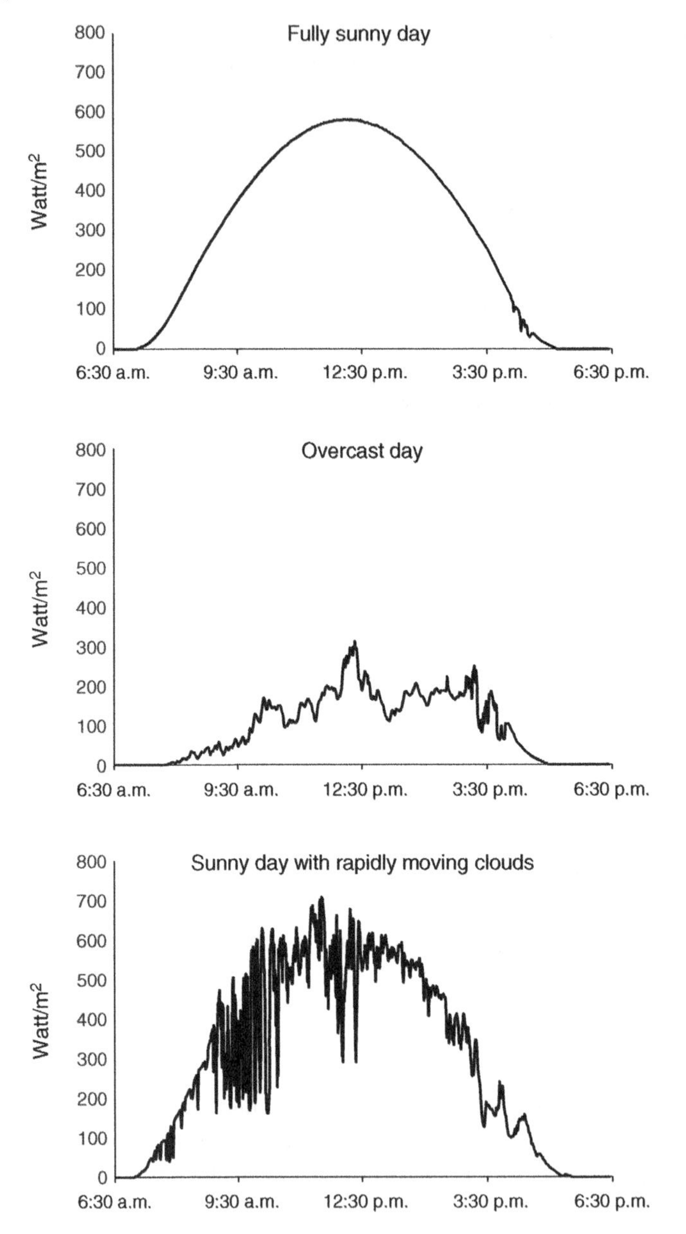

Figure 13.3 Examples of solar irradiation recorded with 30-second resolution in Manhattan, Kansas. Source: Courtesy of Kansas State University.

P^i_G and the reactive power of load is Q^i_G. Hence, $P^i = P^i_G - P^i_L$ and $Q^i = Q^i_G - Q^i_L$. Substituting them into Eq. (13.3) and replacing the voltage in the denominator by the nominal voltage of the system V, we get

$$\Delta V = \frac{R\left(P^i_G - P^i_L\right) + X\left(Q^i_G - Q^i_L\right)}{V} \tag{13.4}$$

For an ideal situation, we can consider ΔV or voltage drop across the line to be zero at a generic time step k, which gives

$$Q^{i(k)}_G = Q^{i(k)}_L - \frac{R}{X}\left(P^{i(k)}_G - P^{i(k)}_L\right) \tag{13.5}$$

Defining a unique R/X ratio for the reactive power control logic is not straightforward due to various network parameters, conductor types (cable or overhead lines), and feeder length. In contrast, voltage sensitivity to active/reactive power variations at each bus can be calculated for each network [4]. Hence, (13.5) can be reformulated as

$$Q^{i(k)}_G = Q^{i(k)}_L - \frac{S^{ii}_{VP}}{S^{ii}_{VQ}}\left(P^{i(k)}_G - P^{i(k)}_L\right) \tag{13.6}$$

where S^{ii}_{VP} and S^{ii}_{VQ} are the voltage sensitivity indices at bus i due to 1 pu active/reactive power change at bus i, respectively. We can write a similar equation for time step $k+1$, or

$$Q^{i(k+1)}_G = Q^{i(k+1)}_L - \frac{S^{ii}_{VP}}{S^{ii}_{VQ}}\left(P^{i(k+1)}_G - P^{i(k+1)}_L\right) \tag{13.7}$$

Now, subtract (13.6) from (13.7), which gives

$$Q^{i(k+1)}_G - Q^{i(k)}_G = Q^{i(k+1)}_L - Q^{i(k)}_L - \frac{S^{ii}_{VP}}{S^{ii}_{VQ}}\left(P^{i(k+1)}_G - P^{i(k)}_G - P^{i(k+1)}_L + P^{i(k)}_L\right) \tag{13.8}$$

Although we are assuming that the solar PV output is fluctuating rapidly, we can assume that the load will not change between the two time steps. Therefore,

$$Q^{i(k+1)}_G - Q^{i(k)}_G = -\frac{S^{ii}_{VP}}{S^{ii}_{VQ}}\left(P^{i(k+1)}_G - P^{i(k)}_G\right) \tag{13.9}$$

or

$$Q^{i(k+1)}_G = Q^{i(k)}_G - \frac{S^{ii}_{VP}}{S^{ii}_{VQ}}\left(P^{i(k+1)}_G - P^{i(k)}_G\right) \tag{13.10}$$

Thus, if we can project change in real power generated by solar PV in the next time step, we can proactively adjust the reactive power to mitigate fluctuations in voltage. While implementing the control, we must ensure that the operation is within the limits specified for the inverters. The details for implementing the

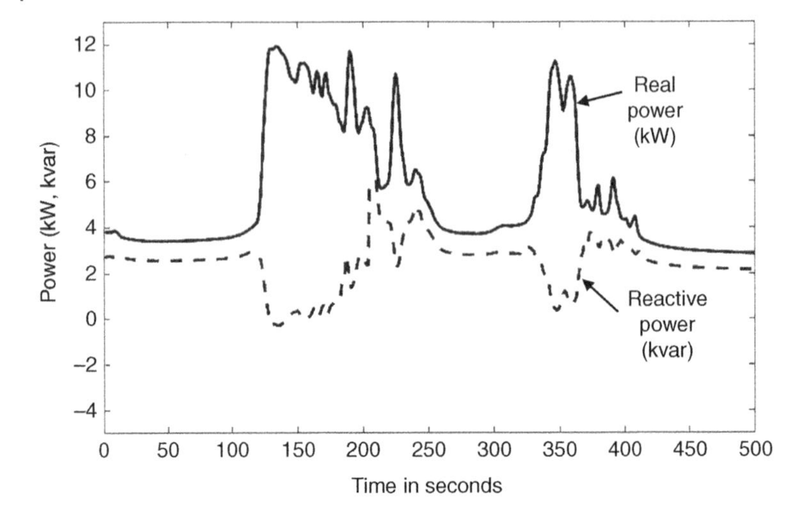

Figure 13.4 Real power variation at a bus due to changing solar irradiation and reactive power injection by inverter to mitigate voltage fluctuations.

control based on this approach are available in a paper [5], and we present some of the results of applying it to a 559-bus test system [6] over a period of 500 seconds. Figure 13.4 shows variation in real power (solid line) at a selected bus in the system due to changing solar irradiation. This figure also shows the reactive power (dashed line) injected by the inverter based on Eq. (13.10). Figure 13.5 shows voltage at the same bus with no reactive power or unity power factor

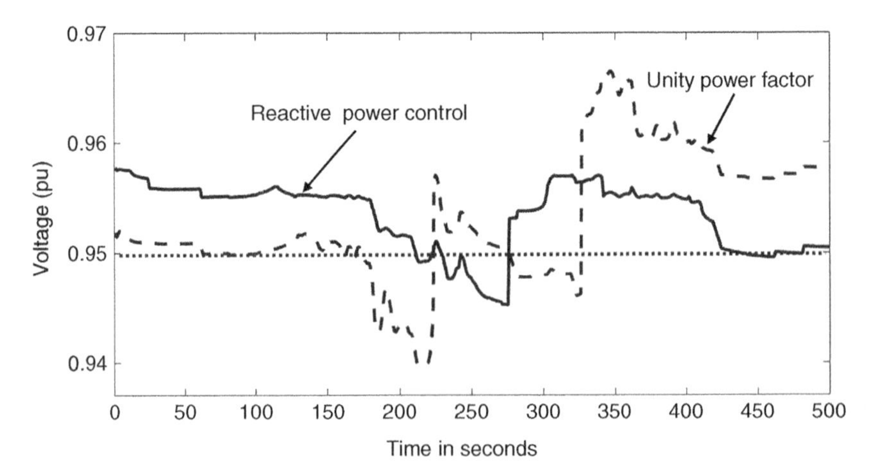

Figure 13.5 Voltage at a bus with unity power factor operation and dynamic control with reactive power injection.

(dashed line) and with reactive power control (solid line). The results clearly show mitigation in voltage fluctuations. Also, the voltage drops below the lower limit of 0.95 pu for a shorter duration with control. In addition, the tap operations at the substation transformer have reduced from two to one with control.

13.5 Microgrids

Over the 150 years of electrification, the power grid has evolved with larger generating resources deployed in centralized locations due to economies of scale. However, due to declining costs, it has become feasible to generate electricity using distributed small-scale generators. In addition to cost, system resiliency due to increased frequency of extreme events causing large-scale disruptions in power supply is becoming a concern. In the past, facilities that required uninterrupted power supply, such as hospitals, deployed a backup diesel or gas generator, which came into action upon loss of power from the grid. However, now small generators can be operated regularly. They supply power to nearby loads but are ready to take over and operate in an islanded mode whenever the connection to the grid is lost. This has led to the concept of MG, which is defined as "a group of interconnected loads and DERs with clearly defined electrical boundaries that acts as a single controllable entity with respect to the grid and can connect and disconnect from the grid to enable it to operate in both grid-connected and island modes" [7].

Microgrids can be operated connected to the main network or autonomously. They have been proposed to integrate high penetrations of distributed generation sources that are becoming more commonplace on the distribution system [8, 9]. Microgrids increase system reliability by reducing customer outage and service restoration time [10]. However, one major issue with implementing MGs is their protection when operating in an islanded mode. This is a result of the fault currents being lower than steady-state currents from voltage-source inverter-connected devices. These devices include battery energy storage systems and PV panels that are often the dominant sources in low- and medium-voltage MGs [11].

13.5.1 Microgrid Types by Supply and Structure

13.5.1.1 ac Microgrids

The best way to address this issue is to consider the example shown in Figure 13.6, which is an 18-bus radial/looped distribution system adapted from [12]. The system has both conventional and renewable generating sources with distinct types. The system also has several single-phase laterals, both overhead and underground.

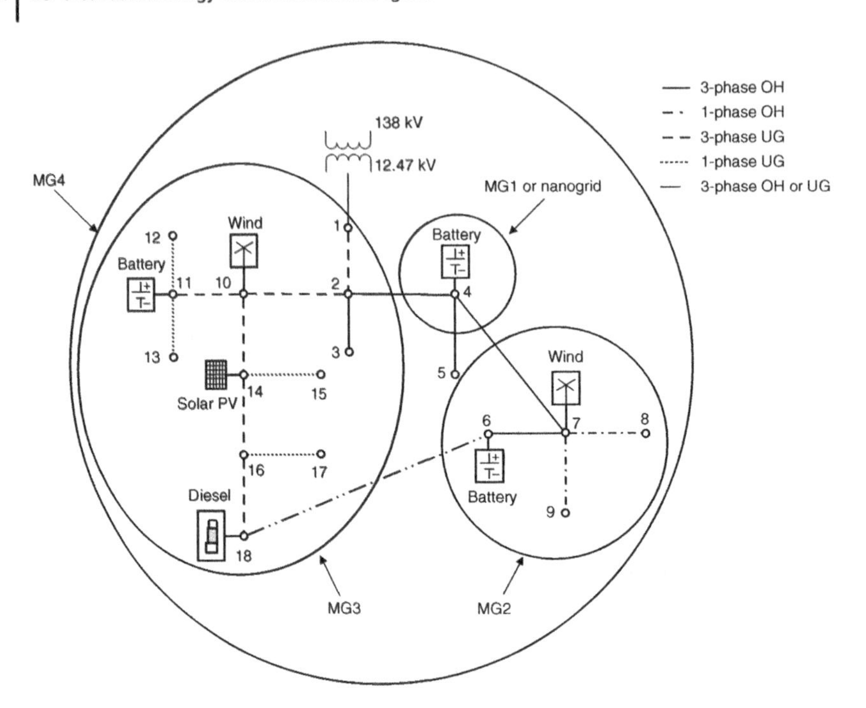

Figure 13.6 An 18-bus radial/looped example system with four microgrids (MGs). Source: Courtesy of Eaton Corporation.

The line connection between buses 6 and 18 is added to close the loop of the otherwise radial system.

Since the system has several sources, we can create different boundaries to define the MGs. Typically, an MG with only one DER is not a true MG. Sometimes it is called a nanogrid. One could conceive of selecting different MGs for the same distribution system, but these must be selected *a priori* so that appropriate real-time actions could be taken when a fault or a disturbance occurs. In this figure, four distinct MGs are identified. The three smaller MGs are nested within the entire substation MG. The selection of MGs is based on matching the available local distributed generation with the local loads. Although in this example we have defined the MGs *a priori*, a flexible approach in which the MG boundaries adapt with changing conditions in the system can be implemented. Such an adaptive system will pose many challenges, which are still under investigation.

13.5.1.2 dc Microgrids

These are slowly evolving to supply dc loads directly from the dc DERs such as solar and batteries without the need for power convertors to ensure higher efficiency. The major problem with a dc MG is the lack of commercial fast breakers or switches. Some of them have been developed for low-voltage applications. This is an area that needs additional research in the development of fast and cost-effective solid-state dc breakers at the medium-voltage levels. Another principal issue is the coordination of dc protective devices, a subject that is yet to be investigated.

13.5.1.3 Hybrid Microgrids

A significant amount of research and development is being carried out in the development of hybrid MGs. From the protection point of view, these are in the infant stage. One could conceive of many topologies which are still being investigated and understood. These MGs involve both ac and dc DER loads, and their protection could become a complex issue which needs further research. In these types of MGs, the role of converters/inverters and their efficient operations including protection need to be well understood. One possible problem is the coordination of a dc breaker with ac protective devices in vogue and future ones to be developed.

13.5.1.4 Networked Microgrids

Tremendous amount of research efforts is underway in several national labs and other institutions on this subject. However, the concept of networked MGs is based on certain assumptions, which are not practical to implement in real life. As the subject of MGs matures, we will get more clarity on the networking aspect.

13.5.2 Microgrid Modes of Operation

The modes of operation of MGs are typically divided into three generalized categories: grid connected, islanded, or mixed mode operation [8, 13]. The modes of operation are influenced by a variety of coordinated dispatch functions and control functions that may be scheduled, autonomous, or responsive to local out-of-specification or abnormal conditions either on the interconnected distribution grid or within the MG. The modes of operation are often much different in that the voltage regulation for the loads is based on either very low impedance power source (the utility) or a higher impedance power source (DER related) with resource-limited energy.

13.5.2.1 Grid-Connected Mode

The grid-connected mode is typically the prevalent operating mode for MGs that are interconnected to the local grid. Energy flow is predominantly from the local grid to local loads within the MG. Other important flows of energy include charging of batteries, conditioning of thermal storage, providing housekeeping for MG controllers, and communications equipment. Transitions to and from the grid-connected mode will often utilize communications for dispatch, synchronization, critical protection data, and data-handling requirements.

13.5.2.2 Islanded Mode

The islanded mode is used for a variety of conditions. Internal supply of power to MG loads during utility outages is used to maintain loads for reliability and resiliency that are typically segregated into critical loads or loads that may be shed according to a schedule of priorities. Other reasons for operating in the islanded mode are becoming increasingly important. A utility may dispatch a MG to curtail the reverse power flow when the distribution system has excess power from the distributed generation such as PVs and wind, and the reverse power may be detrimental to the utility's operation. Such curtailed generation is likely a situation where advanced information is needed to optimize the operations of the MG and the interconnected utility. Another important reason for operation in the islanded mode is to better optimize charges that are being implemented as demand charges or even standby charges.

The scenarios for transitions to islanded mode include the following:

Planned Islanding The process and steps for a planned islanding event include: (i) receive islanding command either as a scheduled event or as dispatch from the DSO; (ii) balance the load and generation (adjust both P and Q to be 0 at the point of interconnection (POI)); (iii) set local controllers and protection devices appropriately; (iv) create the island; and (v) transition to steady-state islanded dispatch mode.

Unplanned Islanding The process and steps for unplanned island events include: (i) detect the need for islanded conditions; (ii) create the island; (iii) set local controllers and protection devices appropriately; (iv) execute the required preplanned actions such as load shedding (and/or implement a black start if required); and (v) transition to steady-state islanded dispatch mode.

Reconnection to Grid The process and steps include: (i) resynchronize, set/match voltage, phase angle, and frequency within prescribed limits specified by applicable grid codes or requirements; (ii) set local controllers and protection devices appropriately; (iii) reconnect; and (iv) transition to steady-state connected dispatch mode and restore noncritical loads as appropriate.

Transition from Grid to Islanded Operation and Vice Versa The modes and methods for transitions with an electrical grid must be well-coordinated processes that involve communications in the form of dispatch requests. A simple scheduled transition is an option if the protocols for the transitions are predetermined. Transitions typically include a verification of a completed process and reporting on the resulting conditions after the transition is completed. The timing and synchronization are often a range of values. The criteria for transitions are followed by the MG

Table 13.1 Typical criteria used for a transition process.

Processes	Parameters	Characterization of parameters	Protection requisites
Transition initiation	Receive transition request or command	Predetermined operational values, limits, and timing needs	Predetermined protection capabilities and limitations.
Load/primary source balancing	Assess microgrid loads, microgrid generation, energy storage status, and data collection systems	Adjust loads and generation to values to assure stable operation	V, f, P, Q, settling time, overshoot, time to island, and steady-state values within contractual requirements and equipment limitations
Transaction to new operating conditions	Transition timing and speed of transfer	Measure and evaluate transients and any oscillations	Communicate with protection equipment on the utility infrastructure and within the microgrid
New settings	Assess stability, generation, and energy storage values	Measure the new stable values	New V, f, P, Q, settling time, overshoot, time to island, and steady-state values within contractual requirements and equipment limitations
Data acquisition and analysis	Assess dynamic values	Record values before, during, and after transition	Store and analyze new data

controller. Table 13.1 shows examples of processes, parameters used before the transition, the needed characterizations of the parameters, and applicable protection requirements.

Metrics are typically specified by the interconnection requirements of the DSO to which the MG is connected, applicable codes and/or standards, or state or local mandates for interconnectivity.

Typical metrics for transitions includes:

- Directly measurable quantities: voltage and current (time-domain waveforms).
- Derived quantities: frequency, root-mean-square (RMS) voltage, RMS current, phase angle, real power (including direction of power flow), reactive power (leading or lagging), energy exchanged at the POI (grid-connected mode),

power quality indices (voltage and current harmonic distortion, individual harmonics, voltage sags, and voltage swells), and reference-tracking errors.

Scenarios are typically defined for the following basic transitions:

- Grid connected to islanded – planned islanding. This transition is initiated upon receipt of an external request, typically sent by the DSO.
- Grid connected to islanded – unplanned islanding. This transition is the result of an event on the distribution grid. It can involve a black start if the control system is designed in this manner.
- Islanded to grid connection. This transition involves resynchronization and reconnection of the MG.

Test scenarios for transitions typically consider the conditions indicated below. They are chosen to allow a complete and comprehensive testing of the transition function, including the required and relevant features of the dispatch function.

- The initial operating conditions of the MG. These include the operating conditions before the transition occurs: the level of local generation and generation mix (dispatchable and nondispatchable), operation of the storage device (mode of operation and SOC), load composition (constant impedance, constant P-Q, and active loads) and load mix (percentage composition), status of breakers, switches, and voltage control devices, and power (P, Q) exchanges between the MG and the grid (prior to a transition from grid connected to island mode).
- The state of the grid at the time of the transition. These include the voltage at the POI and any disturbances occurring on the grid at the time of the transition. In the case of an unplanned islanding event, the nature of the event (typically a fault or an open connection on the feeder connected to the POI initiating the islanding transition) is considered.

Mixed Mode An MG control system set up in mixed mode still must contain the islanded- and grid-connected functions and supports all forms of planned and unplanned islanding and resynchronization. The difference is that the controls of DERs never change regardless of the grid or the island mode connection. The 2018 NREL MG shootout was won using this technique. Massachusetts Institute of Technology (MIT) MG laboratory is using this technique.

13.5.3 Grid-Following vs. Grid-Forming Inverters

In the grid-connected mode, the MG gets the reference frequency from the grid, and the resources in the MG operate in synchronism with the grid while controlling the real and reactive power at the DER. In other words, the IBRs operate in the grid-following mode. However, in the islanded mode, there is no reference signal from the grid, and the resources within the MG must operate in coordination with each other. In the example shown in Figure 13.5, we have included a diesel generator, which could serve as the reference for the MG. However, in the future with waning of diesel generators, we can expect only renewable generation and BESS or IBRs to be deployed in the MGs. In such cases, the MG will have only low-inertia IBRs, which makes operation of the MG in islanded mode challenging. To facilitate operation under such conditions, the concept of grid-forming inverters has come into vogue. The grid-forming inverters have the capability to control frequency and terminal voltage. They also work cooperatively to share the load in the MG in preassigned portion based on the size of the DER they are controlling. Thus, an MG could have a mix of grid-forming and -following inverters. The challenge, however, is that the inverters operating as grid forming in the islanded mode will have to revert to grid following whenever the connection to the grid is made. This subject has only recently started getting attention in the research community. Federal agencies, such as Department of Energy, are investing substantial sums of money on this research, and we are expecting significant developments on this subject in the future.

13.5.4 Microgrid Protection Challenges and Requirements

As innovative technologies and DER resources including renewables penetrate the system, making it complex, it is imperative that the critical issue of protection be identified and ways are paved to achieve efficient and effective means of realizing them. The protection of even a classical, passive, and radial distribution system has always been very challenging. The practices vary widely even within a utility, let alone among utilities.

Protection challenges in the distribution systems abound now with the challenges of DERs including inverter-based generation and storage and new configurations for aggregation and MGs [13–20]. The microprocessor relay of today is a mature and reliable technology. However, its technology must be adapted to the needs of the distribution networks as they have evolved with high penetrations. The starting point for this adaptation is MGs with unique protection requirements.

Protection requirements for MG systems are unique and different from general protection issues, known and applicable to distribution systems, because of the large amount of DERs present and of the MG control systems overseeing the overall operation. Further, the design of protection systems for a MG in the islanded mode is different from that in the grid-connected mode because of the absence of strong voltage sources.

Multifunctional microprocessor relays as intelligent electronic devices (IEDs) are the primary MG control, protection, metering, and monitoring devices for some of the most successful MGs and as such deserve a closer inspection as a solution to these problems. An overview of the protection systems today and the advancements required to meet the challenges for protection in the distribution system, starting with MGs, are discussed in detail in a report [12].

The focus on the modes of operation and transitions is paramount. It is essential to protect an MG for all modes of operation including transitions against all types of faults, and this is a real challenge. The philosophy for MG protection is to have the same protection architecture for both islanded and grid-connected operations. A fast-acting smart switch such as a solid-state circuit breaker is to be developed to open for all faults that could occur in a MG.

13.5.5 Examples of Microgrid in Operation

While the concept of MGs has matured, the technology to support these concepts is still evolving. The idea of operating an existing distribution system as a MG will take several years to mature because the utilities or DSOs do not have much financial incentive to do so. Typically, DERs are owned and installed by customers to increase their own resiliency. The DSOs will have to invest significant sums of money on monitoring and control equipment to operate the system as a MG. While large utilities will not find it advantageous to do so, small municipal DSOs or rural electric cooperatives (RECs) may find it advantageous to do in certain situations. Typically, municipal DSOs and RECs do not have their own generation. They buy electricity from wholesalers and distribute it to their customers. Some such organizations are already contemplating installation of their own solar PV generation and BESS. It could become feasible for them to invest in the MG technology.

The most valuable implementation of MG technology is for institutions or large businesses who have their own campus of large facility. There are several examples of such MGs. Another area where MGs play a crucial role is electrification of remote rural communities. In this section, we look at some MGs that are successfully operating.

13.5.5.1 CERTS Microgrid

The Consortium for Electric Reliability Technology Solutions (CERTS) in the U.S., which includes several national laboratories and universities, introduced their concept of the MG in a white paper [21] to effectively integrate DERs. The CERTS Microgrid Concept provides a novel approach for operation and control of DERs and loads within a MG with minimal communication requirement for safe and stable operation. The CERTS Microgrid Concept especially focuses on automatic and seamless transitions between grid-connected and islanded modes of operation, equipment protection within the MG with low fault currents, and MG control for voltage and frequency stability under islanded conditions without depending on high-speed communications.

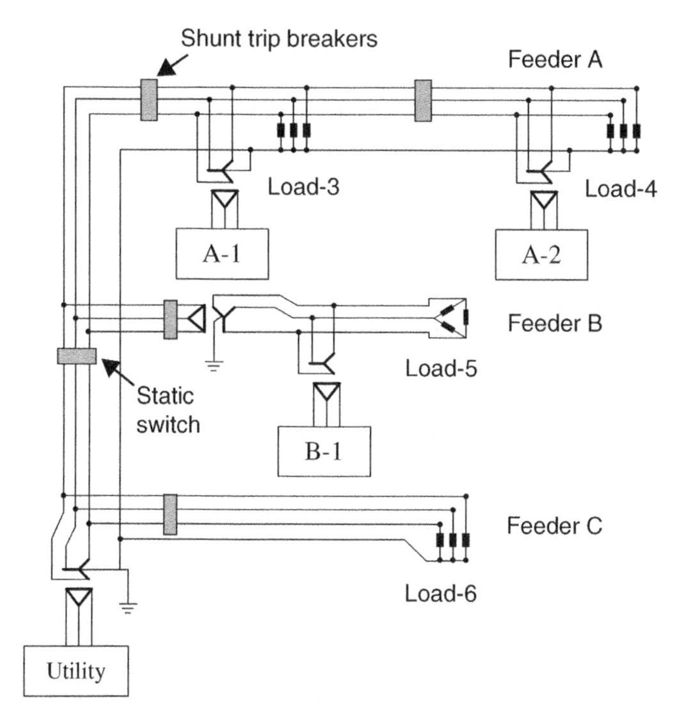

Figure 13.7 CERTS microgrid test bed. Source: Adapted from Lasseter et al. [22].

To demonstrate the concepts, CERTS developed a test bed near Columbus, OH, in collaboration with American Electric Power (AEP). Phase 1 of the test bed shown in Figure 13.7 has three feeders (A, B, and C) with loads and three DERs. One feeder has two DERs (A-1 and A-2), another feeder has one DER (B-1), and the third feeder only has load. The static switch isolates feeders A and B from the grid to allow the MG to operate in the islanded mode.

The four load banks (Load-3 to Load-6) have capabilities of remote control with loads ranging from 0 to 90 kW and 0–45 kvar. Each load bank also has capabilities to simulate faults ranging from bolted faults to high impedance faults. Other equipment include a variable power induction motor (0–20 HP), protection relays, shunt trip breakers, and a digital data acquisition system to capture voltage and current waveforms at different locations. In the later phases of the project, a conventional synchronous generator and a stand-alone electricity storage device were added to the test bed.

13.5.5.2 IIT Microgrid

Following the 12 major power outages, Illinois Institute of Technology (IIT) decided to join the Galvin Electricity Initiative (GEI) for perfect power [23]. While the campus already had two 4-MW combined cycle gas units and a small wind turbine, rooftop PV and a 500-kWh battery were added to boost the generation capacity to meet the needs of the IIT campus with a peak load of around 10 MW. Several system upgrades were implemented, and control equipment have been deployed to operate the system as an MG. The MG can operate both in the grid-connected and islanded modes.

The overall goal of the perfect power MG is to provide real-time reconfiguration of power supply assets, real-time islanding of critical loads, and real-time optimization of power supply resources. In addition to providing reliable power supply to the campus, the MG is also a test bed for testing emerging smart grid technologies.

13.5.5.3 Philadelphia Navy Yard Microgrid

The Philadelphia Navy Yard was abandoned about 25 years ago but now is a vibrant commercial center managed by Philadelphia Industrial Development Corporation (PIDC). A unique feature of the redeveloped Navy Yard is that it is powered by an MG. About 170 employers occupy 7.5 million square feet of space, and 15,000 individuals work there. A motivation to deploy the MG was to provide premium power to the customers while leveraging natural gas and local renewables to enhance the Navy Yard and community grid independence and energy security. In addition, the MG will enhance the reliability, efficiency, capacity, and resilience of the PECO's local distribution grid. The facility is also a test bed for evolving MG technologies. The MG includes an 8-MW natural gas generator, a 600-kW fuel cell plant, 1-MW solar PV, and a 6.2-MW/14.8 MWh Li-ion BESS.

13.6 Off-Grid Electrification

Electrical energy needs are growing globally due to increased dependence on electricity for daily activities in homes and businesses. While much of the world enjoys electricity for conveniences and necessities of daily life, about one billion people have no access to electricity. The absence of electricity creates barriers for family life as well as educational and economic opportunities. Also, people without access to electricity depend on wood for cooking and kerosene for lighting. Emissions from burning these fuels are harmful to people's health and contribute significantly to greenhouse gases.

The extremely prohibitive cost of grid expansion and a lack of available fuel resources to generate electricity have been hurdles for electrification of regions such as sub-Sahara Africa. Extending the grid to meet electricity needs of a small population, especially in rural areas with low population density, is not always viable. Instead, localized electricity generation in stand-alone or MG applications is a suitable option. Small-scale wind and solar power generation are attractive for such applications. A solar panel with a battery can meet the lighting needs of a small house, while a small wind machine can power a few homes. The most technically and economically promising solution for such situations is a combination of different resources for electricity production. The systems for each location must be appropriate for the local economic and social conditions.

13.6.1 Designing Off-Grid Systems

Designing an off-grid system requires consideration of several factors, which are described below [24–26].

13.6.1.1 Load Estimation

The first step is determination of load demand for the community. This starts by counting the number of households, schools, shops, and other entities in the community. The next step is the estimation of loads in each energy-consuming unit and the usage pattern. This can be done by conducting a survey of the communities or by considering a fixed load for each household, such as one or two lights. The usage pattern will require information on weather conditions and lifestyle of the local population. Based on this information, daily hourly aggregate load profile for different seasons can be developed.

13.6.1.2 Resource Assessment

This step requires an evaluation of available resources in the area, such as wind, solar, microhydro, and conventional resources. Wind and solar resource assessment will require yearly wind and solar irradiation data. National Renewable Energy Laboratory (NREL) in the U.S. maintains such data for many parts of the world. Microhydro availability will depend on the local conditions, such as a mountain stream or a nearby river as well as water flow pattern for the year. Conventional resource availability will depend on the local market conditions for specific equipment and fuel.

13.6.1.3 Optimal System Design

The optimal design for the given locality requires cost data for different feasible resources. The data include capital as well as operation and maintenance cost. For conventional resource, the cost of fuel and penalties for emissions, if any, are needed. The cost would also include any specialized equipment such as batteries and other control equipment, such as inverters, meters, and communication layer. We can use these data along with load and resource information to determine an optimal design for a given life span of the MG.

13.6.1.4 Other Factors

The basic design assumes that all the households receive electricity whenever they need without any curtailment. However, if they are willing to accept curtailments, the cost of generating electricity can be reduced significantly because that allows reduction in the size of batteries and other generating resources needed in the system.

References

1 Wu, B., Lang, Y., Zargari, N., and Kouro, S. (2011). *Power Conversion and Control of Wind Energy Systems*. Hoboken, NJ: Wiley-IEEE Press.
2 Messenger, R.A. and Ventre, J. (2017). *Photovoltaic Systems Engineering*, 4e. Boca Raton, FL: Taylor & Francis.
3 Lawder, M.T., Suthar, B., Northrop, P.W.C. et al. (2014). Battery energy storage system (BESS) and battery management system (BMS) for grid-scale applications. *Proceedings of the IEEE* 102 (6): 1014–1030.
4 Tonkoski, R., Lopes, L.A.C., and El-Fouly, T.H.M. (2011). Coordinated active power curtailment of grid connected PV inverters for overvoltage prevention. *IEEE Transactions on Sustainable Energy* 2 (2): 139–147.
5 Malekpour, A.R. and Pahwa, A. (2017). A dynamic operational scheme for residential PV smart inverters. *IEEE Transactions on Smart Grid* 8 (5): 2258–2267.
6 Malekpour, A.R. and Pahwa, A. (2015). Radial test feeder including primary and secondary distribution network. In North American Power Symposium, Charlotte, NC.
7 Ton, D. and Smith, M.A. (2012). The U.S. department of energy's microgrid initiative. *The Electricity Journal* 25 (8): 84–94.
8 Nikkhajoei, H. and Lasseter, R.H. Microgrid protection. In IEEE PES General Meeting in Tampa, Florida (24–28 June 2007).
9 Directorate-General for Research (2006). *European SmartGrid Technology Platform, Vision and Strategy for Europe's Electricity Networks of the Future*. Brussels, Belgium: EU Commission, Information and Communication Unit.
10 Moreira, C.L., Resende, F.O., and Lopes, J.A.P. (Feb. 2007). Using low voltage microgrids for service restoration. *IEEE Transactions on Power Systems* 22 (1): 395–403.
11 Sortomme, E., Mapes, G.J., Foster, B.A., and Venkata, S.S. (2008). Fault analysis and protection of a microgrid. In North American Power Symposium, Calgary, Canada, 1–6.
12 Venkata, S.S., Reno, M.J., Bower, W., Manson, S., and Reilly, J. (2019). Microgrid protection: advancing the state of the art. *Sandia Natl. Lab. Rep.*
13 Barker, P.P. and De Mello, R.W., Determining the impact of distributed generation on power systems. I. Radial power systems. In IEEE Power Engineering Society Summer Meeting (16–20 July 2000), vol. 3, 1645–1656.

14 Brahma, S.M. and Girgis, A.A. (2011) Effect of distributed generation on protective device coordination in distribution system. In Large Engineering Systems Conference on Power Engineering (11–13 July 2001), 115-119.

15 Slaman, S.K. and Rida, I.M. (2001). Investigating the impact of embedded generation on relay setting of utilities. *IEEE Transactions on Power Delivery* 16 (2): 246–251.

16 Brahma, S.M. and Girgis, A.A. (2001). *Impact of Distributed Generation on Fuse and Relay Coordination: Analysis and Remedies*, 384–389. Clearwater, FL: International Association for Science and Technology Development.

17 Doyle, M.T. (2002). Reviewing the impact of distributed generation on distribution system protection. In IEEE Power Engineering Society Summer Meeting (25–28 July 2002), vol. 1, 103–105.

18 Gomez, J.C. and Morcos, M.M. (2002). Coordinating overcurrent protection and voltage sags in distributed generation systems. *IEEE Power Engineering Review* 22 (2): 16–19.

19 Keil, T., Jager, J., Shustov, A., and Degner, T. (2007). Changing network conditions due to distributed generation – systematic review and analysis of their impacts on protection, control and communication systems. In CIRED 19th International Conference on Electrical Distribution (21–24 May 2007).

20 Keil, T., Jager, J., Shustov, A., and Degner, T. (2009). Key requirements for system protection in distribution network with distributed generation – results of holistic analysis. In 3rd International Conference on Advanced Power System Automation and Protection, South Korea.

21 Lasseter, R.H. (1998). Control of distributed resources. *Bulk Power System and Controls IV Conference*. p. 2428.

22 Lasseter, R.H., Akhil, A., Marnay, C., Stephens, J., Dagle, J., and Guttromson, R., et al. (2002). The CERTS microgrid concept. In White Paper for Transmission Reliability Program Office of Power Technologies.

23 Flueck, A. and Li, Z. (2008). Destination perfection: the journey to perfect power at Illinois Institute of Technology. *IEEE Power & Energy Magazine* 6 (November/December): 36–47.

24 Pahwa, A. (2016). Partnerships to facilitate electricity access for the remote rural communities of sub-Sahara Africa. In IEEE PES Power Africa Conference, Livingstone, Zambia.

25 Sadiqi, M., Pahwa, A., and Douglas Miller, R. (2012). Basic design and cost optimization of a hybrid power system for rural communities in Afghanistan. In North American Power Symposium, Urbana-Champaign, IL. 1–6.

26 Louie, H. (2018) *Off-Grid Electrical Systems in Developing Countries*, Springer.

Appendix A

Per-unit Representation

In power systems, many transformers at various (different) voltage levels are involved. Per-unit Representation or System is a normalization procedure that provides a mathematical basis for analyzing power networks with relative ease and convenience. In addition, when various quantities are expressed in per unit (pu) or per cent values, they usually convey a message. For example, if a bus voltage is 0.98 pu, it means that this value is 98% of the nominal or base value which could be at any level in the network. It also immediately conveys a message that the value is an acceptable one. On the contrary, if the voltage value is 1.08 pu, then it immediately conveys that the value is higher than the acceptable level of 1.05 pu. Similar conclusions can be drawn for other quantities such as current, power, and impedance. The idea here is to express various quantities as a fraction of their corresponding base (fixed) values. In general, we can define the pu value of a quantity by the following equation:

$$\text{Quantity in per unit (pu)} = \frac{\text{Actual Value of Quantity}}{\text{Base Value of Quantity}} \tag{A.1}$$

A.1 Single-phase Systems

The basic idea is to select two electrical variables such as power and voltage as independent base values. Then, the base values for the other two variables, namely, current and impedance, are computed using the power relationship and the Ohm's law. We illustrate this procedure for single-phase systems.

Let the base value for power be S_B (VA), and the base value for voltage be V_B (Volts). Then, the base value of current is

$$I_B = \frac{S_B}{V_B} \tag{A.2}$$

Electric Power and Energy Distribution Systems: Models, Methods, and Applications, First Edition.
Subrahmanyam S. Venkata and Anil Pahwa.

Companion website: www.wiley.com/go/Pahwa/ElectricPowerDistributionSystems

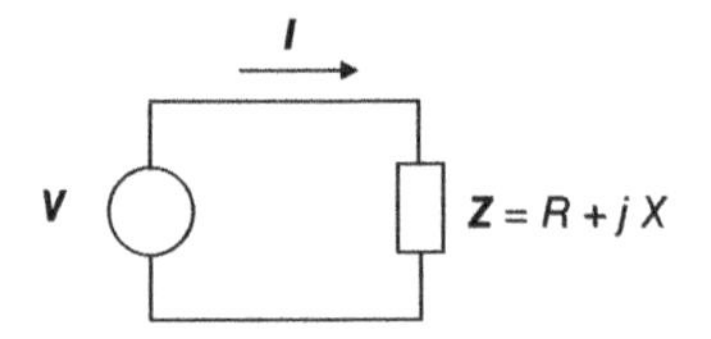

Figure A.1 An example system.

and the base value of impedance is

$$Z_B = \frac{V_B}{I_B} = \frac{V_B^2}{S_B} \tag{A.3}$$

We consider an example shown in Figure A.1 to illustrate the use of pu system.

Note that voltage, current, and impedance are complex values. The base values used for normalization are, however, real values. Therefore, the respective pu values are also complex while they maintain the relative ratio of the real and imaginary parts or the angle of the complex quantities in polar form.

Now,

$$\boldsymbol{V_{pu}} = \frac{\boldsymbol{V}}{V_B} \tag{A.4}$$

$$\boldsymbol{I_{pu}} = \frac{\boldsymbol{I}}{I_B} \tag{A.5}$$

$$\begin{aligned}\boldsymbol{Z_{pu}} &= \frac{\boldsymbol{Z}}{Z_B} = \frac{R + jX}{Z_B} \\ &= R_{pu} + jX_{pu}\end{aligned} \tag{A.6}$$

and

$$\begin{aligned}\boldsymbol{S_{pu}} &= \frac{\boldsymbol{S}}{S_B} = \frac{P + jQ}{S_B} \\ &= P_{pu} + jQ_{pu}\end{aligned} \tag{A.7}$$

A.2 Three-phase Systems

Three-phase systems may be normalized by picking the appropriate three-phase bases. We illustrate the various base choices for both Y and Δ systems on a comparative basis as shown in Table A.1. Note that *LL* and *LN* in parenthesis indicate line to line and line to neutral, respectively. Similarly, *L* signifies line current, and *P* signifies phase current.

A.2.1 Per-unit Values for Δ-Connected Systems

Choose the three-phase apparent power rating as the base value for power (S_{3B}), and the rated line-to-line voltage as the base value for line-to-line voltage ($V_B(LL)$).

Table A.1 Base value selection and relations for wye (Y)- and delta (Δ)-connected three-phase systems.

Wye (Y)	Delta (Δ)
Choose power base: $S_{3B} = 3S_{1B}$	$S_{3B} = 3S_{1B}$
Choose power base: $V_B(LL) = \sqrt{3}V_B(LN)$	$V_B(LL)$
$I_B(L) = \frac{S_{3B}}{\sqrt{3}V_B(LL)}$	$I_B(L) = \frac{S_{3B}}{\sqrt{3}V_B(LL)}$
	$I_B(P) = \frac{S_{3B}/3}{\sqrt{3}V_B(LL)} = \frac{I_B(L)}{\sqrt{3}}$
$Z_B(Y) = \frac{V_B(LN)}{I_B(L)} = \frac{V_B(LL)}{\sqrt{3}I_B(L)} = \frac{V_B^2(LL)}{S_{3B}}$	$Z_B(\Delta) = \frac{V_B(LL)}{I_B(P)} = \frac{3V_B^2(LL)}{S_{3B}} = 3Z_B$

$$I_{Lpu} = \frac{I_L}{I_B(L)} \tag{A.8}$$

$$Z_{pu}(Y) = \frac{Z_Y}{Z_B(Y)} \tag{A.9}$$

A.2.2 Per-unit Values for Δ-Connected Systems

Choose the three-phase apparent power rating as the base value for power (S_{3B}), and the rated line-to-line voltage as the base value for line-to-line voltage ($V_B(LL)$).

$$I_{Lpu} = \frac{I_L}{I_B(L)} \tag{A.10}$$

$$I_{Ppu} = \frac{I_P}{I_B(L)} \tag{A.11}$$

$$Z_{pu}(\Delta) = \frac{Z_\Delta}{Z_B(\Delta)} \tag{A.12}$$

A.3 Base Values for Transformers

Since transformers change voltage from one side to other, we have to select base voltages separately for each side. Proper selection of base voltages allows us to simplify the transformer pu equivalent circuit. The rules for selecting bases values for the transformers are given below. The two sides of the transformer are called side A and side B.

Single-phase transformers:

$$\frac{V_B(A)}{V_B(B)} = \frac{V_{\text{Rated}}(A)}{V_{\text{Rated}}(B)} \tag{A.13}$$

S_B is the same for both sides.

Three-phase transformers:

$$\frac{V_B(LL)(A)}{V_B(LL)(B)} = \frac{V_{\text{Rated}}(A)}{V_{\text{Rated}}(B)} \tag{A.14}$$

S_{3B} is the same for both sides.

With these rules, the pu voltage and current have the same values on both the sides. Also, the pu value of impedance turns out to be the same when computed with respect to either side. Therefore, the transformer is represented by a single reactance in the system when all the quantities are represented in pu.

A.4 Change of Base

It is often necessary to convert the base values of several pieces of equipment connected together to form a power system (or interconnected power system). Usually, the name plate ratings of these individual devices are different, and hence, their respective individual base values will also be different. In order to refer all pu values to a common system base, it is necessary to change all device pu values to pu values computed with the common system bases. The key equations for the conversion are given below:

$$\boldsymbol{Z}_{\textbf{actual}} = \boldsymbol{Z}_{\boldsymbol{pu}}^{\textbf{old}}\, Z_B^{\text{old}} = \boldsymbol{Z}_{\boldsymbol{pu}}^{\textbf{new}}\, Z_B^{\text{new}} \tag{A.15}$$

or

$$\boldsymbol{Z}_{\boldsymbol{pu}}^{\textbf{new}} = \boldsymbol{Z}_{\boldsymbol{pu}}^{\textbf{old}} \left[\frac{Z_B^{\text{old}}}{Z_B^{\text{new}}}\right] = \boldsymbol{Z}_{\boldsymbol{pu}}^{\textbf{old}} \left[\frac{V_B^{\text{old}}}{V_B^{\text{new}}}\right]^2 \left[\frac{S_B^{\text{new}}}{S_B^{\text{old}}}\right] \tag{A.16}$$

For three-phase systems, the total three-phase power for S_B and the line-to-line voltage for V_B must be used.

A.5 Advantages of Per-unit Representation

1. pu representation results in a more meaningful and correlated data. It gives the relative magnitude information.
2. There will be less chance of mixing up between single- and three-phase powers or between line-to-line and line-to-neutral (phase) voltages.
3. The pu representation is very useful in simulating machine systems on analog, digital, and hybrid computers for steady-state and dynamic analysis.
4. Manufacturers usually specify the impedance of a piece of apparatus in pu (or percent) on the base of the name plate rating of power and voltage. Hence, it can be used directly if the bases chosen are the same as the name plate ratings.

5. The pu impedance values of the various apparatus lie in a narrow range, though the actual values vary widely.
6. The pu equivalent impedance of any transformer is the same referred to either primary or secondary side. For complicated systems involving many transformers or different turns ratio, this advantage is a significant one for large systems on computers.
7. Though transformers have different connections (Y–Y, Δ–Δ, Y–Δ, Δ–Y) in three-phase systems, the pu impedance is the same irrespective of the connection.
8. pu method allows the same basic arithmetic operation resulting in per-phase end values, without having to worry about the factor "100," which occurs in percent representation.

Appendix B

Symmetrical Components

C.L. Fortescue introduced the symmetrical components in 1918. They have been used for the analysis of unbalanced power systems for over 100 years. According to Fortescue, unbalanced voltages and currents are resolved into three sets of sequence components, which are

1. **Positive Sequence:** They consist of three phasors with equal magnitude, +120° phase displacement, and *a–b–c* (positive) phase sequence.
2. **Negative Sequence:** They consist of three phasors with equal magnitude, −120° phase displacement, and *a–c–b* (negative) phase sequence.
3. **Zero Sequence:** They consist of three phasors with equal magnitude and zero phase displacement.

Figure B.1 shows an example of sequence components of voltages along with the voltages in the phase domain.

Now we define an operator $\boldsymbol{a} = 1\angle 120°$. Using this operator, we can write the components of the phase voltages as follows:

$$\begin{aligned} \boldsymbol{V}_1^b &= \boldsymbol{a}^2\boldsymbol{V}_1^a \\ \boldsymbol{V}_1^c &= \boldsymbol{a}\boldsymbol{V}_1^a \\ \boldsymbol{V}_2^b &= \boldsymbol{a}\boldsymbol{V}_2^a \\ \boldsymbol{V}_2^c &= \boldsymbol{a}^2\boldsymbol{V}_2^a, \end{aligned} \tag{B.1}$$

and

$$\boldsymbol{V}_0^b = \boldsymbol{V}_0^c = \boldsymbol{V}_0^a$$

Since phase voltages are the sum of their respective components, we can express the relationship between the phase voltages and their sequence components in

Electric Power and Energy Distribution Systems: Models, Methods, and Applications, First Edition.
Subrahmanyam S. Venkata and Anil Pahwa.

Companion website: www.wiley.com/go/Pahwa/ElectricPowerDistributionSystems

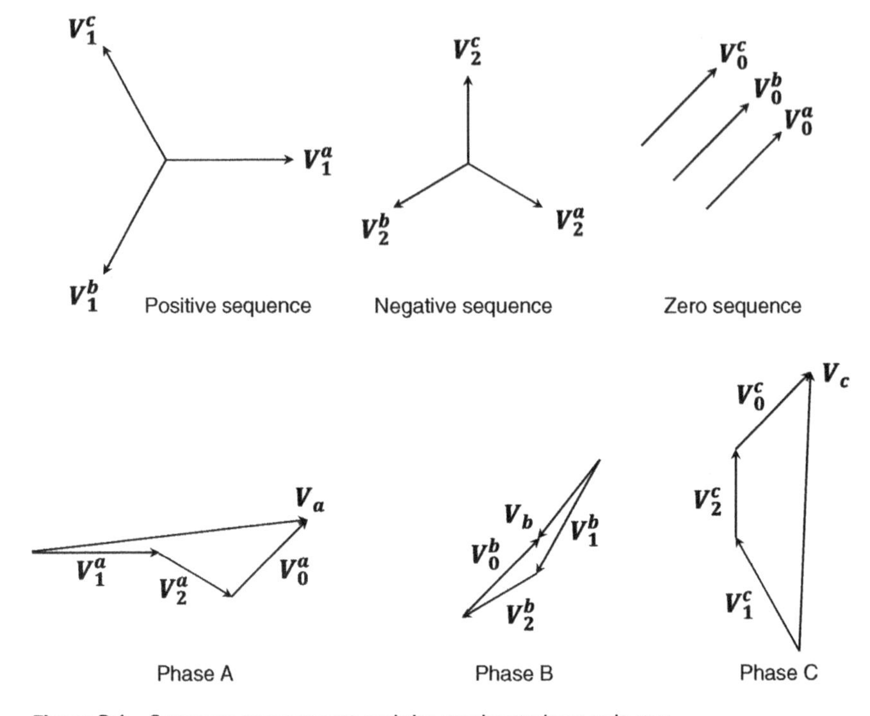

Figure B.1 Sequence components and the resultant phase voltages.

terms of the components of phase a in matrix form. We have dropped the superscript a for simplicity.

$$\begin{bmatrix} V_a \\ V_b \\ V_c \end{bmatrix} = \begin{bmatrix} 1 & 1 & 1 \\ 1 & a^2 & a \\ 1 & a & a^2 \end{bmatrix} \begin{bmatrix} V_0 \\ V_1 \\ V_2 \end{bmatrix} \tag{B.2}$$

or

$$V_p = AV_s \tag{B.3}$$

where

$$V_p = \begin{bmatrix} V_a \\ V_b \\ V_c \end{bmatrix}$$

$$V_s = \begin{bmatrix} V_0 \\ V_1 \\ V_2 \end{bmatrix} \tag{B.4}$$

and

$$A = \begin{bmatrix} 1 & 1 & 1 \\ 1 & a^2 & a \\ 1 & a & a^2 \end{bmatrix}$$

We can compute the inverse of this matrix, which is

$$A^{-1} = \frac{1}{3}\begin{bmatrix} 1 & 1 & 1 \\ 1 & a & a^2 \\ 1 & a^2 & a \end{bmatrix} \tag{B.5}$$

We can build similar relation for currents in the system. Note that if we are dealing with current, the zero-sequence current in terms of phase domain currents is

$$I_0 = \frac{1}{3}(I_a + I_b + I_c) \tag{B.6}$$

For a Y-connected grounded system

$$(I_a + I_b + I_c) = I_n \tag{B.7}$$

Therefore,

$$I_n = 3\,I_0 \tag{B.8}$$

For loads and series-connected impedances, we can develop a relationship to determine the impedance in the sequence domain with the given the impedances in the phase domain. Consider a general case where the following relationship can be written in the phase domain

$$V_p = Z_p I_p \tag{B.9}$$

Converting it to sequence domain gives

$$AV_s = Z_p A I_s \tag{B.10}$$

Therefore,

$$V_s = (A^{-1} Z_p A) I_s \tag{B.11}$$

Now, let

$$(A^{-1} Z_p A) = Z_s \tag{B.12}$$

Then, (B.11) becomes

$$V_s = Z_s I_s \tag{B.13}$$

Consider a general formulation for Z_p

$$Z_p = \begin{bmatrix} Z_{aa} & Z_{ab} & Z_{ac} \\ Z_{ab} & Z_{bb} & Z_{bc} \\ Z_{ac} & Z_{bc} & Z_{cc} \end{bmatrix} \tag{B.14}$$

If $\boldsymbol{Z}_p$ is symmetric or

$$\boldsymbol{Z}_{aa} = \boldsymbol{Z}_{bb} = \boldsymbol{Z}_{cc} \tag{B.15}$$

and

$$\boldsymbol{Z}_{ab} = \boldsymbol{Z}_{ac} = \boldsymbol{Z}_{bc} \tag{B.16}$$

the off-diagonal elements of $\boldsymbol{Z}_s$ become zero, and the diagonal components are given by

$$Z_0 = \boldsymbol{Z}_{aa} + 2\boldsymbol{Z}_{ab}$$

and

$$Z_1 = Z_2 = \boldsymbol{Z}_{aa} - \boldsymbol{Z}_{ab} \tag{B.17}$$

Hence,

$$\boldsymbol{Z}_s = \begin{bmatrix} Z_0 & 0 & 0 \\ 0 & Z_1 & 0 \\ 0 & 0 & Z_2 \end{bmatrix} \tag{B.18}$$

Index

Electric Power and Energy Distribution Systems: Models, Methods, and Applications, First Edition.
Subrahmanyam S. Venkata and Anil Pahwa.

Companion website: www.wiley.com/go/Pahwa/ElectricPowerDistributionSystems

d

n

o

r

t

Printed and bound by CPI Group (UK) Ltd, Croydon, CR0 4YY
16/04/2025
14658345-0002